T0192675

Weak Neutral Currents

Weak Neutral Currents

THE DISCOVERY OF THE ELECTRO-WEAK FORCE

David B. Cline
with classic papers

Routledge
Taylor & Francis Group

LONDON AND NEW YORK

First published 1997 by Westview Press

Published 2019 by Routledge
52 Vanderbilt Avenue, New York, NY 10017
2 Park Square, Milton Park, Abingdon, Oxon OX14 4RN

Routledge is an imprint of the Taylor & Francis Group, an informa business

A Cataloging-in-Publication data record for this book is available from the Library of Congress.

ISBN 13: 978-0-367-21332-9 (hbk)

ISBN 13: 978-0-367-21613-9 (pbk)

Frontiers in Physics
David Pines, Editor

Volumes of the Series published from 1961 to 1973 are not officially numbered. The parenthetical numbers shown are designed to aid librarians and bibliographers to check the completeness of their holdings.

Titles published in this series prior to 1987 appear under either the W. A. Benjamin or the Benjamin/Cummings imprint; titles published since 1986 appear under the Westview Press imprint.

Editor's Foreword

The problem of communicating in a coherent fashion recent developments in the most exciting and active fields of physics continues to be with us. The enormous growth in the number of physicists has tended to make the familiar channels of communication considerably less effective. It has become increasingly difficult for experts in a given field to keep up with the current literature; the novice can only be confused. What is needed is both a consistent account of a field and the presentation of a definite "point of view" concerning it. Formal monographs cannot meet such a need in a rapidly developing field, while the review article seems to have fallen into disfavor. Indeed, it would seem that the people who are most actively engaged in developing a given field are the people least likely to write at length about it.

Frontiers in Physics was conceived in 1961 in an effort to improve the situation in several ways. Leading physicists frequently give a series of lectures, a graduate seminar, or a graduate course in their special fields of interest. Such lectures serve to summarize the present status of a rapidly developing field and may well constitute the only coherent account available at the time. One of the principal purposes of the *Frontiers in Physics* series is to make notes on such lectures available to the wider physics community.

A second way to improve communication in very active fields of physics is by the publication of collections of reprints of key articles relevant to our present understanding. Such collections are themselves useful to people working in the field or to students considering entering it. The value of the reprints is, however, considerably enhanced when the collection is accompanied by a commentary which might, in itself, constitute a brief survey of the present status of the field.

The present volume represents just such a collection, with reprints selected and annotated by a leading contributor to the field, David B. Cline. Taken together, his introductory material and the reprints of both key scientific contributions and personal reminiscences by a number of the leading researchers provide the reader with a detailed account of the remarkable developments during the past forty years in the physics of the weak interactions. I am pleased to welcome Professor Cline and his fellow contributors to *Frontiers in Physics*.

David Pines
Urbana, IL
June, 1997

PREFACE

This book attempts to trace the key experimental developments that led to the discovery of weak neutral currents in 1973 and the W, Z bosons in 1983 – all of the results of which culminated in the identification of the unified-electroweak force. As much as possible, we use the primary scientific papers and some individual accounts by pioneers in the field to attempt to recreate the excitement of this series of experimental discoveries. The individual Chapters are written by the author. In addition, several overview papers from *Physics Today* and elsewhere are reprinted. The individual acounts come from the proceedings of the meeting, "50 Years of Weak Interactions," which was held at Frank Lloyd Wright's "Wingspread" (near Racine, Wisconsin) in 1984 and published by the University of Wisconsin in that same year. They were then reprinted for the proceedings of the conference, "30 Years of Weak Neutral Currents," held in Santa Monica, California in 1993 (AIP Conference Proceedings 300, *Discovery of Weak Neutral Currents: The Weak Interaction Before and After*), edited by A. K. Mann and the author. Interested readers may consult this reference for a much wider account of the subject.

It is rare that scientists get to participate in and witness a development as exciting as the formation of the Standard Model of Elementary Particles, and the discovery of weak neutral currents and of the electroweak force, which were at the core of this revolution. I have attempted to present these developments largely from the experimental viewpoint, and I hope that I have been fair to the many beautiful experiments performed during this period that bear directly on the theme of this book. I also have attempted to predict the future (in a modest way) concerning the eventual complete search for the Higgs boson at the Large Hadron Collider at CERN in the next decade.

First and foremost, I would like to acknowledge the U.S. Department of Energy, Division of High Energy Physics, for all of the support it gave us over the many years of experiments that have been required to bring us to this point. Truly, without that support, this discovery would not have been possible.

I also would like to acknowledge here all of the great people I enjoyed working with during the period covered by this book: Jack Fry, Ugo Camerini, Don Reeder, Alberto Benvenuti, Carlo Rubbia, Al Mann, Peter McIntyre, Vernon Barger, Francis Halzen, Jim Rohlf, and many many more.

In addition, I thank Ms. Joan George and her UCLA student workers for their excellent help in preparing this work.

In the time between the completion and the printing of this book, Professor C-S. Wu passed away. I had the great privilege to interact with her on several occasions but, unfortunately, never to work with her. In my opinion, Professor Wu was the best experimental weak-interaction scientist since E. Fermi and Madame Curie. After the W was discovered, she was so complimentary, but actually it took a cast of thousands to find the W and only a few people to prove parity violation.

I had so hoped to send her a copy of this book to get her reaction, but that is now not possible. Instead, I can only offer these few words in appreciation of her accomplishments and contributions.

David B. Cline
Los Angeles, 1997

CONTENTS

CHAPTER 3
The Search for Other Forms of the Weak Interaction
Changing Flavors (1963 – 1970)

CHAPTER 4
The Electroweak Interaction Picture Emerges
Everyone Missed It (1962 – 1973)

CHAPTER 5
The Discovery of Weak Neutral Currents
Nothing In – Nothing Out (1973 – 1974)

CHAPTER 6
Other Weak-Neutral Current Processes, Parity Violation in Weak Neutral Currents, and $\sin^2 \theta_W$
A Force in the Mirror (1975 – 1978)

CHAPTER 7
On to the *W* and *Z* Particles
The Electroweak Force Rises (1976 – 1983)

CHAPTER 8
High Precision Studies of the Electroweak Force
From the Top Down (1983 – 1995)

CHAPTER 9
Back to the Future with the Higgs Boson
The God Particle Again (1995 – 2010)

NOMENCLATURE

A	Axial vector
AGS	Alternating Gradient Synchrotron
ALEPH	LEP detector
ANL	Argonne National Laboratory
ATLAS	A Toroidal LHC ApparatuS at CERN
BNL	Brookhaven National Laboratory
Caltech	California Institute of Technology
CDF	Collider Detector at Fermilab
CDHS	CERN-Dortmund-Heidelberg-Saclay
CERN	European Organization for Nuclear Research
CL	Confidence Level
CLEO	CLEOpatra (name for the collaboration that was to design, build and operate the general purpose detector using the storage ring, CaESaR, developed by the Cornell Laboratory of Nuclear Studies in the mid-1970s)†
CMR	D. Cline, A. Mann, and C. Rubbia
CMS	Compact Muon Solenoid, detector at CERN's LHC experiment
DELPHI	DEtector with Lepton, Photon and Hadron Identification at CERN
FCNC	Flavor-Changing Neutral Current
FNAL	Fermi National Accelerator Laboratory; also known as Fermilab (at Batavia, IL)
GGM	GarGaMelle, bubble chamber at CERN
GIM	Glashow-Iliopoulos-Maiani
GWS	Glashow-Weinberg-Salam
HPW	Harvard-Penn-Wisconsin
HPWF	Harvard-Penn-Wisconsin-FNAL
ISABELLE	Intersecting Storage Accelerator at Brookhaven
ISR	Intersecting Storage Ring
IVB	Intermediate Vector Boson
KEK	Koh-Enerugii Butsurigaku Kenkyuusho

---

†Thanks for the memories, Ed Thorndike.

L3	LEP experiment at CERN
LBL	Lawrence Berkeley Laboratory, now known as
LBNL	Lawrence Berkeley National Laboratory
LEP	Large Electron-Positron collider at CERN
LHC	Large Hadron Collider at CERN
LLNL	Lawrence Livermore National Laboratory
MIT	Massachusetts Institute of Technology
MURA	Midwestern University Research Association
NC	Neutral Current
OPAL	Omni-Purpose Apparatus at LEP
PDF	Particle Data Group
$\bar{p}p$	Proton–anti-proton collider at CERN
PS	Photon Synchrotron
QCD	Quantum ChromoDynamics
QED	Quantum ElectroDynamics
RMC	C. Rubbia, P. McIntyre, D. Cline
SLAC	Stanford Linear Accelerator Center
SLC	Stanford Linear Collider
SPEAR	Stanford Positron–Electron Accelerator Ring
S$\bar{p}p$S	Super Proton–anti-proton Synchrotron collider at CERN
SPS	Super Proton Synchrotron collider at CERN
UA1, UA2	Underground Areas 1, 2 (detectors at the $\bar{p}p$ collider, CERN)
V	Vector
V-A	Vector–Axial vector
W	W boson (weakon)
WNC	Weak Neutral Current

ACKNOWLEDGEMENT AND PERMISSIONS

I wish to acknowledge the role that publishers played during the exciting period in the history of high-energy physics that is covered in this book and to thank them for generously granting us permission to use reprints of many of the papers that they published. The publishers involved are listed below in alphabetical order. Complete information on these papers may be found in the Contents of this book.

American Institute of Physics (500 Sunnyside Blvd., Woodbury, NY 11797), *AIP Conference Proceedings 300*, 1994:

2(A), (B), (C)	8(A)
4(C)	9(A)
5(E)	6(E) *JETP Letters*: v. 27 (1978)

Physics Today (editorial office: American Center for Physics, One Physics Ellipse, College Park, MD 20740-3843):

2(D) March (1975)	7(B) August (1980)

American Physical Society (One Physics Ellipse, College Park, MD 20740-3844), *Physics Review of Letters*:

2(E) v. 9 (1962)	5(D) v. 32 (1974)
3(B) v. 13 (1964)	6(A) v. 33 (1974)
3(C) v. 24 (1970)	6(B) v. 37 (1976)
4(A) v. 30 (1973)	6(C) v. 41 (1978)
4(B) v. 35 (1975)	6(D) v. 37 (1976)
5(C) v. 32 (1974)	

Elsevier Science N.L. (Sara Burgerhartstraat 25, 1055 KV Amsterdam, The Netherlands), *Physics Letters*:

5(A) v. 46B (1973)	7(D) v. 122B (1983)
5(B) v. 46B (1973)	7(E) v. 122B (1983)
6(F) v. 77B (1978)	7(F) v. 126B (1983)
6(G) v. 71B (1977)	7(G) v. 129 (1983)
7(C) v. 107B (1981)	

Gordon and Breach Scientific Publishers (820 Town Center Dr., Langhorne, PA 19047):

3(A) *Methods in Subnuclear Physics* (1969)

8(B) *Comments on Nuclear and Particle Physics*, v. 16 (1986)

Friedr. Vieweg & Sohn Verlagsgesellschaft mbH (Postfach 5829, D-65048 Wiesbaden, Germany):

7(A) *Proceedings of the International Neutrino Conference Aachen 1976* (1977)

CHAPTER 1
THE WEAK INTERACTION UNCOVERED
(H. Becquerel to W. Pauli)

In the Beginning (1896 – 1933)

1.1. INTRODUCTION

The standard model of elementary particles, in a broad sense, represents a body of knowledge of the physical world. It synthesizes classical mechanics, classical electromagnetism, special relativity, quantum mechanics, quantum electrodynamics (QED), quantum chromodynamics (QCD) and, together with general relativity, serves as a basis for fields from chemistry to astrophysics and cosmology. Gravity is yet to be incorporated into this picture. The final stages of the standard model were formulated through the interplay of experimental observations, theoretical insight, and phenomenology during the 1960s, '70s, and '80s. Among the key experimental developments were the search for the flavor-changing neutral current (FCNC) process in the 1960s (~ 1963), the discovery of the weak neutral current (WNC) in 1973, and the discovery of some of the elements of the three families of quarks (strange, charm, and beauty) and of leptons (v_μ and τ), as well as the discovery of the W and Z particles in 1983. (The quark is a fundamental point-like constituent of protons and other hadrons.) Thus, over a period of approximately two decades, the key experiments were carried out that covered a center-of-mass energy range of approximately ½ GeV (K mass) to 630 GeV – three orders of magnitude. These remarkable experimental developments were matched with the brilliant theories of Weinberg [1.1], Salam [1.2], Glashow [1.3], J.C. Ward, B. W. Lee, Higgs [1.4], Veltman [1.5], 't Hooft [1.6], S. Bludman and others, to formulate the SU(2) × U(1) model of the electroweak interaction (to be discussed in Chapter 4).

It can now be said that there have been three major advances in physics during the past 100 or so years. The theory of relativity can be dated as being formulated during 1905-1919; the development of quantum mechanics during 1913-1930; and the experimental and theoretical formulations of the standard model of elementary particles (called "rise of the standard model"), which began around 1960 and continued until approximately 1983. Of course, there have been many other important advances in nuclear physics, condensed matter physics, cosmology, and so forth. However, the three advances mentioned above are distinguished by having led to a nearly complete picture of a major area of fundamental science that has stood the test of time and verification by many experiments. The period of the rise of the standard model was full of major theoretical and experimental discoveries. Among the most important events during this period was the discovery of the WNC in 1973.

All three of the breakthroughs mentioned above have one thing in common–the basic theory was tested intensively by experiment very early. For general relativity, the observed deflection of starlight by the Sun in 1919 was a crucial experiment; for quantum theory, it was the study of various atomic-physics systems, as well as the observation of the particle-like nature of light (Compton effect) and the wave-like nature of matter (2-slit electron-interference pattern). For the standard model of elementary particles, the key period of experimentation was between 1974 and 1978. It was during this period that the newly discovered WNC was quickly confirmed and studied in numerous experiments around the world. Given the breadth of the physical phenomena involved, this was a truly remarkable period.

1.2.   THE EARLY OBSERVATIONS OF WEAK INTERACTIONS

Radioactivity and the weak interactions were discovered almost by accident in 1896 when H. Becquerel placed a sample of uranium ore and a photographic plate in the same drawer. He quickly recognized that the ore produced the fogging and later that β particles were emitted (electrons). During the period of 1896 to about 1930, the major studies were of the energy spectrum of the β rays that were emitted in the process. In 1914, J. Chadwick showed that the β spectrum was continuous [1.7]. We now know that the charged current interaction was being observed by the production of an electron and antineutrino pair. During the 1920s, the major question was whether the β rays were emitted with a unique energy or with a continuous spectrum of energies. This was not a trivial issue, since all of the experiments had some degree of material, and the energy loss of the β rays in this material was not really known. Finally, by the end of the 1920s, the β spectrum was determined to be continuous, which set off another controversy about the physical origin of this spectrum. Many people, including N. Bohr, believed that energy conservation was violated for these processes. In a now famous letter, W. Pauli proposed that the explanation was due to the emission of a neutral particle, named "neutrino" (little neutral one) by E. Fermi, in the process. During the period from 1933 to 1948, the importance of both the neutrino and the weak interaction was realized by Fermi, Pauli, and others [1.8],[1.9].

Note that none of the experiments from 1896 to 1960 were sensitive to the WNC interaction (final states with neutrino–antineutrino pairs, for example). That required either the use of strange particles or intense neutrino beams, which only became available 60 years after the initial discovery of the weak interaction, as we shall discuss in detail in Chapters 3 and 5.

1.3.   BRIEF HISTORY OF THE CONCEPT OF WNC

Let us trace the early history of the WNC concept. The earliest ideas concerning the unity of weak interactions and WNC seems to have been described by the Swedish physicist, Oscar Klein, in 1938 in a very obscure paper presented to the League of Nations meeting in France and in pre-wartime Warsaw, Poland [1.10]. In the 1940s, T. D. Lee of Columbia University, M. Rosenbluth of the University of California, San Diego, and C. N. Yang of Stony Brook Laboratory, invented the concept of the W, the carrier of the weak force [1.11]. It was not until the 1960s that S. Glashow of Harvard University [1.3], S. Weinberg of the University of Texas [1.1], and A. Salam of the Theoretical Center at Trieste, Italy [1.2] developed the modern concept (the GWS model) of the W and Z bosons, implying the existence of WNC in Nature.

By 1973, when the Fermi National Accelerator Laboratory (FNAL or Fermilab) neutrino beam was just being commissioned, and the very large bubble chamber, Gargamelle (GGM), was taking copious data at CERN, the issue of the existence of the WNC was coming to a boil! It was in that year, starting with the GGM and the Harvard–Penn–Wisconsin (HPW) experiments at FNAL, that WNCs were observed. (This will be discussed in detail in Chapter 5.)

The search for FCNC was carried out with strange particles, mainly in the period 1963–1970. The first definitive search for FCNCs that change the quark family was undertaken by this author and his colleagues using the strange particle beam at Lawrence Berkeley Laboratory (LBL) [1.12]. In the mid-'70s, the absence of FCNC prompted the concept of "natural family conservation" for FCNCs [1.13] and the Glashow–Iliopoulos–Maiani (GIM) mechanism [1.14]. This has come to be a key development in the rise of the standard model and strongly limits the types of quarks that exist in Nature. (This is an example of how null experiments can strongly influence the direction of a field of science.)

During the period 1974–1978, a remarkable synthesis occurred: the WNC parameters were found to fit precisely the expectation of the GWS model! And the key parameter, $\sin^2\theta_W$, of the standard model was born (discussed in Chapter 6).

1.4. THE STANDARD MODEL, THE INTERMEDIATE VECTOR BOSON, AND THE ELECTROWEAK FORCE

The standard model of elementary particles was created in the 1960s when the quark model was invented, the SU(2) × U(1) theory of GWS formulated, and the first evidence for the concept of quarks discussed (the first evidence for partons came from the SLAC–MIT deep-inelastic scattering experiments) [1.15]. The concept of the intermediate vector boson (IVB) dates back to the late 1930s and '40s and, starting in the 1950s, it was the subject of intense speculation and experimental search. By the end of the '50s, the limit on the IVB mass was of order 2 GeV. The earliest neutrino experiments at Brookhaven National Laboratory (BNL), and later at CERN, put slightly better limits. The 200-GeV machine at Batavia, Illinois at FNAL was constructed largely to discover the IVB and, in principle, it could have been at the unitary limit of the electroweak scale of ~300 GeV. In Fig. 1.1, we attempt to indicate some of the most important developments in the rise of the standard model and the discovery of the electroweak force.

While the IVB was not directly observed, the phenomenology of weak interactions strongly pointed to such an object. (We will discuss the discovery of the $W^\pm$ and Z^0 in Chapter 7.) Thus, by 1970, there was a feeble limit on the W mass,

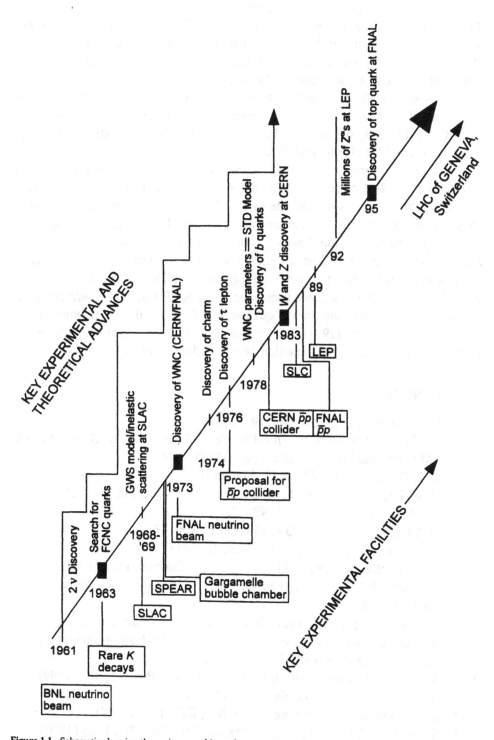

Figure 1.1. Schematic showing the various machines, detectors, theoretical advances, and discoveries that framed the rise of the standard model.

no evidence for WNC, strong constraints on FCNC processes, and general excitement about the new high-energy machines at Batavia (Fermilab) and CERN. No one could guess the great revolution that lay three years ahead. By 1996 more than 10 million Z^0's have been observed, and the model of GWS fits the data so well that it seems safe to say that this revolution is here to stay (Chapter 8).

However, the key idea of this theory – the Higgs mechanism – has yet to be really tested [1.4] and must await the completion of new machines. Thus, it will only be with the discovery of the Higgs boson that we will be able to state that the full understanding of the electroweak force is here (Chapter 9).

The earliest proposal (~1933) for the weak interaction form was made by Fermi, the so-called current–current four-fermion interaction. This was a very radical concept at the time, since he included five different types of interactions (axial vector, vector, tensor, scalar, and pseudoscalar). It included, in principle, the possibility of parity violation if two types of interactions [e.g., vector (V) and axial-vector (A)] were present. It is assumed that Fermi believed that only the vector interaction would be important, since the form of the weak interaction was made as an analogy to the electromagnetic interaction, on which Fermi had become an expert in the early days after the Dirac equation was formulated. Later when parity violation was discovered (1957), the essential form of the interaction was determined to be the V-A four-fermion interaction 24 years after the initial work!

REFERENCES

1.1. S. Weinberg, *Phys. Rev. Lett.*, **19**, 1264 (1967).

1.2. A. Salam, *Proceedings of the VIII Nobel Symposium*, N. Svartholm, ed. (Almquist and Wiksells, Stockholm, 1968), p. 367.

1.3. S. L. Glashow, *Nucl. Phys.*, **22**, 579 (1961).

1.4. P. W. Higgs, *Phys. Lett.*, **12**, 132 (1964); *Phys. Rev.*, **145**, 1156 (1966).

1.5. M. Veltman, *Acta Phys. Pol.*, **8B**, 475 (1977).

1.6. G. 't Hooft, *Nucl. Phys.*, **33B**, 173 (1971); **35B**, 167 (1971).

1.7. J. Chadwick, *Verh. Dtsch. Phys. Ges.*, **16**, 383 (1914).

1.8. E. Fermi, *Ric. Sci.*, **2**(2), 491 (1933); *Nuovo Cimento*, **11**, 1 (1934).

1.9. W. Pauli, *VII Congrès de Physique Solvay 1933* (Gauthier-Villars, Paris, 1934), p. 324.

1.10. O. Klein, in *Les Nouvelles Théories de la Physique, Proceedings of a Symposium held in Warsaw, May 30–June 3, 1938* (Institut International de Coopération Intellectuelle, Paris, 1939), p. 6.

1.11. T. D. Lee, M. Rosenbluth, and C. N. Yang, *Phys. Rev.*, **75**, 905 (1949).

1.12. V. Carmerini *et al.*, *Phys. Rev. Lett.*, **13**, 318 (1964).

1.13. S. Glashow and S. Weinberg., *Phys. Rev.*, **D15**, 1958 (1977).

1.14. S. Glashow, J. Iliopoulos, and L. Maiani, *Phys. Rev.*, **D2**, 1285 (1970).

1.15. M. Breidenbach *et al.*, *Phys. Rev Lett.*, **23**, 935 (1969).

CHAPTER 2
FORTY YEARS OF WEAK INTERACTIONS
(CHARGED CURRENTS)

Fermi to Fermilab (1933 – 1973)

2.1. THE STUDY OF β DECAY AND THE SPACE-TIME NATURE OF THE WEAK INTERACTION

The initial formulation by E. Fermi of the so-called "four-fermion" theory (or point-like theory) of the weak interaction was remarkably successful [2.1]. The theory, however, allowed for many kinds of space-time interactions (*e.g.*, tensor, scalar, pseudoscalar, vector, and axial vector). While it was known that there were problems with this model (later cured by the IVB and the GWS theory [2.2]), they were overlooked during the period from about 1940 to 1956, when the study of β decays of many nuclei were used to attempt to determine the space-time interaction. The history of this period is very complex with many false starts, poor experiments, and bad guesses. Nevertheless, there was progress. The account by C-S. Wu, reprinted here [paper (A)], starts with the work of J. Chadwick in 1914 [1.7] and gives a brief history of this period. As she points out, the whole enterprise took almost three decades. One of the key missing measurements at the time was the use of polarized targets, which only became available in the mid-1950s. Had these been available earlier, the correct interaction structure might have been determined at that time.

2.2. THE *V-A* THEORY IS CHOSEN BY THE DISCOVERY OF PARITY VIOLATION IN 1957

In the early 1950s, a puzzle with strange particle decays led D. Dalitz to question some of the basic assumptions of the weak interaction – invariance under parity transformations [2.3]. T. D. Lee and C. N. Yang, in a beautiful and classic paper, proposed that the weak interaction might not be invariant under parity transformation (parity violation) and suggested in detail the experiments that might reveal this violation [2.4]. In less than a year, they were proven correct [2.5]–[2.8], a remarkable achievement. In a very short time, this discovery was used to identify the space-time structure of the weak interaction. Reprinted here (B) is an account of this development by E. Sudarshan and R. Marshak, two of the pioneers of this activity along with others [2.9]–[2.12].

The most interesting aspect of this development was the discovery that the weak interaction is a mixture of *V* and *A* (*V-A*), which leads to maximum parity violation [2.9]–[2.11]. Thus, parity violation is a key indicator of the weak force and not a tiny side effect. In a very short time, the work of four to five decades was completed, and large numbers of weak decays were now understood – a completion of the initial work of Fermi 25 years later!

2.3. THE DISCOVERY OF THE ELECTRON NEUTRINO AND OF OTHER NEUTRINO FAMILIES

In 1955, the key ideas of the weak force were the space-time structure and the existence of an as-yet undiscovered neutral particle that was presumably emitted in every β-decay event. The neutrino concept was perhaps the boldest idea ever put forward in science: to invent a whole new, unseen object to explain the continuous nature of the β spectrum [2.13]. We now know that neutrinos are the most abundant particles in the Universe. The discovery of antineutrinos was a landmark in experimental science [2.14]. The account of this discovery by Fred Reines is reprinted here (C). In a long overdue reward, he received the 1995 Nobel Prize for this discovery.

Neutrino studies took a completely new turn in the late 1950s, when M. Schwartz [2.15] and B. Pontecorvo [2.16] suggested producing intense neutrino beams at the new high-energy proton synchrotrons. [Some of the references to this development can be found in the reprinted *Physics Today* article (D) by Cline, Mann, and Rubbia (CMR).] Much of the rest of this book concerns the use of high energy neutrinos to uncover the electroweak force. The first major discovery with these neutrinos was the muon neutrino at BNL, and we reprint this discovery paper here (E) [2.17].

The experiment of Schwartz, Sternberger, and Lederman that led to the discovery of the ν_μ was very important in many ways to the future of particle physics because:

1. It was the first acceleration neutrino experiment, setting the stage for future neutrino experiments (*e.g.*, the discovery of the WNC in 1973).
2. It established methods to detect muons from neutrino interactions; and it was also very important for the WNC discovery, since the absence of muons was involved.
3. The discovery of the muon neutrino led to the design of the ν_μ beams, which have been the workhorses of neutrino physics ever since!

This was a seminal experiment in elementary particle physics.

2.4. HIGH-ENERGY NEUTRINO INTERACTIONS AND THE WEAK INTERACTION

By 1960, the space-time structure of the weak interactions was known, and there was little more to learn from β decay. What was needed was a completely new tool to study the weak interaction in the high-energy region, where it was suspected that the Fermi theory no longer held.

Two major breakthroughs were needed to move forward:

1. The development of intense high-energy neutrino beams at Fermilab and CERN, and
2. The discovery of point-like structures in the nucleon, allowing for large, weak cross sections at high energy [2.18].

These advances will be recounted later in the book; a concise summary of both developments can be found in the reprinted *Physics Today* article (D) by CMR, which gives some insight into the thinking at that time.

By 1973, the $V-A$ interaction was established, the muon neutrino confirmed, and the point-like structure in the nucleon established. The stage was set for the major discoveries about the weak force that had eluded scientists for 76 years – the weak neutral current!

REFERENCES

2.1. E. Fermi, *Ric. Sci.*, **2**(2), 491 (1933); *Nuovo Cimento*, **11**, 1 (1934).

2.2. S. L. Glashow, *Nucl. Phys.*, **22**, 579 (1961); S. Weinberg, *Phys. Rev. Lett.*, **19**, 1264 (1967); A. Salam, in *Proceedings of the VIII Nobel Symposium*, N. Svartholm, ed. (Almquist and Wiksell, Stockholm, 1968), p. 367.

2.3. R. Dalitz, *Philos. Mag.*, **44**, 1068 (1953); *Phys. Rev.*, **94**, 1046 (1954).

2.4. T. D. Lee and C. N. Yang, *Phys. Rev.*, **104**, 254 (1956).

2.5. C-S. Wu, E. Ambler, R. W. Hayward *et al.*, *Phys. Rev.*, **105**, 1413 (1957); also see article (A).

2.6. R. L. Garwin, L. M. Lederman, and M. Weinrich, *Phys. Rev.*, **105**, 415 (1957).

2.7. J. I. Friedman and V. L. Telegdi, *Phys. Rev.*, **105**, 1681 (1957).

2.8. M. Goldhaber, L. Grodzins, and A. Sunyar, *Phys. Rev.*, **109**, 1015 (1958).

2.9. R. P. Feynman and M. Gell-Mann, *Phys. Rev.*, **109**, 193 (1958).

2.10. E. C. G. Sudarshan and R. E. Marshak, *Phys. Rev.*, **109**, 1860 (1958); also see article (B).

2.11. J. J. Sakurai, *Nuovo Cimento*, **7**, 649 (1958).

2.12. S. S. Gerschtein and J. B. Zeldovick, *Zh. Eksp. Teor. Fiz.*, **29**, 698 (1955).

2.13. W. Pauli, *VII Congrès de Physique Solvay 1933* (Gauthier-Villars, Paris, 1934), p. 324.

2.14. F. Reines and C. L. Cowan, Jr., *Phys. Rev.*, **114**, 273 (1959); also see article (C).

2.15. M. Schwartz, *Phys. Rev. Lett.*, **4**, 306 (1960).

2.16. B. Pontecorvo, *Sov. Phys.-JETP*, **10**, 1236 (1960).

2.17. G. Danby *et al.*, Phys. Rev. Lett., **9**(1), 36 (1962) [article (E)].

2.18. M. Breidenbach *et al.*, Phys. Rev. Lett., **23**, 935 (1969).

CONTRIBUTIONS OF β-DECAYS TO THE DEVELOPMENTS
OF ELECTRO-WEAK INTERACTION

Chien-Shiung Wu
Columbia University, New York, N.Y. 10027

INTRODUCTION

It is indeed exciting to gather in this famous Wingspread Hall
in the prairie, built by the famed American architect Frank Lloyd
Wright, to reminisce with so many pioneer physicists from different
lands; physicists who have played major roles in the developments
of the weak interactions which have now been finally unified
through a gauge theory, with the electromagnetic force to establish
the fundamental natural interaction known as the "Electro-Weak"
Interaction.

The most exciting conversations permeating the conference
were mainly on the spectacular discoveries of the charged vector
bosons $W^{\pm}$ and neutral Vector boson Z° (UA1 and UA2) in the $(p\bar{p})$
collider of CERN in 1983. The collider had reached, at that time,
into a new domain of particle energies at $\sqrt{S} = 540$ Gev. The two
important discoveries were:

$$\bar{P} + P \rightarrow W^{\pm} + X \qquad \bar{P} + P \rightarrow Z^{\circ} + X$$
$$\quad\hookrightarrow e^{\pm} + (\bar{\nu}) \quad \text{and} \quad \hookrightarrow e^{+} + e^{-} \text{ or } \mu^{+} + \mu^{-}$$

The observed properties of these bosons $W^{\pm}$ and Z° turned out
to be in excellent agreement with that predicted by the unified
gauge theory. The comparison between experimental results and
theoretical expectations can be seen in Table 1.

Contents   Page

Table I Comparison of measured and predicted
properties of $W^{\pm}$ and Z°

	Phys.Lett. UA1 126B,139 122B,108	Phys.Lett. UA2 129B,130 122B,476	Standard Model with $Sin^2\theta_w=0.217\pm0.014$
M_w(Gev)	80.9±1.5±2.5	81.0±2.5±1.3	$83.0^{+2.9}_{-2.7}$
M_z(Gev)	95.6±1.5±2.8	91.9±1.3±1.4	$93.8^{+2.4}_{-2.2}$
M_z-M_w(Gev)	14.7±2.1±0.4	10.9±2.8±0.2	10.8±0.5
$Sin^2\theta=\left(\dfrac{38.65Gev^2}{M_w}\right)$	0.228±0.008±0.014	0.228±0.014±0.007	0.217±0.014
ρ	0.928±0.038±0.016	1.004±0.052±0.010	

The unified gauge theories seemed to work extremely well for
the unified electromagnetic and weak interactions as shown in
Table 1. This fine agreement may also imply that there are
possible existences of deeper symmetries for all elementary forces.
At present, however, only further persistent tests on the CERN
$P\bar{P}$ collider or on other future higher energy accelerators (such
as SSC) may reveal what the future holds for the deeper unified
guage theories for all subnuclear phenomena. It reminded us that
at the very beginning of the development of nuclear beta decay,
the β-decay picture was rather confusing and controversial. By
the time of the late forties, the correct types of β-energy
spectrum for the allowed and forbidden transitions were observed
which could be precisely interpreted by the Fermi theory of β-decay
Furthermore, the concept of the Universal Fermi Interaction
was suggested and well-received in the late fifties to cover all
weak interactions involving four Fermion particles. Soon after
the parity non-conservation in Weak Interactions was discovered in
1957, it was clarified that the interaction forms in weak interactions are (V-A). Please refer to the genetic tree, Fig.(6), and
the extremely close equality between $g_v^\beta \cong g_\mu$ could be reasonably
interpreted by the renormalizability of the proposed Conserved
Vector Current theory (CVC theory).
It might be appropriate to emphasize here that a series of
experimental verifications, such as the β-decay of ^{12}B and

$^{12}N^{(Wu-77)}$, the rate of Pion β-decay (Ba-63, DeP-63) and the foremENTIONED equality between $g_v^\beta \cong g_\mu$ were the first quantitative tests linking the electromagnetic observables with the weak interaction observables. Their satisfactory agreements with the CVC predictions provided a strong incentive and stimulus to a later persistent search for a unifying theory for the Electroweak Interaction. There was, however, a short period between 1965 to 1977 when the field of β-decay was plagued by the question whether the second class current$^{(We-58)}$ existed in the weak interactions or not, but it became quite clear in 1977, after some detailed and critical survey in experimental results, that the evidences were consistent with the absence of the second class current$^{(UN1-77)}$ and in strong support of the CVC theory.$^{(UN1-77)}$

The basic strategy of developing the unified gauge theories for electromagnetic and weak interactions progressed, at first rather slowly and could be attributed to Glashow$^{(Gl-61)}$, Weinberg$^{(We-67)}$ and Salam$^{(sa-68)}$. Their construction of a renormalizable theory was based on the notion of spontaneously broken gauge symmetry. The resulting theory was shown by 'tHooft$^{('tHo-71)}$ to be a renormalizable quantum field theory. Finally, the inclusion of quarks in this theory was achieved by Glashow, Illopoulos and Maiani$^{(Gl-70)}$. As a result, the electomagnetic and weak interactions are unified into a gauge theory, with three intermediate vector bosons and the photon as gauge particles which were confirmed in the CERN's p$\bar{p}$ collider's beam in 1983 by two gigantic collaborations known as UA1 and UA2 with a total number of close to 200 physicists working together.

Now let's look back to see what developments in β-decays have contributed to the development of the Electroweak Interaction

THE CORRECT β-ENERGY SPECTRUM OBSERVED AT MORE THAN FIVE DECADES AFTER THE DISCOVERY OF β-DECAY

From the very beginning, the development of nuclear beta decay is full of surprises and subtleties. Several seemingly convincing and essential experimental resutls of β-decays turned out to be wrong and led down to a tortuous path. For example, that the discrete electron-line spectrum seen in β-spectrum had been mistakenly interpreted as electron transitions in β-decays was eventually refuted. By using an electron counter as electron-detector in a magnetic spectrometer, it was possible to show in 1914 by Chadwick$^{(Ch-14)}$ see Fig.(1) that despite the prominent appearance of the sharp lines on a photographic plate, it was the continuous distribution of electrons that contributed the

major portion of the disintegration electrons. This finding
raised a deeper question in 1922 by Lise Meitner[Me-22]. If the
states of a nucleus are quantized, why should the electrons have
a continuous energy distribution? What, then, caused the
continuous energy distribution? Could it be introduced after
the actual expulsion from the nucleus by loss of energy in
collisions with electrons? If this energy variation was somehow
introduced after the electron's emission from the nucleus, it
would imply that the electrons all had the same disintegration
energy, equal to the upper limit of the spectrum. Most physicists
at that time were inclined to believe in this explanation. But
Ellis[El-27] held his ground and believed in the continuous
energy distribution of the primary β-rays. This idea was eventually
put to a test by determining the average energy of disintegration
of a β-emitter in a microcalorimeter in 1927, when Ellis and
Wooster obtained their result,[El-27] which was in striking
agreement with the mean energy of the RaE β spectrum. In reacting
to this startling result, they remarked[El-27] on how contrary this
might appear to the general principles of quantum theory. Three
years later, in 1930, Meitner and Orthmann[Me-30] published their
results which were the same as those of Ellis and Wooster. Pauli
was pleased because the Polemics was happily resolved by an
experiment. To this experiment, Pauli also commented: "It's
Meitner's Experimental Masterpiece, because she also reported on
additional important information which showed no continuous
γ-rays were present in RaE decay." This episode clearly illustrated
that the fine competitive spirit and the great emphasis in
experimental results were already deeply rooted in the early stage
of nuclear physics.

    Undistorted β-energy spectra[Wu-50] which were in good
agreements with the theoretical predictions (Fermi's theory of
β-decay)[Fe-34] were not observed until four or five decades after
the discovery of the β-decay as the β-particles, being fast
moving electrons, light and charged, therefore easily scattered
in their sources or on their flight paths. Fortunately, the
development of the nuclear reactor was successfully completed soon
after the second World War and many kinds of high specific
β-sources became plentiful from reactors. The required experimental
conditions for making good β-spectrum measurements were
dramatically improved. There were (1) thin and uniform
β-sources fabricated possibly by evaporation; (2) Iron-free
magnetic spectrometers were employed; (3) cylindrical-type
magnetic spectrometers with large solid angles were designed and
built and (4) Improved thin window and energy sensitive
β-detectors were used.

Eliminating the excess low-energy electrons of the allowed
β-spectrum in the early fifties removed the obvious discrepancies
standing in the way of acceptance of the Fermi theory of β-decay.
By then, the proposed Konopinski-Uhlenbeck interpretation[KO-41] was
definitely ruled out. The absence of the Fierz interference terms
in the spectrum shows that the β-interaction must be either
(V,A) or (S,T).

However, the above conclusions could hardly be considered a
crucial test of the Fermi theory, because the shape of an allowed
spectrum is given by a statistical factor $PE(E_0-E)^2$, which is the
calculated phase volume corresponding to the sharing of the
disintegration energy between the electron and the neutrino.
Nothing about this shape is very sophisticated!

UNIQUE FORBIDDEN SPECTRA

On the other hand, there also must be some "forbidden"
spectra radically different from the allowed shape. Such spectra
were indeed being uncovered in rapid succession in various
laboratories[La-49b; Wu-50,55] starting in 1949, and they looked
altogether different from the familiar allowed shape.

It was good to see a real forbidden spectrum after seeing
so many allowed ones! And the maxim "It never rains but it
pours" seemed to apply in this case too. Not only were abundant
examples of unique first forbidden spectra observed, but even
the rare types of unique second and third forbidden spectra were
soon identified. The experimental observation of the three unique
first, second and third forbidden spectra of Y^{91} [La-49,Wu-50],
Be^{10} (Fe-49, Fel-50) and K^{40}(Fe-49,52) respectively as theoretically predicted, is a triumphant proof for the theory of
β decay. See Fig.(2) → Fig.(4).

"UNIVERSAL FERMI INTERACTION"

Begun in 1948, the μ-decay, μ-capture etc. were discovered
and investigated one by one. All of them can be described by
approximately the same coupling constant "g" as that of β-decay.
The similarity in strength among them gave rise to the notion of a
"Universal Fermi Interaction"
which should be on the same footing as the strong nuclear, the
electromagnetic and the gravitational force. That the significance
of this similarity among the β-decay, μ-decay and μ-capture
was conceived but not apparently evident at first, can be found in
Fermi's Sillman Lecture Series [Fe-50] at Yale University in 1950
in which Fermi repeatedly emphasized by saying:

> "The similarity among these three
> coupling constants is not an
> accident, but has some deep
> meaning which however is not
> understood at present."

Then he again stressed:

> "That this fact was noticed
> independently by Klein[K1-48];
> Tiomno and Wheeler[T1-49];
> Lee, Rosenbluth and Yang[Le-49]
> and other authors, probably is
> not a coincidence, although at
> present, its significance is
> unknown."

Of course, there was also Puppi[Pu-49] whose name was
associated with the Puppi-Triangle illustrating the similarity
relationship between the three interactions.

FIG. 5 PUPPI-TRIANGLE

The exciting reminiscences about how the notion of the
Universal Fermi Interaction occured and developed among various
theoretical groups were among the most discussed topics of this
conference in the first two days. It showed that the idea of
the 'UFI" was rooted early and deep in the developments of Weak
Interactions.

THE (V-A) INTERACTION

Any universality should imply not only the equality of the
coupling constants but also the similarity of the structure of the
interactions. For a long time, there was no conclusive
evidence to substantiate the latter. In the classical studies of
β-decays, the two possibilities of identifying β-interactions
either V,A or S,T from the absence of the Fierz term in the
spectrum could be resolved by the (β-ν) angular correlation
experiments. In an allowed (Gamow-Teller)β-transition,
($\Delta I=1$, No), the angle between (β,ν) is quite different depending
on the β-interaction and whether it is Tensor or Axial-Vector.

Unfortunately, in 1952, Rustad and Ruby[Ru-53] published their experimental results on (β-ν) relations in ^{6}He decay and gave the distribution for the angle θ between (β-ν) yielding*

$$\lambda_{exp} = +0.34 \pm 0.09$$

while theoretically, one expects

$$\lambda_{theo} = \begin{cases} +1/3 \text{ for T} \\ -1/3 \text{ for A} \end{cases}$$

The results indicated definitely that the β-interaction in (G-T) type transition should be T. The acceptance of this result ruled out the theoretical idea of intermediate vector boson as the Tensor Coupling cannot be transmitted by a spin 1 field; because of the absence of derivative coupling and it cannot be transmitted by a spin 2 field. Fortunately, this misrepresented situation was soon cleared away under the tremendous activities that followed right after the discovery of non-conservation of P and C in β-decay as shown in Fig.(6). Re-examining the anlaysis of Rustad's ^{6}He experiment, it was found that due to a very small amount of residue ^{6}He gas in the recoil ion chamber, the proper conclusion of the experiment[Wu-58] should have been A instead of T. The revised conclusion was written up and forwarded to Rustad and Ruby and also Feynman (he was scheduled to speak at an APS meeting in New York City).

OVERTHROW OF THE LAW OF PARITY AND CHARGE CONJUGATION IN WEAK INTERACTIONS[Le-56,Wu-57]

The overthrow of the law of parity and charge conjugation in the weak interactions occured at the end of 1956 and the beginning of 1957, first in nuclear β-decay, then in rapid succession in ν and μ decays[Fr-57] and also in the decay of the charged K mesons; it was like a bandwagon at a political convention. It prompted me to say during a talk at the International Conference on High Energy Physics at Rochester in 1957 that "I am here, on the strength of the Weak Interactions." How true it was then, and still is. The story of the overthrow of parity in 1957 is now a familiar one with physicists, so I will not bore you with it here. However, with the help of the theory of two components of neutrino, i.e. the handedness of the neutrino, together with all of the traditional methods that classic β-decay can provide, the form of the β-interaction was conclusively determined as (V-A) as shown by the genetic tree (Fig. 6).

*see Appendix

THEORETICAL FORMULATION OF THE UNIVERSAL
(V-A) FERMI INTERACTION

If the interaction were pure (V-A), it would be interesting to explore some possible theoretic arguments that would lead to such a linear combination. The (V-A) form was indeed reached independently by three different theoretic approaches, all based on the principle of representing the four component spinor ψ in terms of two component spinors ϕ_+ and ϕ_-. To allow only one of the two-component spinors to appear in the reaction, different hypothetic principles were proposed to justify its restriction.

Table II Hypothetic Principles for (V-A) Interaction

Authors	Conjectures
Sudarshan and Marshak	Chirality Invariance
Feynman and Gell-Mann	The Two-Component Formulation of Dirac Spinors
Sakarai	Mass-Reversal Invariance

The interesting consequences of these assumptions are that the interaction is now uniquely determined to be (V-A).

IS $g_v^\beta \cong g_\mu$ AGAIN A PUZZLE?

However, the vector-coupling constant g_v^β in β-decay is found to be not only remarkable constant among all superallowed $0^+ \rightarrow 0^+$ β-transitions but also very nearly equal to that of the Fermi constant g_μ of μ-decay within 1-2%. However, this excellent agreement turns out to be by no means a blessing, but rather a puzzle! Why? A proton is not just a simple Dirac proton, but, rather, it is dressed by a meson cloud! Then, why should the meson cloud have the same weak interaction as the bare Dirac particle ."μ"? In other workds, there is no a priori reason why the coupling constants g_v^β and g_μ should be identical!

THE CONSERVED VECTOR CURRENT (CVC) HYPOTHESIS

To explain the unexpectedly good agreement between the vector coupling constant g_v^β in β-decay and the Fermi constant g_μ of μ-decay within 1 or 2%, Feynman and Gell-Mann$^{(Fe-58)}$ and earlier, Gershtein and Zeldovich$^{(Ge-55)}$ proposed the conserved vector current theory (CVC Theory) based on the analogy in electromagnetism, where the observed coupling strength e with electromagnetic field is the same for all particles coupled. The universality of electric charge follows because the <u>electromagnetic current is conserved</u>. If the weak vector current is similarly conserved, the vector coupling constant will be a universal constant. Let us look at Fig. (7). The Feynman-Gell-Mann hypothesis amounts to the assumption that the total isotopic spin current, including both nucleonic and pionic terms, is conserved.

WEAK MAGNETISM TERM

Although the effective coupling strength is not renormalized by pionic corrections in the CVC hypothesis, a nucleon also possesses a magnetic moment that is greatly altered by the pion cloud. Physically, for a given charge, the bare pion carried a large magnetic moment, due to its smaller mass. The CVC hypothesis implies that such anomalous magnetic moment terms (the weak magnetism terms) must also appear in β decay. As an example, Gell-Mann proposed to measure the $\beta^\pm$ spectra of ^{12}B and $^{12}N^{(Ge-58,59)}$. The ground state of ^{12}B, the 15-Mev state of ^{12}C, and the ground state of ^{12}N all have spin I=1, positive parity, and comprise the three members of the isotroplet (T=1). They emit β^-, γ, and β^+, respectively to reach the ground state of ^{12}C, which has isospin 0, spin 0 and positive parity as in Fig.(8). The β spectrum for (G-T) transitions is given by

$$P(E)dE=[f_A^2|\int\vec{\sigma}|^2/2\pi^3]pE(E_0-E)^2F_0(\pm Z,E)[1+R(E,E_0)]$$

$$X[1\pm(8/3)"a"E] \text{ for } \beta^\pm$$

$$\text{with } "a"=-(f_v/f_A)(1+\mu_p-\mu_n)/2M$$

The coefficient "a" in the spectral shape factor was measured by the Columbia group Lee, Mo and Wu$^{(Le-63)}$ in 1963 in an iron free magnetic spectrometer and the shape factor "a" was in fair agreement with that anticipated from the weak-magnetism term from (CVC). In 1977, the Heidelberg group,Kaina et.al.$^{(Ka-77)}$ reinvestigated the ^{12}B and ^{12}N spectra by using a NaI spectrometer and found the shape factors in fair agreements. At that time, our attention was called (Ca-76) to the numerical errors introduced into the calculated Fermi Functions by Bhalla and Rose (F_{B-R}) which we used in 1963

in our analysis of the β-spectrum. Since the end points energies branching ratios and calculated f-values (the integrated F functions), had all been revised by that time, Wu, Lee and Mo[Wu-77] decided in 1977 to completely reanalyze the original experimental data (1963). It turned out that although replacement of the erroneous Fermi Functions (F_{B-R}) by (F_{B-J}), indeed greatly reduces the slope of the

shape factors of ^{12}N and ^{12}B, the presently accepted values of the branching ratios and the integrated F functions "f" actually affect the slopes of the shape factors considerably but in the direction opposite to that of change of (F_{B-J}). The final

comparison between experimental and theoretical results are satisfactory as shown in Fig.(9). Furthermore, the theoretical group[KO-79] with Professor M. Morita at the University of Osaka,

restudied this weak magnetism problem in ^{12}B and ^{12}N by taking into account all higher order contributions for evaluation of the spectral shape factors. Theoretical values of the shape factors

for ^{12}B and ^{12}N were plotted and compared with the experimental results of Columbia and Heidelberg respectively as in Fig. (10a) and Fig.(10b). The overall agreements support strongly the validity of the   Conserved Vector Current Theory.

CONCLUSION

In the past fifty years, many exciting and beautiful revelations have taken place in the β-decay field. First, the surprising revelation of continuous distribution of β-energy spectrum was demonstrated, as a result of which, an outlandish zero restmass particle, later known as neutrino was proposed. *
When parity puzzle occured in the "τ-θ" problem, we found a way out by the startling discovery that the parity and charge conjugation both are not conserved, not only in nuclear beta decay

such as in ^{60}CO but in all weak interactions! That manifested all weak-interactions into a fundamental interaction.

To understand the close equality between $g_v^\beta \approx g_\mu$, it was

proposed that the vector current in the weak interaction is conserved as that of the electromagnetic current and the weak magnetism term is also observable. Isotriplet states T=1 in the ^{12}B and ^{12}N

decay by e^- and e^+ to ^{12}C, the $\beta^\pm$ spectra should show definite spectrum shape "a".

The theoretical branching ratio of π -decay according to CVC theory, should be

$$\text{Ratio}_{th} = \frac{\text{Rate}(\pi^{\pm} \to \pi^{\circ} e^{\pm} \nu)}{\text{Rate}(\pi^{\pm} \to \mu^{\pm} \nu)} = 1.07 \times 10^{-8} \; ;$$

The observed $\text{Ratio}_{exp} = (1.02 \pm .07) \times 10^{-8}$

The agreement is good.

It can be seen that we have been trotting through the β-decay field by carrying out pertinent tests linking the electromagnetic observables and the weak observables continuously and found the experimental verifications most encouraging and satisfactory. The spectacular production and identification of $W^{\pm}$ and Z° by the CERN collider in such a short period and with such an elegant and grand manner, has our highest admiration and warmest congratulations.

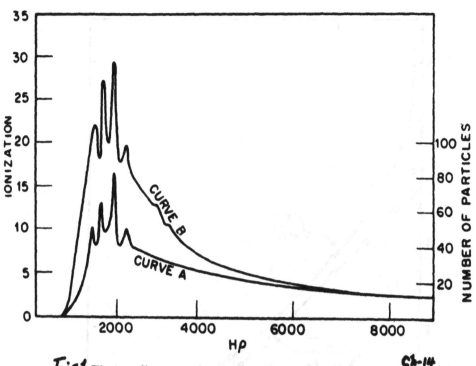

Fig1 Electron lines superimposed on the continuous β-spectrum $^{Ch-14}$
Curve *A* by counter method
Curve *B* by ionization chamber method

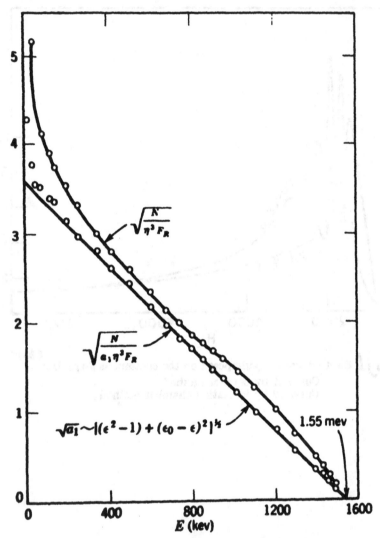

Fig 2 The Kurie plot of Y^{91} using the unique first forbidden correction factor. From (Wu-50).

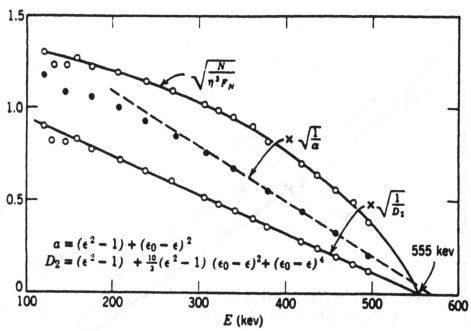

$a = (\epsilon^2 - 1) + (\epsilon_0 - \epsilon)^2$

$D_2 = (\epsilon^2 - 1) + \frac{10}{3}(\epsilon^2 - 1)(\epsilon_0 - \epsilon)^2 + (\epsilon_0 - \epsilon)^4$

$\sqrt{\frac{N}{\eta^3 F_N}}$

$\times \sqrt{\frac{1}{a}}$

$\times \sqrt{\frac{1}{D_2}}$

555 kev

E (kev)

Fig3 Kurie plots of the Be^{10} β spectrum. From (Wu-50).

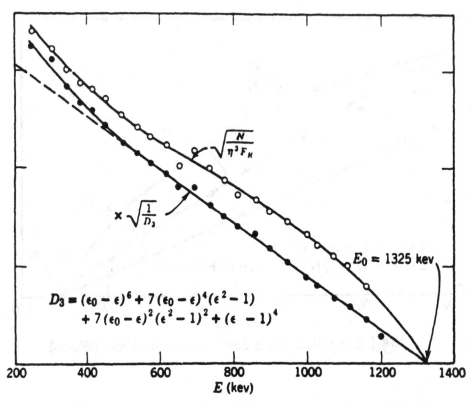

$$D_3 = (\epsilon_0 - \epsilon)^6 + 7(\epsilon_0 - \epsilon)^4(\epsilon^2 - 1)$$
$$+ 7(\epsilon_0 - \epsilon)^2(\epsilon^2 - 1)^2 + (\epsilon - 1)^4$$

Fig 4 Kurie plots of the K^{40} β spectrum. From (Wu-50).

Fig 5 The Puppi Triangle $^{Pu-48}$

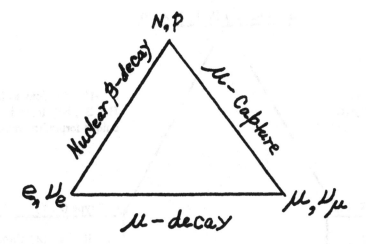

Fig 6
III The Form of the β Interaction

The genetic tree illustrating the important developments in determining the form of the β interaction can be represented as shown:

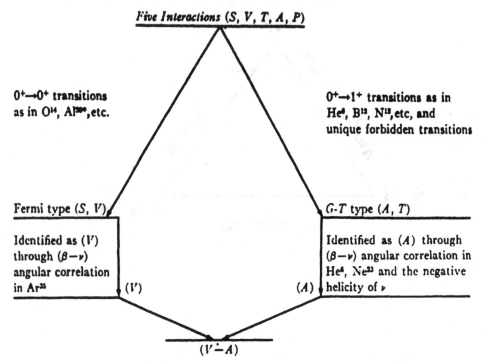

Five Interactions (S, V, T, A, P)

$0^+\rightarrow0^+$ transitions as in O^{14}, Al^{20+}, etc.

$0^+\rightarrow1^+$ transitions as in He^4, B^{12}, N^{12}, etc, and unique forbidden transitions

Fermi type (S, V)

G-T type (A, T)

Identified as (V) through $(\beta-\nu)$ angular correlation in Ar^{35}

(V)

Identified as (A) through $(\beta-\nu)$ angular correlation in He^4, Ne^{23} and the negative helicity of ν

(A)

$(V\overset{.}{-}A)$

The phase factor "$-$" is determined from the polarized neutron experiment and the factor $|C_A/C_V| = 1.18$ is derived from the ft values of the neutron and the average ft value of $0^+\rightarrow0^+$ transitions.

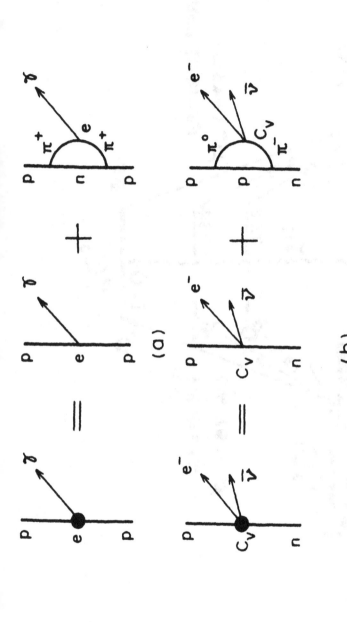

Fig 7 Gamma decay and beta decay in the conserved vector current (CVC) theory.
(a) Gamma decay of the physical proton is the sum of contributions from the bare proton
and the pion cloud. (b) Beta decay of the physical neutron is the sum of contributions
from the bare neutron and the pion cloud.

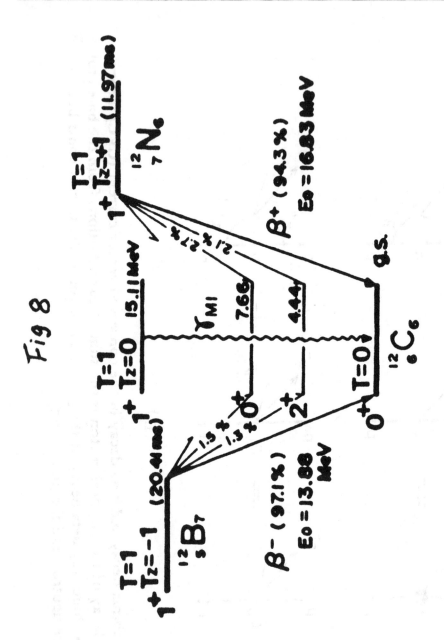

Fig 8

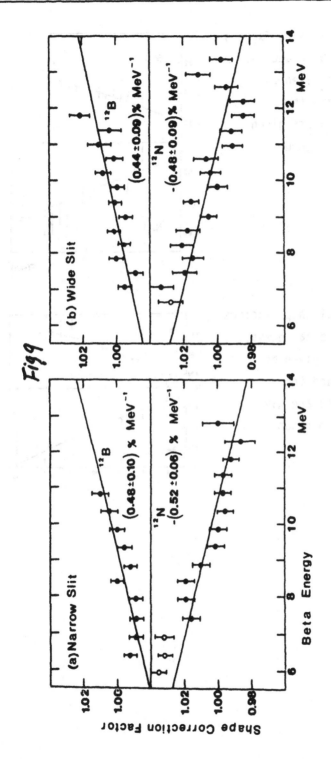

Fig 9

Fig. *10a* Spectral shape factors
for ^{12}B and ^{12}N beta decays.
Solid curves are given by
the theory,   and the
experimental data are given
by the Columbia   group.

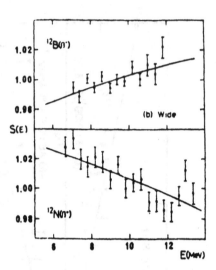

Fig. *10b* Spectral shape factors
for ^{12}B and ^{12}N beta decays.
Solid curves are given by
the theory,   and the
experimental data are given
by the Heidelberg group.

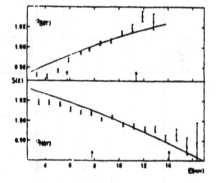

REFERENCES

UA1, Collaborations on CERN's P$\bar{\text{P}}$ Collider, Phys. Lett. 122B, 108;
126B,139 (1983).
UA2, Collaborations on CERN's P$\bar{\text{P}}$ Collider, Phys. Lett. 122B,476;
129B,130 (1983).
Wu-77, Columbia Group, Wu, C.S., Y.K. Lee and L.W. Mo Phys. Lett.
39, 72 (1977).
Ba-62, Bacastow, R. et.al. Phys. Rev. Lett. 9,400 (1962).
Dep-63, DePommier, P. et.al. Phys. Lett. 5,61 (1963).
We-58, Weinberg, S., Phys. Rev. 112, 1375 (1958).
UN1-77, C.S. Wu, Proceedings of the Symposium on Unification of
Elementary Forces and Gauge Theories, p.549-592, Edited by D.B.
Cline and F.E. Mills published by London, Harwood Academic Publishers
1978).
G1-61, S.L. Glashow, Nucl. Phys. 22, 579 (1961).
We-67, S. Weinberg, Phys. Rev. Lett. 19, 1264(1967).
Sa-68, A. Salam, Proceedings of the VIII Nobel Symposium, p.367,
Edited by N. Svartholm.
'tHo-71, G.'tHooft, Nucl. Phys. B.33, 173(1971); 35, 167 (1971).
G1-70, S.L. Glashow, J. Iliopoulos and L. Maiani, Phys. Rev. D.
2, 1285 (1970).
Ch-14, V. Chadwick, Phys. Gens. 16, 383 (1914).
El-27, C.D. Ellis and W.A. Wooster, Proc. Roy. Soc. (London) A117,
109 (1927).
Me-22, L.Meitner Z. Phys. 9, 131,145 (1922).
Me-30, L.Meitner, W. Orthmann, Z. Phys. 60, 143 (1930).
Wu-50, C.S. Wu, Rev. Mod. Phys. 22, 386 (1950).
Fe-34, E. Fermi, Z. Phys. 88, 161 (1934).
Ko-41, E.J. Konopinski and G.E. Uhlenbeck, Phys. Rev. 60, 308 (1941).
La-49, L.M.Langer, H.C. Price, Jr. Phys. Rev. 76, 641 (1949).
Fe-49,52, L.Feldman and C.S.Wu, Phys. Rev. 76, 698(1949); 76, 697
(1949).
Fe1-50,L.Feldman and C.S.Wu, Phys. Rev. 78, 318 (1950).
K1-48, O.Klein, Nature, 161, 897 (1948).
Ti-49, J.Tiomno and J.A. Wheeler, Rev. Mod. Phys. 21, 153 (1949).
Le-49, T.D. Lee, M. Rosenbluth and C.N.Yang, Phys. Rev. 75,905
(1949).
Pu-49,G.Puppi, Nuovo Cimento 6, 194 (1949).
Ru-53, B.Rustad and S.Ruby, Phys. Rev.
Wu-58, Written up as a Pupin-Report; it is also included in this
paper as an appendix.
Le-56, T.D.Lee and C.N.Yang, Phys. Rev. 104, 254(1956).
Wu-57, C.S.Wu, E.Ambler, R.W.Hayward, D.D. Hoppes and R.P. Hudson,
Phys. Rev. 105, 1413 (1957).
Ga-57, R.L.Garwin, L.M.Lederman and M.Weinrich, Phys. Rev. 105, 1415
(1957).
Fr-57,J.J.Friedman and V.L.Telegdi, Phys. Rev. 105, 1681 (1957).
Ma-57,R.E. Marshak, E.C.G. Sudarshan, Proc. Int. Conf. Elem.
Particles, Padua, Venice (1957), Phys. Rev. 109, 1860 (1958).

Sa-58, J.J. Sakurai, Nuovo Cimento 7, 649 (1958).

Fe-58, R.P. Feynman and M. Gell-Man, Phys. Rev. 109, 193 (1958).

Ge-55, S.S. Gershtein and I.A.B. Zel'Dovich, JETP. 29, 698 (1955);
Sov.Phys. JETP 2, 576 (Eng. transl).

Ge-58, M. Gell-Man, Phys. Rev. III, 362 (1958).

Ge-59, M. Gell-Mann and S.M. Berman, Phys. Rev. Lett. 3,99 (1959).

Le-63, Y.K. Lee, L.W.Mo, and C.S.Wu, Phys. Rev. Lett. 10, 253 (1963).

Ka-77, W. Kaina, V.Soergel, H. Thies and W.Trost, Phys. Lett. 70B,
411(1977).

Ko-79, K.Koshigiri, M.Nishimura, H.Ohtsubo and M.Morita, Nucl.
Phys. A319, 301 (1979).

Ca-76, F.P.Chalaprice and B.R. Holstein, Nucl. Phys. A273, 301(1976).

CU-173
AT 30-1 GEN 72

PUPIN CYCLOTRON LABORATORY
AND PEGRAM LABORATORY

A Critical Examination of the He^6 Recoil Experiment

of Rustad and Ruby[*]

by

C. S. Wu and A. Schwarzschild

April 21, 1958

COLUMBIA UNIVERSITY

DEPARTMENT OF PHYSICS

NEW YORK 27, NEW YORK

[*] Insert this paper aa an appendix.

A CRITICAL EXAMINATION OF THE He^6 RECOIL EXPERIMENT OF RUSTAD AND RUBY

C. S. Wu and A. Schwarzschild

Status of Parity-Nonconservation Experiments in β-Decay

By the early part of August 1957, the conflicting and confusing evidence on β-decay, which appeared earlier in the summer on the various types of parity experiments, was gradually discarded and cleared away. It was generally agreed then that there was no strong evidence against the two component theory of the neutrino.[1] (1) The asymmetrical distribution of β$^-$ particles from polarized nuclei undergoing G-T transitions, such as in Co^{60}, indicates a maximum possible asymmetry. $I(\theta)d\theta = C(1+\alpha \cos \theta) \sin \theta \, d\theta; \quad \alpha = -\frac{\tilde{v}}{c} \frac{\langle J_Z \rangle}{J}$. (2) β-γ (circularly polarized) angular correlation experiments have more or less definitely established that the interference term $\sim(C_S C_T'^* + C_S' C_T^* - C_V C_A'^* - C_V' C_A^*)$ between Fermi and G-T interactions does exist, and that in fact it approaches a maximum possible value in the cases of Sc^{46} and Au^{198}. (3) Furthermore, the longitudinal polarization of electrons and positrons from G-T transition, Fermi transition or Fermi and G-T mixed transition all exhibit an amount of polarization equal to $\mp v/c$ where the - sign is for electrons and the + sign is for positrons.

From this experimental evidence one can conclude that the relationships between the coupling constants are:

---

[1] C. S. Wu, in Proceedings of the Rehovoth Conference on Nuclear Structure, pp. 346-366

$$C_T = -C_T' \qquad C_S = -C_S' \qquad\qquad C_T^L = 0 \qquad C_A^R = 0$$

$$\text{or}$$

$$C_A = +C_A' \qquad C_V = +C_V' \qquad\qquad C_S^L = 0 \qquad C_V^R = 0$$

Incidentally, these relationships between the coupling constants as shown above give the Fierz interference term identically zero.

However, the existence of the interference term between Fermi and G-T transitions in β-asymmetry (Mn^{52}) and β-γ (circular polarization) angular correlation experiments (Sc^{46} and Au^{198}) shows that the β-interaction contains either S and T or V and A or all four V-A-S-T.

Information from β-α Angular Correlations

To gain more information on the type of interactions in β-decay prior to the parity-nonconservation era, one must turn to the results of β-ν angular correlations. The correlation depends on the value of λ_1.

$$I_{\beta\nu} \sim (1 + \lambda_1 \tfrac{v}{c} \cos \theta_{\beta\nu})$$

	S	V	T	A
λ_1	-1	+1	$+\frac{1}{3}$	$\frac{1}{3}$

So far, there are only five β-decays whose β-ν angular correlations have been investigated. They are n, He^6, Ne^{19}, and Ne^{23} and A^{35}. The choices of the interactions[2] are:

---

[2] E. J. Konopinski, in Proceedings of the Rehovoth Conference on Nuclear Structure, pp. 328-330.

He^6, n, Ne^{19} in agreement with S-T

A^{35}, n, Ne^{19} in agreement with V-A

The result of $_2Ne^{23}$ is preliminary and agrees with neither. In the case of A^{35}, $\dfrac{(H_F)^2}{(H_{GT})^2} \sim 95\%$. The latest results of A^{35} strongly indicate a vector-interaction in Fermi transition. The evidence of the recoil experiment of He^6 by Rustad and Ruby has been considered rather conclusively for tensor interaction. So, here we are in a predicament! It is therefore highly desirable to look carefully into the He^5 experiment and decide whether any systematic errors due to oversight could have affected their interpretation. The following is a report of such an investigation, and the conclusion is that unfortunately some corrections which the authors neglected in their calculations turned out to be quite substantial and rendered the conclusion which they originally made definitely questionable.

Angular Variation of Detecting Efficiency of the β-Counter

The β-ν angular correlation experiments can be carried out in two distinctly different manners. In the first place, to detect the neutrino (ν) directly is out of the question. Therefore, one can obtain the desired information indirectly either by measuring the energy spectrum of the recoil nuclei or by investigating the angular correlation between β and recoil nuclei. The sensitivity of the various arrangements depends on the degree of limitations allowed for the geometrical conditions and energy discriminations. In all the recoil experiments mentioned above, only in the He^6 case investigated by Rustad and Ruby was the angular correlation between the β and the

recoil ion actually measured. In this experiment the active He^6 gas
was continuously flowing into the source volume and then evacuated by
differential pumping through a diaphragm with eight small holes which
also served as the collimators for the recoil ions (Figure 1). It is
quite obvious that the detection efficiency of the β-counter for He^6
decays in the collimating "chimney" or hole varies with the position
of the β-counter. When the β-detector is at right angles to the recoil detector, it detects only those very energetic electrons which
are capable of penetrating through the diaphragm and reaching the
counter. But the β-detector looks down directly through the diaphragm
holes when it is opposite to the recoil detector. This implies that
the angular variation of the detection efficiency of the β-detector
for those decays in the chimney will be quite considerable. This is
particularly important for correlation between β and the recoil ion
because the geometric efficiency of the recoil detector is much larger
for decays in the chimney than for decays in the supposed source
volume. In order to establish this angular variation effect on the
measured angular correlation due to the decay of He^6 in the hole, one
must know the pressure gradient of the gas in the hole.

The Pressure Gradient of the Gas in the Diaphragm Hole

The diaphragm holes are extremely small. They are 4.5 mm
in diameter and 7 mm in length. It is difficult to measure the gas
pressure gradient by any known pressure gauges. However, since the
gas pressure inside of the gas volume is only 0.4 micron, the mean
free path will be of the order of 50 to 100 cm, which is large com-

pared with the dimensions of the apparatus. Under these conditions we
have free molecular flow, and the gas pressure gradient in the hole can
be actually calculated. Nevertheless, the calculations will be quite
involved, because each diaphragm hole was further equipped with a set
of four aluminum rings equally spaced to reduce the scattering of recoil ions from the wall. A very rough estimate by the well known
formula[3] for effusion from long channels

$$Q = (1/K) \; 1/4 \; n \; \bar{v} \; A_s \quad \text{where } 1/K \cong \frac{1}{1 + 3/8 \; \ell/a} \cong 0.5$$

indicates that the gas density near the lower end of the first diaphragm
must be not less than 12% of that in the source volume.

Optical Analog Study of Gas Pressure Gradient

Since the mean free path is much larger than the dimensions
of the apparatus, the gas molecules will not collide with each other
but only impinge on the wall and then be reflected. The situation is
very much like the case of optical diffusion from a perfect reflecting
wall. Measurement of the variation of light intensity inside the
source volume and in the collimating chimney will give some information
on the pressure gradient throughout the system. A scale model ten
times larger than the actual size of the apparatus was constructed.
The inside wall of the source volume and the chimney was smoked with a
thick layer of MgO to get high reflecting surface. A diffused light
source was set up in a side arm of the source volume. The surfaces

[3]N. F. Ramsey, Molecular Beams (Oxford Press), and P. Claussing,
Physica 9, 65 (1929).

between the upper and lower diaphragms were painted black to simulate
perfect pumping conditions and therefore gave a lower light intensity
than that of the actual situation. The loss of light intensity due
to multipole reflections on the MgO wall was corrected by auxiliary
experiments. A small probe of ground glass of 5 mm diameter connected
to a small photocell (Clairex C13) through a long light pipe (10 cm in
length) was used to measure the light intensity at various locations.
It is important to make the sensitivity of the ground glass probe isotropic in all directions. This is accomplished by smoking the probe
with a thin uniform layer of MgO. The measured relative light intensities are shown in Figure 2. It can be seen that the intensity is
constant inside of the sensitive volume and drops approximately linearly throughout the chimney. Between the two diaphragms where the pumping region is, the light intensity drops quickly and then it remains
more or less constant in the second collimating hole. This demon-
strates the predicted collimating action of the first chimney.

The relative coincidence counts at 180° due to decay of He^6
at different levels of the chimney is proportional to the product of
gas density and the solid angle subtended by the recoil detector and
the effective volume V. (The solid angle of the β detector varies very
little in this region.) In Figure 3, curve P represents the relative
gas pressure, curve Ω represents the relative solid angle, and $V \cdot P \cdot \Omega$
is the product of the three. It is quite obvious that the areas under
the curve $V \cdot P \cdot \Omega$ for the region of the source volume and the area of
the collimating chimney are comparable, and therefore the decay due to
the He^6 gas in the chimney is not insignificant in the correlation
results.

Semitransparency of the Diaphragm

If the diaphragm were completely opaque to the β-particles, correction of the original correlation results due to He^6 in the chimney might be applied with great confidence. Unfortunately, this was not the case, the collimating structure being composed of lucite and very thin aluminum plates.

An attempt was made to determine empirically the effects of the transparency of the collimator. The β-counting rates at different angular positions $\theta = 180°$, $165°$, $150°$, $135°$, $120°$ and $105°$ were measured by using a plastic scintillation spectrometer with a 256-channel analyzer and a Rh^{106} source ($E_{max} = 3.53$ Mev) on a tip of a pin point inserted to the different positions E, A, B, C, D, F and G along the chimney. The curves from different holes are slightly different. A few curves are shown in Figure 4. Then these measurements were repeated again under similar conditions except that a brass plate 2 mm thick with eight tapered holes drilled in it was laid on top of the polystyrene diaphragm. The later measurements should represent a completely opaque case. Figure 5 shows that in the actual case the diaphragm is semitransparent. It implies that many β-particles have suffered energy loss and multiple scattering on their way out and therefore destroyed their correlation relations.

Scattering from the Wall of the Chimney

Measurements also show that the backscattering of β-particles from the diaphragm walls is considerable. Figure 6 shows two β-spectra of Rh^{106} taken with the source position at the center of the source volume (curve a) and near the mouth of the chimney (curve b). The one

taken with the source near the mouth of the chimney shows the increase
of low energy electrons due to backscattering effect. This observed
effect may also constitute a plausible explanation for the observed
electron-recoil ion coincidence spectrum (Figure 13 of Rustad and
Ruby's paper) taken at fixed position of 180°. The excess of low
energy β-particles due to the backscattering effect would distort the
spectrum from axial vector curve to tensor curve.

Interpretation and Conclusion

          With the semitransparency and the multiple scattering effect
of the diaphragm holes, a precise correction of the measured coincidence results is obviously impossible. To attempt to evaluate the
extent that these corrections would alter the measured correlation
and its conclusion, we can inspect two extreme cases. One is to apply
the correction due to the angular variation of detecting efficiency of
the β-counter as measured with the polystyrene diaphragm alone.
ly, the corrections in this case will not be sufficient due to the
semitransparency of the diaphragm. On the other hand, if one con-
siders that those β-particles coming out from the opaque diaphragm
are unaffected and retain the correlation relationship with the recoil ions, but that the excessive number of β-particles from the polystyrene diaphragm over the opaque diaphragm have lost their original
correlation with the recoil ions and exhibit isotropic correlation,
then the correction is much closer to the true situation. Figure 7a,
          show the curves corresponding to these   extreme cases
and the results are more in favor of axial vector than tensor, contradictory to the original conclusion.

The purpose of this re-examination is to point out the detri-
mental effects on the correlation results due to the presence of even
a small quantity of He^6 gas in the diaphragm holes. Furthermore, by
using an optical analog it has been possible to infer directly the
pressure gradient of He^6 along the chimney hole and therefore substantiate our supposition. This is, unfortunately, what happened.
The coincidence investigations of β-ν angular correlation of He^6 and
Ne^{23} are, nevertheless, still of paramount importance. The investigation of the energy distribution of recoil ions in coincidence with β-
particles of a selected energy interval and at nearly opposite directions still offers the most sensitive method to determine the relative
amounts of A and T. What one has learned through this re-examination
may be of some help to future investigations.

Acknowledgments

We wish to thank Dr. Richard Garwin and Dr. Luke Yuan for
many valuable discussions concerning the measurement of the gas pressure
gradient. We particularly want to thank Mr. K. Lattemann for his
patient help in carrying out the investigation of the angular variation
effect. To Mr. B. Rustad and Mr. S. Ruby we express our thanks for
their cooperation.

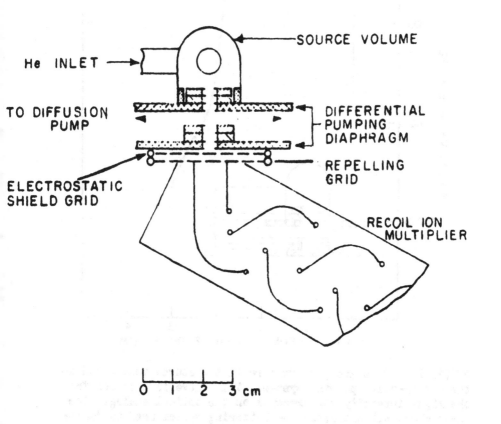

FIGURE 1.   Scale diagram of the angular correlation equipment of Rustad and Ruby (Phys. Rev. 97, 991 (1955)) showing the source volume, diaphragm, recoil detector and one of the positions for the β-detector.

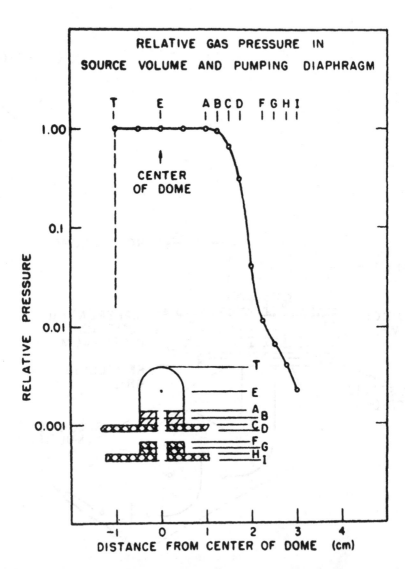

FIGURE 2. Relative gas pressure in the source volume and in the differential pumping system. This curve is inferred from the light intensity measurements on the optical analog. The inset shows schematically the lettering system used to denote the different regions of the apparatus.

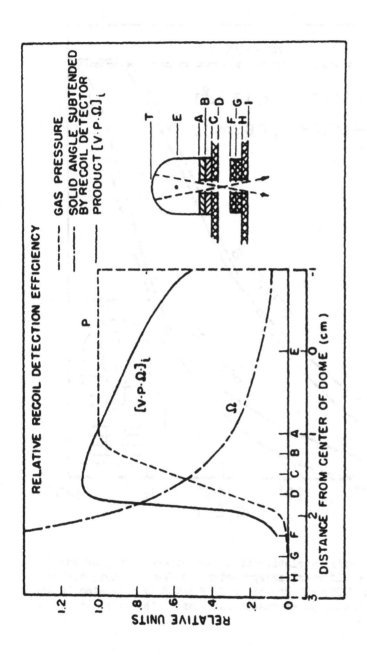

FIGURE 3. The three curves of P, Ω and V x P x Ω, respectively the gas pressure, the geometric solid angle from the position to the recoil detector, and the product of V, P and Ω where V is the effective source volume of the region plotted vs the position in the apparatus.

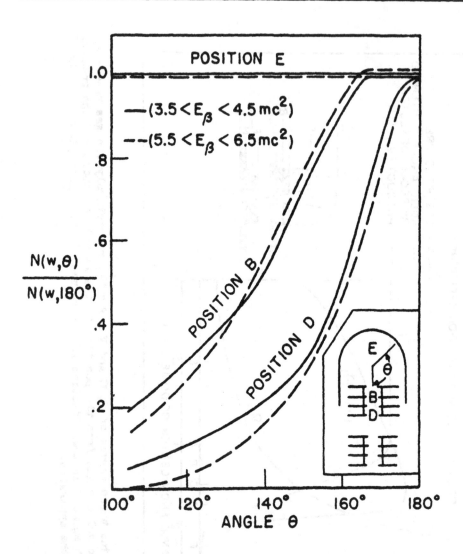

FIGURE 4. The angular variation of detection efficiency of the β-detector for two different energy intervals from β-decays occurring in the three different positions E, B and D as indicated in the inset. The β-intensity is expressed in terms of the intensity at the 180° position.

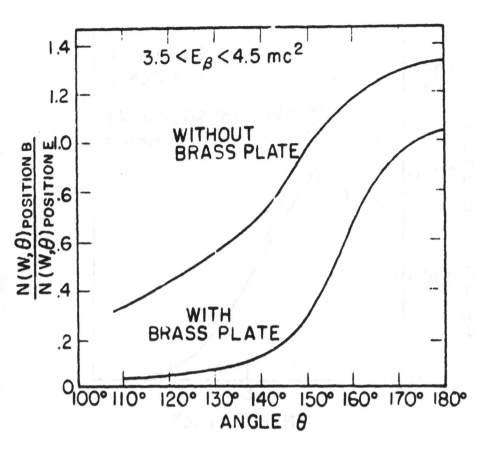

FIGURE 5. The effect of semi-transparency of the polystyrene diaphragm. The curve with brass plate represents the geometric collimation of the diaphragm holes on the electron angular distribution. The large number of electrons at low angles shown by the curve without brass plate indicates the semi-transparency of the polystyrene. It should be noted that the ordinate represents the ratio of counts obtained from a constant β-source in the position B divided by the counts obtained with the same source at position E – the center of the nominal source volume. The ratio 1.4 at 180° for the curve without brass plate is indicative of the large distortion of the spectrum and in-scattering due to the polystyrene diaphragm.

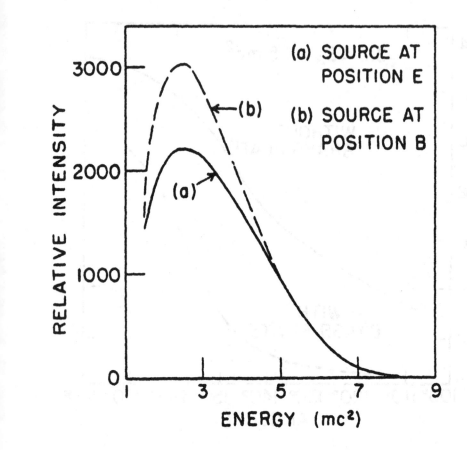

FIGURE 6. The β-spectra of the same Rh^{106} source at positions E and B. The two curves have been normalized over the energy region 5-7 mc^2. The large excess of low energy electrons in the curve (b) is indicative of the unfavorable scattering effects.

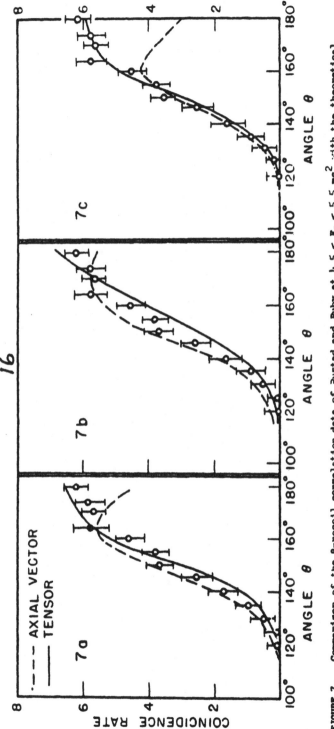

16

FIGURE 7. Comparison of the β-recoil correlation data of Rustad and Ruby at $4.5 < E_\beta < 5.5 \; mc^2$ with the theoretical distribution curves. Figure 7c is without correction for the He^6 gas decaying in the diaphragm and is identical to Figure 11 of Rustad and Ruby's paper. In Figure 7a and 7b the theoretical distributions have been corrected for the semi-transparent and the opaque plus semi-transparent diaphragm effects respectively as described in the text. It should be noted that these corrections tend strongly to remove the expected dip in the axial vector distribution at high angles. These curves of Figure 7a and 7c have been normalized to the data in the angular region $160°-180°$ where the statistical errors are smallest. Correction to the theoretical curves for the energy region $2.5 < E_\beta < 4.0 \; mc^2$ have also been performed and a similar merging of the tensor and axial vector distributions and the experimental points occurs. However, due to the serious scattering situation it is felt that a rather large portion of the electrons detect-ed in this low energy region have been scattered through large angles, and therefore correction of the theoretical curves

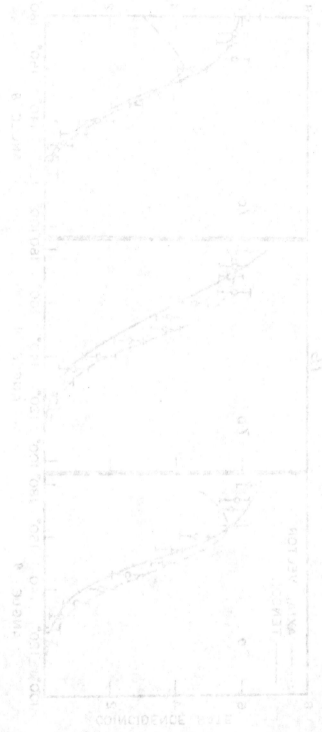

ORIGIN OF THE UNIVERSAL V-A THEORY

E. C. G. Sudarshan
University of Texas, Austin, TX 78712

R. E. Marshak
Virginia Polytechnic Institute and S.U., Blacksburg, VA 24061

§1. FERMI'S THEORY OF β DECAY (1934-47)

When David Cline invited one of us (R.E.M.) to give the first talk at this conference, it provided an opportunity to pay tribute again to one of the great physicists of the twentieth century, Enrico Fermi. The "50 years" in the title of this conference refers, of course, to the first explicit formulation of a theory of weak interactions by Fermi in 1934[1]. In these seminal papers, Fermi applied the methods of second quantized field theory to the β decay process: $n \rightarrow p + e^- + \bar{\nu}$ (accepting Pauli's neutrino hypothesis) and worked out the consequences of his postulated vector interaction among the four spin 1/2 particles. With an eye for the physically relevant, Fermi used non-relativistic wave functions for the nucleons to exhibit the selection rules for β transitions in nuclei. He deduced the important features of forbidden transitions and calculated the effect of a non-vanishing neutrino mass on the β spectrum.

Fermi selected the vector (V) interaction out of five possible choices (the others being scalar (S), pseudoscalar (P), axial vector (A) and tensor (T)) in analogy to the electromagnetic interaction even though the analogous "current" in β decay was charged and not neutral. The analogy with the electromagnetic interaction was not pursued further - either in the direction of extending gauge invariance in some fashion or invoking boson mediation of the weak interaction (let us recall that a meson theory of nuclear forces was only put forward by Yukawa[2] a year later). Parity conservation, as well as baryon and lepton conservation, were implicitly assumed. Fermi's four-fermion vector theory of β decay was a splendid beginning and provided the basic framework for the chiral (V-A) interaction proposed by us in 1957. A decade later, the chiral (V-A) interaction was used as the starting point for the electroweak gauge theory that has passed its first major test recently with the detection of the W and Z weak bosons.

But let us return to the extensions and refinements that followed Fermi's initial papers. While Fermi's selection rule for an allowed transition (ΔJ = 0, no - no refers to parity change) was later confirmed, Gamow and Teller[3] pointed out quite early that the β interaction can depend on the spin of the nucleon and, in that case, the selection rule is ΔJ = 0, ± 1 (no 0 → 0), no for an allowed transition. A prime example supporting the Gamow-Teller conjecture was He^6 (J = 0^+) → Li^6 (J = 1^+) + e^- + $\bar{\nu}$, which decay played such a crucial role in the quest for a universal β interaction in later years. Hence the distinction between Gamow-Teller selection rules (corresponding to the A or T interaction) and Fermi selection rules (corresponding to the V or S interaction).

The observation of Gamow-Teller β transitions implied that Fermi's V interaction could not be the sole β interaction and might even be absent. The precise structure of the β interaction became a burning question and a variety of methods was suggested to determine its form. Thus, Feriz[4] pointed out that the presence of A and V or T and A in the β interaction leads to an interference term in the allowed β spectrum which vanishes in the absence of an admixture of S and V or T and A. Möller[5] suggested that additional information could be obtained from K electron capture in nuclei and several authors[6] noted that a study of forbidden β spectra could reveal a great deal about the form of the interaction. Other possible experiments to shed light on the structure of the β interaction, e.g. electron-neutrino angular correlation and β-γ angular correlation experiments, were proposed. However, World War II interrupted the implementation of this program and, in 1947, the discovery of the second generation lepton, the muon, broadened the scope of beta interaction physics to weak interaction physics.

§2. BEGINNINGS OF A UNIVERSAL THEORY OF WEAK INTERACTIONS WITHOUT PARITY VIOLATION (1947-56)

It was known before 1947 that the cosmic ray meson underwent electron decay with a long lifetime but the decay products were unknown and no plausible connection had been established between this phenomenon and β decay. Indeed, Yukawa[2] had failed to establish this connection within the framework of his meson theory. But events beginning in 1957 altered the entire situation.

There was first the Italian experiment[7], published in February 1947, which disclosed that a substantial fraction of negative sea level mesons decayed in a carbon plate, but were absorbed in an iron plate. Analysis of this experiment by Fermi, Teller and Weisskopf[8] led to the startling conclusion that there was a factor of 10^{12} discrepancy between the meson's production and absorption cross sections. At the Shelter Island Conference in early June of 1947, Marshak and Bethe put forward the two-meson hypothesis[9] to explain the factor of 10^{12} by having the strongly interacting (Yukawa) meson produced in the upper atmosphere decay into the lighter weakly interacting meson at sea level. By mid-June, Nature had arrived in the U.S. with the beautiful photographs of $\pi \rightarrow \mu$ decay discovered by the Bristol group[10]. And by the end of June, Pontecorvo[11] made the brilliant observation: "We notice that the probability ($\sim 10^6$ sec^{-1}) of capture of a bound negative meson [in carbon] is of the order of the probability of ordinary K capture processes, when allowance is made for the difference in the disintegration energy and the difference in the volumes of the K shell and of the meson orbit. We assume that this is significant and wish to discuss the possibility of a fundamental analogy between β processes and processes of emission or absorption or charged mesons...". Pontecorvo did not know about the two-meson theory nor the $\pi \to \mu$ decays but he was really asserting that the muon was a "heavy electron" ("second generation lepton" in modern parlance) - an identification that has withstood the test of time. By the end of 1947, one was poised for the extension of Fermi's theory of β decay to other weak processes like muon decay and muon capture.

By the beginning of 1948, the production of pions in the Berkeley synchrocyclotron and further work on muon decay and muon capture had made it clear that theory was called upon to explain the interrelationship of the various processes shown diagramatically in Fig. 1. It is true that by mid-1947, Pontecorvo[11] had already noted the rough equality $g_3 \sim g_1$ (see Fig. 1), and that Marshak and Bethe[9] had related $g_{\pi\mu\nu}$ to g_3 and $g_{\pi NN}$; but muon decay had not been brought into the discussion and serious calculations of muon capture still had to be made. Through 1948 and the early part of 1949, a number of authors[12] examined the relationship of the various weak processes implied by Fig. 1 with or without the mediation of the strong pion-nucleon interaction. The record shows that the first and most comprehensive attempt to relate g_1, g_2 and g_3 (the legs of the large triangle in Fig. 1) was made by Tiomno and Wheeler[12] who found $g_1 \sim g_2 \sim g_3$ as long as the β interaction was not predominantly P. This work and that of the others in Ref. 12 gave great impetus to the concept of a universal Fermi interaction (UFI) even though the structures of the three weak interactions were as yet undetermined. Ruderman and Finkelstein[13] immediately latched on the UFI and pointed out that if UFI is accepted, the ratio of the decay rates of the pseudo-scalar π into (e, ν) and (μ, ν) is independent of the strong pion-nucleon interaction and depends only on the form of the weak coupling, in the following fashion: $\pi_{e2}/\pi_{\mu2} \sim 10^{-4}$ for the A weak interaction, ~ 1 for the P weak interaction and $=0$ for the S, V or T weak interaction. The measurement of the $\pi_{e2}/\pi_{\mu2}$ ratio was important in checking the universal V-A theory, as we shall see below.

The period 1947-56 also saw considerable progress in the unraveling of the structure of the β interaction although it ended on an inconclusive note. During this period, further experiments on muon decay fixed the decay products and value of the Michel parameter for the electron spectrum[14], and new muon capture experiments were consistent with UFI. However, the greatest effort went into β spectra (requiring A or T), and the occurrence of J = 0 → 0, no transitions (requiring the presence of S or V). The measurement of the electron-neutrino angular correlation coefficient $\lambda = 0.33 \pm 0.08$ in the decay of He[6] favored T as the Gamow-Teller contribution to the β interaction by a rather wide margin[16]. Since the P interaction is very elusive in β decay measurements (because it does not contribute in the non-relativistic limit), one entered the crucial year 1956 with only two allowable combinations of the β interaction,

namely S, T <u>or</u> V, T, except for a possible admixture of P[17].
Whether either preferred choice of the β interaction could be
reconciled with the as-yet-undetermined forms of the muon decay
and muon capture interactions remained to be seen. It was also
unclear whether UFI could be extended to the strange particle
decay processes that were coming under close scrutiny.

§3. PARITY VIOLATION AND THE DILEMMA FOR UFI (1956-57)

In one of those fortunate circumstances that leads to a quantum
jump in scientific understanding, the θ-τ puzzle triggered a series
of rapid-fire developments (both theoretical and experimental) that
first created an impasse for UFI but which was soon resolved by
the universal V-A theory. By 1956, the θ-τ puzzle (wherein two
strange mesons, θ and τ, decaying respectively into two and three
π's, were observed to have the same masses and lifetimes) was
becoming increasingly troublesome. The subject came under intense
discussion at the Sixth Rochester Conference (April 1956) and
parity violation was one of the possible explanations suggested[18].
It devolved upon Lee and Yang[19] to delineate with great care
other weak decay processes in which parity violation would manifest
itself other than through the 2π and 3π decay modes of strange
mesons. The parity violation hypothesis was spectacularly confirmed within months of the publication of the Lee-Yang paper by
Wu and collaborators[20] who looked for an electron asymmetry from
polarized CO^{60}. The backward asymmetry which was found gave
unequivocal evidence for parity violation and could be explained
(since the decay of Co^{60} was a Gamow-Teller transition) by a T
interaction with a right-handed neutrino (ν_R) or an A interaction
with a left-handed neutrino (ν_L). Consequently, the combination
of the Co^{60} parity-violation experiment, the Fierz interference
experiments and the (e-ν) correlation experiment in He^6 mandated
the choice S, T for the β interaction (with Fermi's V interaction
lost in the shuffle!).

During the hectic year from the Spring of 1956 to the Spring
of 1957, the parity-violation hypothesis was also tested in muon
decay through a measurement of the backward electron asymmetry
with respect to the muon momentum[21]. If one assumed the two-
component neutrino[22] and the conservation of leptons - which
was consistent with all other experiments - this results required
the V, A interaction for muon decay. Apparently, UFI was in
deep trouble. At the Seventh Rochester Conference in April 1957,
T. D. Lee acknowledged (in his introductory talk at the session
on weak interactions[23]) that: "Beta decay information tells us
that the interaction between (p, n) and (e, ν) is scalar and
tensor, while the two-component neutrino theory plus the law of
conservation of leptons implies that the coupling between (e, ν)
and (μ, ν) is vector. This means that the Universal Fermi Interaction cannot be realized in the way we have expressed it....at
this moment it is very desirable to recheck even the old beta

interactions to see whether the coupling is really <u>scalar</u>...". The
T contribution to the β interaction was still not being questioned
because of He[6]!

The dilemma became more acute after C. S. Wu's talk at the
conference wherein she reported on her unpublished measurement
of the e$^+$ asymmetry from Co^{58} (undergoing a mixed-Fermi plus Gamow-
Teller-transition) which was giving a smaller value than the e$^-$
asymmetry from Co^{60} and of opposite sign. This result could be
explained if Co^{58} decay was primarily Gamow-Teller; however, if
one inserted the accepted ratio of Fermi to Gamow-Teller matrix
elements, the interference term between S and T produced disagreement with the experimental result. This discrepancy led Wu to
remark[24] that: "The evidence on the relative strengths of scalar
and vector components in the Fermi interaction is no longer so
convincing as we previously had thought...The decay of A^{35} would
furnish a much more sensitive test...". The implication was that
an appreciable amount of V in the β interaction would help to
explain the measured positron asymmetry in Co^{58}. However, if the
β interaction was predominantly V, T (despite the evidence of
some old parity-conserving β experiments[17]), one would be forced
to assign opposite helicities to the neutrinos emitted in Fermi-
and Gamow-Teller-type β transitions, a very displeasing prospect
indeed. To add to the confusion, the possibility of a V, T beta
interaction was reinforced by two rumors circulating at the
Seventh Rochester Conference (β experiments were being performed
at an incredible rate!): one rumor was that Boehm and Wapstra[25]
had obtained a similar result to that of Wu in measuring the β-γ
(circularly polarized) correlation in Co^{58}. The second rumor was
that an Illinois group[26] had measured the electron-neutrino
angular correlation coefficient from A^{35} (a dominantly Fermi
transition) and was finding λ = -1 (as required by the V interaction) instead of λ = +1 (as required by the S interaction).
Could the beta interaction be V, T after all[27] and UFI have to
be abandoned?

It was our original intention to make a brief report at the
Seventh Rochester Conference on the universal V-A theory. We had
identified the problems with reconciling all the known β decay
experiments with a unique β interaction and had recognized that
some experiments must be wrong. But since ECGS was a graduate
student at the time and since REM was making a major presentation
on nuclear forces (the Signell-Marshak potential), it was decided
that P. T. Matthews, then a Visiting Professor at Rochester (who
was conversant with our work) would report on the V-A theory in
place of ECGS. For reasons unconnected with the V-A theory,
Matthews never made the presentation. During the conference, REM
would have stepped into the fray but for the specter of a V, T
interaction in β decay (requiring opposite helicities for the
neutrino); he was reluctant to argue for V-A as the UFI option
as long as a consistent picture did not emerge from the <u>parity-
violating</u> experiments in weak interactions.

It was essential to clarify as soon as possible whether the
V, T combination was a mirage insofar as the parity-violating β
decay experiments were concerned. This clarification came during
the first week of July (1957) as the result of a meeting with
F. Boehm[28] where we presented our arguments for the universal
V-A theory and asked for an updating on the β-γ (circulary
polarized) correlation program in which he was engaged. Boehm
informed us that his latest experiment on Sc^{46} [25] gave a much
larger correlation coefficient than Co^{58}, implying that the choice
V, T (or S, A) for the β interaction was excluded; presumably,
the estimate for the ratio of Fermi to Gamow-Teller matrix elements
was in error for Co^{58}. With this assurance, and the benefit of
several additional experimental numbers (see §4), we were able to
complete our paper within a matter of days and to send off an
abstract to the organizers of the Padua-Venice Conference where
we expected to present our work in September.

§4. UNIVERSAL V-A THEORY AND ITS RAPID CONFIRMATION (1957-59)

The several months' delay - from April to July 1957 - in
putting the finishing touches on our paper was most useful since
it allowed time for certain key β experiments to pass from the
rumor to completion stage and thereby to consolidate the experimental underpinning of our theory. Thus, we were able to discuss
not only the electron asymmetry experiment in Co^{60}, the "Fierz
interference" experiments, and the electron-neutrino angular
correlation experiment in He^6, but also the electron polarization
experiment on the Fermi decay of Ga^{66} [29] and the electron-neutrino
angular correlation experiment in A^{35} [26], in addition to the
β-γ correlation experiment in Sc^{46} [25].
This comprehensive analysis of β processes led us to conclude
in our Padua-Venice paper (entitled "Nature of the Four-Fermion
Interaction") that: "The present β decay data, while still somewhat contradictory from an experimental point of view, seem to
suggest some definite choices for the coupling types...the simplest
inference would be that the β decay coupling is either AV or ST....
The AV (or ST) combination has the added merit that the neutral
particle emitted in electron decays is then right-handed (or left-
handed) both for the Fermi and the G-T interactions [the neutral
particle emitted in electron days is the antineutrino]...In the
case of both AV and ST, the Fierz interference terms in allowed
spectra and first forbidden spectra vanish identically. The
choice between AV and ST thus hinges essentially on the electron-
neutrino angular correlations or, equivalently, on the determination
of the spirality of the neutral particle emitted in β decay. As
regards the electron-neutrino angular correlations, this implies
a choice between the A^{35} and He^6 experiments...". [The term
"spirality" was used interchangeably with "helicity" in the early
days of parity violation.]
We then proceeded to consider the evidence from other weak
interactions. Our analysis of muon decay was, of course, in

accord with T. D. Lee's, and we stated that "The muon decay data
thus suggest A, V interaction irrespective of the spirality of the
neutrino field. The latter can be unambiguously determined if
one measures the longitudinal polarization of the positron from μ^+
decay. The positron would be expected to be right- or left-
polarized, according as the Co^{60} transition proceeds via axial
vector or tensor interactions, provided the Law of Conservation
of Leptons is valid...".

   We continued with an analysis of the evidence from $\pi_{e2}/\pi_{\mu2}$
and $K_{e2}/K_{\mu2}$ and finally concluded that "the only possibility for
a Universal Fermi Interaction is to choose a vector + axial vector
coupling [the nomenclature V-A was adopted later] between every
two of the pairs of fields $\mu\nu$, $e\nu$, np, $\Lambda^\circ p$, np leading to the
τ and θ modes of the K meson. In the framework of our hypothesis,
the β decay interaction is defined uniquely by the sign of the
electron asymmetry in the decay of oriented Co^{60}. This unique
form is: $g \, \bar{P} \, \gamma_\mu \, (1 + \gamma_5) \, N \, \bar{e} \, \gamma_\mu \, (1 + \gamma_5) \, \nu$ + h.c. The hypothesis
of Universal Interaction generalizes this β coupling to a coupling
of four Dirac fields A, B, C, D in the form: $g \, \bar{A} \, \gamma_\mu \, (1 + \gamma_5) \, B \, \bar{C}$
$\gamma_\mu \, (1 + \gamma_5) \, D$. Since γ_5 and γ_μ anticommute, one can rewrite the
interaction of the four field A, B, C, D, in the form:

$$g \, \bar{A} \, \gamma_\mu \, (1 + \gamma_5) \, B \, \bar{C} \, \gamma_\mu \, (1 + \gamma_5) \, D = g \, \bar{A}' \gamma_\mu \, B' \, \bar{C}' \, \gamma_\mu \, D' \qquad (I)$$

where A', B', C', D' are the "two component" fields:

$$A' = (1/\sqrt{2}) \, (1 + \gamma_5)A, \; \bar{A}' = (1/\sqrt{2}) \, \bar{A} \, (1 - \gamma_5), \text{ etc.}$$

Now the "two-component" field $(1/\sqrt{2}) \, (1 \pm \gamma_5)$ A is an eigenstate
of the chirality operator[30] with eigenvalue ± 1. Thus the
Universal Fermi Interaction, while not preserving parity, preserves chirality and the maximal violation of parity is brought
about by the requirement of chirality invariance. This is an
elegant formal principle, which can now replace the Lee-Yang
requirement of a two-component neutrino field coupling (or
equivalently the Salam postulate of vanishing bare mass and
self mass for the neutrino)...Thus our scheme of Fermi interactions
is such that if one switches off all mesonic interactions, the
gauge-invariant electromagnetic interactions (with Pauli couplings
omitted) and Fermi couplings retain chirality as a good quantum
number...".

   We ended our paper with: "While it is clear that a mixture
of vector and axial vector is the only universal four-fermion
interaction which is possible and possesses many elegant
features, it appears that one published and several unpublished
experiments cannot be reconciled with this hypothesis. These
experiments are:
   (a) The electron-neutrino angular correlation in He^6...
   (b) The sign of the electron polarization from muon decay...
   (c) The frequency of the electron mode in pion decay...
   (d) The asymmetry from polarized neutral decay...

All of these experiments should be redone, particularly since some of them contradict the results of other recent experiments on the weak interactions. If any of the above four experiments stands, it will be necessary to abandon the hypothesis of a universal V+A four-fermion interaction or either or both of the assumptions of a two-component neutrino and/or the law of conservation of leptons."

The quotations are all from the paper presented to the Padua-Venice Conference on "Mesons and Recently Discovered Particles" held September 22-28, 1957. Our paper was published in the proceedings of this conference in late Spring 1958[31] and reprinted in P. K. Kabir's book on "History of Weak Interaction Theory"[32]. In those halcyon days of collegiality, it never occurred to us to republish the Padua-Venice paper in a journal; we did send out a preprint dated September 16, 1957 (a date we remember because it happened to be ECGS's 26th birthday!). Several months later, we decided to publish a short note on "Chirality Invariance and the Universal Fermi Interaction"[33] to make some new points and to take stock of experimental developments following the Padua-Venice Conference. Thus, we remarked in that note: "since the conference, the validity of the He^6 experiment has been questioned[34], the polarized neutron experiment has come down to a value consistent with the V-A theory[35] and the helicity of the positron from μ^+ decay has turned out[36] to be +1, as it should. There has been no change in the experimental situation with regard to the electron decay of the pion but it is clear that this very difficult experiment should be redone...". Our note (sent to the Phys. Rev. on Jan. 10, 1958) was not intended as a substitute for our 1957 Padua-Venice paper but, unfortunately, it was treated by all too many physicists in later years as the sole publication of our universal V-A theory[37].

Apart from the priority question (which will be discussed in the next section), the fact is that within a year and a half of the Padua-Venice Conference, the four experiments, whose demise was required by the universal V-A theory, had all been redone and the new results were in complete accord with the theory. Not only had the electron asymmetry from polarized neutrons come down and the polarization of e from μ decay acquired the correct sign and magnitude but also the electron-neutrino angular correlation coefficient in He^6 had become $-0.39 + 0.02$[38] and the $\pi_{e2}/\pi_{\mu2}$ ratio had changed to $0.93 + 0.37 \times 10^{-4}$[39]. The most striking confirmation of the V-A theory was the direct measurement of the neutrino helicity as -1 in an ingenious experiment on K capture in Eu^{152} performed by Goldhaber and collaborators[40]. Experimental support for the universal V-A theory (for charged currents, of course, and with the Cabibbo or, shall we say, Kobayashi-Maskawa modification) has continued to pile up ever since 1959 - with one experimental success following another. Indeed, within the past year, the measurement of the angular distribution of decay leptons from the very massive charge W boson has confirmed the V-A theory up to 80 GeV!

And so it came to pass - only three years after parity violation in weak interactions was hypothesized - that the pieces fell into place and that we not only had confirmation of the UFI concept but we also knew the basic (V-A) structure of the charged currents in the weak interactions for both baryons and leptons. Let us remind ourselves that in 1959 - at the midpoint in time between the date when Fermi's β decay theory was formulated and the present - there was only one neutrino, only two charged leptons, no quarks, no Cabibbo mixing, no neutral currents, no CP violation, no role for Yang-Mills fields (proposed five years earlier[41], and, of course, no electroweak model. The past quarter of a century has seen enormous progress in the theory of weak interactions but it is fair to say that the form of UFI that we wrote down in our Padua-Venice paper (Eq. (I) above) recognized for the first time the importance of chiral fields, irrespective of the fermion mass[42]. Chiral fermion fields are not only crucial for the electroweak theory but appear to be essential to progress with grand unification and composite models as well.

§5. CONCLUDING HISTORICAL REMARKS

In a perfect world we could have ended our story at this point but, in the imperfect world which we inhabit, it is incumbent upon us to make some historical comments. The very success of the universal V-A theory has led to claims that other work either anticipated our work or was conceived independently of it. In order to contribute to the historical record on the origin of the universal V-A theory, we shall briefly evaluate these claims as objectively as we can.

In the early 1950's - after UFI was proposed and before parity violation was confirmed in weak interactions - a number of authors attempted to deduce the form of UFI from some type of symmetry principle. The papers that were closest in spirit to the chirality invariance underlying the V-A theory were written by Tiomno and by Stech and Jensen. Tiomno[43] invented the idea of "mass reversal invariance" (the idea that the Dirac equation is invariant under the transformation $\psi \rightarrow \gamma_5\psi$, $m \rightarrow -m$) and postulated the invariance of the weak current $\psi_1 O_\mu \psi_2$ (where O_μ is the S, V, P, A or T operator) under the simultaneous transformation (to conserve parity): $\psi_1 \rightarrow \pm \gamma_5\psi_1$, $\psi_2 \rightarrow \pm \gamma_5\psi_2$. Tiomno found that if the signs in the γ_5 transformations are the same, O_μ has to be a combination of V and A whereas if the signs are different, the combination has to be S, P, T. This clearcut separation into two classes of Fermi interactions was interesting but still quite different from the idea of applying separate chirality invariance to each Dirac field which led to parity violation and the V-A interaction. Stech and Jensen[44] proposed to consider the limiting case $m \rightarrow 0$ for the Dirac particles (since $m \rightarrow -m$ has no field-theoretic meaning), applied simultaneous chirality transformation to the Dirac spinors and therefore arrived at the same bifurcation of interactions into (S,P,T) or (V,A). They did go a step further and argued that the four-fermion interaction should be invariant under Fierz rearrangement and ended up with the two combinations: (S+P-T) or V-A. They favored the S+P-T beta interaction for the usual reasons. We were aware of these papers at the time that we wrote ours but we chose not to refer to them because of their limitation to the parity conservation case. In hindsight, we consider these papers as valuable contributions to the chirality invariance approach and wish to correct the record on this score.

Two papers which bear on chirality invariance in relation to parity violation, of which we were not aware when we wrote our paper, are those by Salam and Tiomno. Salam brought his unpublished paper (dated February 1957)[46] to the attention of one of us (REM) in 1968, with the consequence that it was acknowledged in the book by Marshak, Riazuddin and Ryan[46]. In his Nobel address[47], Salam mentions his contribution to the development of the V-A theory as follows: "The idea of chiral symmetry leading to a V-A theory. In those early days my suggestion of this was limited to neutrinos, electrons, and muons; shortly after that, Sudarshan and Marshak, Gell-Mann and Feynman, and Sakurai had the courage to postulate γ_5 symmetry for baryons as well as leptons, making it into a universal principle of physics...". In his unpublished paper[45], Salam examined muon decay, wrote down the four-fermion interaction in charge retention order, adopted the two-component neutrino hypothesis, and applied Tiomno's mass reversal invariance to the e and μ spinors; he thereby deduced a combination of V and A interaction (not necessarily V-A) for muon decay. As Salam implies in his Nobel address, he did not question the conventional wisdom at that time that the β interaction was a combination of S and T. Unbeknown to us, Tiomno's paper on "Nonconservation of Parity and the Universal Fermi Interaction" was sent to Nuovo Cimento[48,49] in early July 1957 and published in October. He went beyond Salam in trying to reconcile the accepted (S,T,P) combination for the β interaction with the (V, A) muon interaction by postulating opposite helicities for the neutrino and thus ended up with a somewhat inelegant and incorrect UFI.

We now come to the Feynman-Gell-Mann and Sakurai papers. It is clear from the record that Feynman was toying with the idea of using the 2-component Klein-Gordon equation in place of the 4-component Dirac equation to express parity violation in weak interactions as early as April 1957[50]. It is a fact that Gell-Mann was informed of our work on the universal V-A theory not later than the first week of July, at which time our paper was completed and an abstract sent off to Padua. It also seems clear from Tiomno's paper at this Racine conference that the Feynman-Gell-Mann paper was written during the Summer of 1957 (with the help of amateur radio between Rio de Janiero and Pasadena!) with the result that the paper was dispatched to the Physical Review by September 16, precisely the date on which our preprint was circulated. The first public presentation of our work was made during the Padua-Venice Conference September 22-28, 1957 and several months later, the Feynman-Gell-Mann paper was published in the Physical Review (January 1, 1958)[51]. Our followup note

on the universal V-A theory was published in the March 1, 1958
issue of the Physical Review while the publication of our first
paper in the Padua-Venice conference proceedings was unexpectedly
delayed[52] to May 1958. With this complicated set of facts, how
does one settle the priority question in which historians of
science are interested? In this instance, perhaps the simplest
solution is to quote Feynman[53], who said a decade ago: "We have
a conventional theory of weak interactions invented by Marshak
and Sudarshan, published by Feynman and Gell-Mann and completed
by Cabibbo - I call it the conventional theory of weak interactions - the one which is described as the V-A theory."

For purposes of the historical record, it may also be worthwhile to compare the approaches of the V-A papers by ourselves
and Feynman and Gell-Mann. Our paper adopted the "inductive"
approach - after a thoroughgoing analysis of all key parity-
violating and parity-conserving weak interaction experiments
then extant, we reached the unequivocal conclusion that the only
possible UFI was the V-A interaction, at the expense of a certain
number of explicitly identified contradictory experiments. We
noted that the V-A interaction possessed a number of interesting
properties, chief among them was the invariance of the V-A
interaction under separate chirality transformations of the Dirac
spinors. The Feynman-Gell-Mann paper adopted the "deductive"
approach, purporting to derive the V-A interaction by using
half of the solutions of the 2-component Klein-Gordon equation
without gradient coupling. Their "derivation" is no more
perspicuous than our "derivation" based on chirality invariance
and has been less successful in withstanding the test of time[54].
Feynman and Gell-Mann then proceed to confront the V-A theory
with experiment, using pretty much the same empirical findings
as we do and, of course, come to similar conclusions. The novel
feature of the Feynman-Gell-Mann paper is a rather extensive
discussion of the conserved vector current hypothesis as a
further argument for the universality of V-A; apparently, the
authors were not aware of the earlier work of Gershtein and
Zeldovich[55] on the subject but, in any case, examined the
consequences in greater depth. All in all, the Feynman-Gell-Mann
paper was a most valuable contribution to the theory of weak
interactions.

We conclude with a brief comment concerning Sakurai's work
on the universal V-A theory. In the acknowledgement to his
paper[56], Sakurai states: "The present investigation is directly
stimulated by conversations the author had with Professor R. E.
Marshak, to whom he wishes to extend his sincere thanks...".
It is true that Sakurai did meet with one of us (REM) in Rochester
at the beginning of October 1959 to be briefed concerning the
status of the universal V-A theory; he also received copies of
the preprints of our paper and that of Feynman and Gell-Mann.
He prepared a paper, upon his return to Cornell, in which he
pointed out that separate chirality invariance of the four-
fermion interaction could be restated in terms of separate

mass reversal invariance with the same resulting V-A interaction.
He then argued that the use of mass reversal invariance to
"derive" the V-A interaction was justified by the fact that the
relationship between momentum and energy for a particle, as well
as the 2 component Klein-Gordon equation used by Feynman and
Gell-Mann, depend on m^2 (not on m). Sakurai then repeats some
of the experimental discussion contained in our paper and that
of Feynman and Gell-Mann, paying somewhat greater attention to
the compatibility of the V-A interaction with the experimental
results on the non-leptonic decays of the strange particles.
Sakurai's paper was sent to Nuovo Cimento on October 31, 1957
and was published March 1, 1958, several months before the
publication of our Padua-Venice paper. Apart from the priority
question - which seems easy to resolve - it is difficult to see
how the mass reversal invariance argument improves upon chirality
invariance in "deriving" the universal V-A intearction.

REFERENCES

1. E. Fermi, Nuovo Cimento 11, 1 (1934) and Zeits. f. Phys. 88,
    161 (1934).
2. H. Yukawa, Proc. Phys. Math. Soc., (Japan) 17, 49 (1935).
3. G. Gamow and E. Teller, Phys. Rev. 49, 895 (1936).
4. M. Fierz, Zeits. f. Phys. 104, 553 (1937).
5. C. Möller, Physik, Zeits. Sovijetunion 11, 9 (1937).
6. E. J. Konopinski and G. E. Uhlenbeck, Phys. Rev. 60, 308
    (1941) and R. E. Marshak, Phys. Rev. 61, 431 (1942).
7. M. Conversi, E. Pancini and O. Piccioni, Phys. Rev. 71, 209
    (1947).
8. E. Fermi, E. Teller and V. F. Weisskopf, Phys. Rev. 71, 314
    (1947).
9. R. E. Marshak and H. A. Bethe, Phys. Rev. 72, 506 (1947); see
    also S. Sakata and T. Inoue, Prog. Theor. Phys. 1, 143 (1946).
10. C. M. Lattes, H. Muirhead, G.P.S. Occhialini and C.F. Powell,
    Nature 159, 694 (1947).
11. B. Pontecorvo, Phys. Rev. 72, 246 (1947).
12. J. Tiomno and J. A. Wheeler, Rev. Mod. Phys. 21, 144 & 153 (1949);
    O. Klein, Nature 161, 897 (1948); G. Puppi, Nuovo Cim. 5, 587
    (1948) and ibid 6, 194 (1949); T. D. Lee, M. Rosenbluth and
    C. N. Yang, Phys. Rev. 75, 905 (1949).
13. M. Ruderman and R. Finkelstein, Phys. Rev. 76, 1458 (1949);
    see also S. Sasaki, S. Oneda and S. Ozaki, Sci. Reports of
    Tohoku Univ. XXIII, 77 (1949).
14. L. Michel, Nature 163, 959 (1949).
15. See C. S. Wu and S. A. Moszkowski, "Beta Decay", John Wiley
    (1966).
16. B. M. Rustad and S. L. Ruby, Phys. Rev. 89, 880 (1953); ibid
    97, 991 (1955).
17. Cf. H. M. Mahmoud and E. J. Konopinski, Phys. Rev. 88, 1266
    (1952).
18. E.g. in the general discussion at the session on "Theoretical
    Interpretation of New Particles" (Proc. of Sixth Rochester

Conf. (1956), p. VIII-27) there is the following exchange:
"Feynman brought up a question of [Martin] Block's: Could
it be that the θ and τ are different parity states of the
same particle which has no definite parity, i.e., that
parity is not conserved. That is, does nature have a way
of defining right or left-handedness uniquely? Yang stated
that he and Lee looked into this matter without arriving
at any definite conclusions...".

19. T. D. Lee and C. N. Yang, Phys. Rev. 104, 254 (1956).
20. C. S. Wu, E. Ambler, R. Hayward, D. Hoppes and R. Hudson,
    Phys. Rev. 105, 1413 (1957).
21. R. Garwin, L. Lederman and M. Weinrich, Phys. Rev. 105, 1415
    (1957); J. Friedman and V. Telegdi, Phys. Rev. 105, 1681
    (1957).
22. A. Salam, Nuovo Cim. 5, 29 (1957); L. Landau, JETP (U.S.S.R.)
    32, 407 (1957); T. D. Lee and C. N. Yang, Phys. Rev. 105,
    1671 (1957).
23. T. D. Lee, Proc. of Seventh Rochester Conf., p. VII-7.
24. C. S. Wu, Proc. of Seventh Rochester Conf., p. VII-22.
25. F. Boehm and A. H. Wapstra, Phys. Rev. 109, 456 (1958)
    (paper received Sept. 25, 1956); see also C. S. Wu and
    A. Schwarzschild, Columbia University preprint April 21,
    1958 (unpublished).
26. W. Hermannsfeldt, D. Maxson, P. Stähelin and J. Allen,
    Phys. Rev. 107, 641 (1957) (paper received May 28, 1957).
27. It is interesting to note that J. Schwinger [Annals of
    Phys. 2, 407 (1957) - paper received July 31, 1957] was
    wrestling at this time with a V, T beta interaction in con-
    nection with his intermediate vector boson hypothesis.
28. The meeting with Boehm was attended by Murray Gell-Mann,
    who actually arranged it, by B. Stech, one of Boehm's
    collaborators at that time, and by R. A. Bryan, a Rochester
    graduate student who had driven ECGS to California to see
    the sights. One of us (REM) had encountered Gell-Mann in
    late June at the Rand Corporation in Santa Monica, California
    (where we both happened to be consultants) and, after a
    briefing concerning V-A the the Co^{58} problem, Gell-Mann
    graciously consented to set up a luncheon meeting with his
    Cal Tech colleague, Felix Boehm.
29. M. Deutsch, B. Gittleman, R. Bauer, L. Grodzins and A.
    Sunyar, Phys. Rev. 107, 1733 (1957).
30. S. Watanabe, Phys. Rev. 106, 1307 (1957).
31. E. C. G. Sudarshan and R. E. Marshak, Proc. Padua-Venice
    Conf. on "Mesons and Recently Discovered Particles,"
    (1957), p. V-14.
32. P. K. Kabir, "Development of Weak Intearction Theory",
    p. 118, Gordon and Breach (1963).
33. E. Sudarshan and R. Marshak, Phys. Rev. 109, 1860 (1958).
    (References 34-36 are identical with references 10-12
    of our paper[33])
34. C. S. Wu (private communication); the experiment must be
    redone.

35. Burgy, Krohn, Novey, Ringo, and Telegdi find for the asymmetry 0.15 + 0.08 (private communication from V. Telegdi).
36. Culligan, Frank, Holt, Kluyver, and Massam, Nature (to be published).
37. The inaccessibility of the Padua-Venice conference proceedings does not explain this phenomenon since the Kabir book was available after 1963!
38. W. Hermannsfeldt, R. Burman, P. Stähelin, J. Allen and T. Braid, Phys. Rev. Lett. 1, 61 (1958).
39. G. Impeduglia, R. Plano, A. Prodell, N. Samios, M. Schwartz and J. Steinberger, Phys. Rev. Lett. 1, 249 (1958); see also T. Fazzini. G. Fidecaro, A. Merrison, H. Paul and A. Tollestrup, ibid 1, 247 (1958) had earlier obtained the value $> 2 \times 10^{-5}$ for the $\pi_{e2}/\pi_{\mu2}$ ratio.
40. M. Goldhaber, L. Grodzins and A. Sunyar, Phys. Rev. 109, 1015 (1958).
41. C. N. Yang and R. L. Mills, Phys. Rev. 96, 191 (1954).
42. See also E.C.G. Sudershan, Proc. Indian Acad. Sci. A49, 66 (1959).
43. J. Tiomno, Nuovo Cim. 1, 226 (1955); see also his Princeton Ph.D. thesis (1950).
44. B. Stech and J. D. Jensen, Seits. f. Phys. 141, 175 (1955).
45. A. Salam, Imperial College preprint dated Feb. 1957 (unpublished).
46. R. Marshak, Riazuddin and C. Ryan, "Theory of Weak Interactions in Particle Physics", Wiley-Interscience (1969), p. 89, footnote.
47. A. Salam, Science 210, 723 (1980).
48. J. Tiomno, Nuovo Cim. 6, 912 (1957).
49. See also B. F. Touschek, ibid. 5, 1281 (1957).
50. R.P. Feynman, Proc. of Seventh Rochester Conf. (1957), p. IX-44.
51. R.P. Feynman and M. Gell-Mann, Phys. Rev. 109, 193 (1958).
52. "Owing to the unexpectedly large number and completeness of the papers..., publication has been considerably delayed...The complete version which will consist of some 1000 pages...will be ready by the beginning of May." (letter from N. Dallaporta to R. Marshak, dated April 28, 1958).
53. R.P. Feynman, Proc. of "Neutrino-1974", U. of Pennsylvania, p. 300.
54. Cf. R.E. Marshak, Prod. of Conf. on "Weak Interactions as Probes of Unification" (A.I.P. No. 72, 1981), p. 665.
55. S.S. Gershtein and Y.B. Zeldovich, JETP (U.S.S.R) 2, 576 (1956).
56. J.J. Sakurai, Nuovo Cim. 7, 649 (1958).

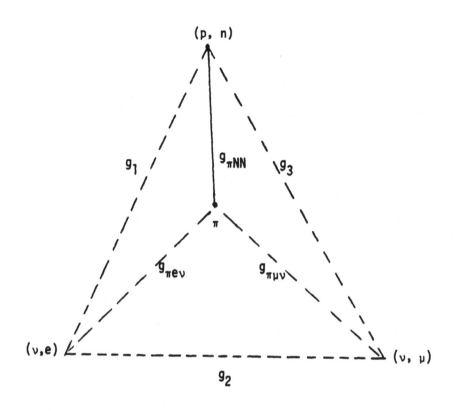

Fig. 1. Diagrammatic sketch showing the
weak interactions (dotted lines) and the
strong interaction (solid line).

SEARCH FOR THE FREE NEUTRINO

F. Reines, University of California, Irvine

The Second World War had a great influence on the lives and careers of very many of us for whom those were formative years. I was involved during, and then subsequent to the war, in the testing of nuclear bombs, and several of us wondered whether this man-made star could be used to advance our knowledge of physics. For one thing this unusual object certainly had lots of fissions in it, and hence, was a very intense neutrino source. I mulled this over somewhat but took no action. Then about 1951, I decided I really would like to do some fundamental physics. Why not, I thought, pick an important but "impossible" problem and solve it-- that would really be something! This brash approach, which so often characterizes the young and naive, is an invitation to failure, but there it was. In any event the question raised was, why not use the bomb as a source of neutrinos? After all, it is extraordinarily intense and so signals produced by neutrinos might be distinguishable from background. Some handwaving and rough calculations led me to conclude that the bomb was the best source. All that was needed was a detector measuring a cubic meter or so. I thought, well I must check this with a real expert. It happened during the summer of 1951 that Fermi was at Los Alamos, and so I went down the hall, knocked timidly on the door and said, "I'd like to talk to you a few minutes about the possibility of neutrino detection." He was very pleasant, and said, "Well, tell me what's on your mind?" I said, "First off as to the source, I think that the bomb is best." After a moment's thought he said, "Yes, the bomb is the best source." So far, so good! Then I said, "But one needs a detector which is so big. I don't know how to make such a detector." He thought about it some and said he didn't either. Coming from the master that was very crushing. I put it on the back burner until a chance conversation with Clyde Cowan. We were on our way to Princeton to talk with Lyman Spitzer about controlled fusion when the airplane was grounded in Kansas City because of engine trouble. At loose ends we wandered around the place, and started to discuss what to do that's interesting in physics. "Let's do a real challenging problem," I said. He said, "Let's work on positronium." I said, "No, positronium is a very good thing but Martin Deutsch has that sewed-up. So let's not work on positronium." Then I said, "Clyde let's work on the neutrino." His immediate response was, "GREAT IDEA". He knew as little about the neutrino as I did but he was a good experimentalist with a sense of derring do. So we shook hands and got off to working on neutrinos.

Why detect the free neutrino? By 1951 essentially everybody had accepted the idea of the neutrino's existence. The Pauli-Fermi Theory was beautiful. It explained an enormous number of phenomena and who needed to <u>see</u> this thing? What did that mean anyway? Why were we interested in the neutrino? Because everybody said, you couldn't see it. Not very sensible, but we were attracted by the challenge. After all, we had a bomb which constituted an excellent intense neutrino source. So, maybe we had an edge on others. But there is a more profound reason which occurred to us as we worked on the problem--in order to determine that the particle existed, you had to see it do something at a point remote from its place of origin. Otherwise, you were not proving it existed. In rebuttal to this point of view, people said "But after all, recoils during K-capture, would accomplish the equivalent." The counter rebuttal is that if one did not see a recoil, then the neutrino hypothesis would be in trouble. But even if you did see the recoil, you still hadn't proved the existence of the neutrino--it was not ruled out because it might be that energy-momentum is not conserved in beta decay. In fact this was a possibility pointed out by Bohr, Kramers and Slater. So, if you didn't have conservation at the "scene of the crime"-- so to speak--no fair using that lack to attribute to the neutrino the responsibility for the recoil in K-capture. That simply was an exercise in circular reasoning. Well, once again being brash, but having a certain respect for certain authorities, I commented in this vein to Fermi who agreed. And in fact, if you look in the famous Chicago notes of Fermi's nuclear-physics course, you will see his description of the convincing way to go and detect the neutrino. A formal way to make some of these comments is to say that, if you demonstrate the existence of the neutrino in the free state i.e. by an observation at a remote location you extend the range of applicability of these fundamental conservation laws to the nuclear realm. On the other hand, if you didn't see this particle in the predicted range then you have a very real problem. To quote an unnamed sage "If you find something you thought was correct to be wrong, then you have made great progress."

As Bohr is reputed to have said "A deep question is one where either a yes or no answer is interesting." So I guess this question of the existence of the "free" neutrino might be construed to be deep. Alright, what about the problem of detection? We fumbled around a great deal before we got to it. Finally we chose to look for the reaction $\bar{\nu} + p \rightarrow n + e^+$. If the free neutrino exists this inverse beta decay reaction has to be there as Bethe and Peierls recognized, and as I'm sure did Fermi, but they had no occasion to write it down in the early days. Further, it was not known at that time whether ν_e and $\bar{\nu}_e$ were different. We chose to consider this reaction because if you believe in, microscopic reversibility, (or is it detailed balancing?) and use the measured value of the neutron half life

then you know what the cross section has to be--a nice clean
result. (In fact, as we learned some years later from Lee and
Yang the cross section is a factor of two greater because of
parity non-conservation). Well we set about to assess the problem
of neutrino detection. How big a detector is required? How many
counts do we expect? What features of the interaction do we use
for signals? Bethe and Peierls in 1934, almost immediately after
the Fermi paper, estimated that if you are in the few MeV range
the cross section with which you have to deal would be ~ 10^{-44}
cm^2. To appreciate how minuscule this interaction is we note that
it is ~ 1000 light years of liquid hydrogen. No wonder that Bethe
and Peierls concluded in the article which appeared in Nature
(April, 1934), "there is no practically possible way of observing
the neutrino." I confronted Bethe, with this pronouncement some
20 years later and with his characteristic good humor he said,
"Well you shouldn't believe everything you read in the papers."

    What in the face of this extraordinarly small interaction
changed the prospects for neutrino detection? Two developments:
the discovery of fission and the discovery by Kallmann and others
of organic liquid scintillators. This meant to us that a large
detector was possible although at the time Cowan and I got into
the act, a "big" detector was only a liter or so in volume.
Despite the large (> 3 orders of magnitude) extrapolation in
detector size we were envisioning it seemed to us an interesting
approach worth pursuing. The detection idea, then, was to use the
$\bar{\nu}$ + p reaction with protons in the scintillator as the target.
The bomb neutrinos would be emitted in a great pulse lasting a few
tens of seconds--until the mushroom cloud left the vicinity of the
detector. "But how do you identify a signal as due to
neutrinos?" Bethe asked me this question when I was describing
all this to him and I said, "Oh that's easy. You use the delayed
coincidence and that identifies it." Clyde and I were well aware
of the delayed coincidence between the positron and neutron
capture pulses but it hadn't yet occurred to us that we were
describing a way to cut the background and, more importantly, that
such a distinctive signature might mean maybe you don't have to
use a bomb. We realized the significance of this somewhat later,
fortunately before we actually carried out the experiment in which
we were set to use the bomb.(Fig. 1) But how sensitive would the
bomb experiment have been? It's interesting that when we
presented the idea for approval by the Los Alamos Director, Norris
Bradbury, we could not see how to detect the predicted cross
section for bomb neutrinos of ~ 10^{-43} cm^2. In fact we figured
that the best we could do was a cross section of 10^{-39} cm^2, four
orders of magnitude short of our goal. Yet it was an improvement
by a factor of a thousand or so from the previous limits and the
Director, in his wisdom, decided to let us try it. So we worked,
and while we were working it occurred to us that you could use the
delayed coincidence and search in the quieter circumstances
associated with a nuclear pile. And so we told Fermi, who as you

can well imagine, was interested in all these things. He wrote
the letter which some of you may have seen.(Fig. 2)

    Well that kind of reinforcement was great stuff.
Incidentally, the detector we designed turned out to be big enough
so that a person bent up, could fit in an insert placed in it.
Intrigued, we proceeded to measure the total K^{40} radio
radioactivity in a couple of humans. Prior to this detector
development if you wanted to measure the K^{40} in a human being you
had to ash the specimen or reduce backgrounds by putting geiger
counters deep underground. Incidentally it was an excellent
neutron as well as gamma ray detector but we resisted the
temptation to be sidetracked and harvest these characteristics for
anything other than the neutrino search. After some preliminary
tests we took the detector to the Hanford plutonium production
reactor. The picture illustrates the arrangement.(Fig. 3) Notice
the sign which reads "Project Poltergeist." What results did we
get from this particular (1953) reactor experiment? We had a 300
liter liquid scintillator viewed by 90, 2 inch photomultiplier
tubes. Backgrounds were very troublesome and we found it
necessary to pile and unpile hundreds of tons of lead to optimize
the shielding. We took the data with reactor on and off and
labored until we were absolutely exhausted. Finally after we had
done the best we could we took the train back to New Mexico. On
the way home we analyzed the data with the results shown in the
Table.(Fig. 4) We had checked by means of neutron sources and
shielding tests that the trace of a signal 0.4±0.2/min. wasn't
just reactor neutrons leaking into the detector. These marginal
results merely served to whet our appetites--we figured that we
had to do better than that. So we designed a detector (1954)
which was rather more sophisticated. The idea is shown in Fig.
5a. The anti-neutrino comes from the reactor, interacts in the
water, produces a positron which annihilates, throwing gammas into
the counters. The neutron moderates and then in 5.5 microseconds
it is captured in the cadmium compound dissolved in the water
target emitting gamma rays, which are again seen by the counters
above and below. This sequence of coincidences is extraordinarily
distinctive and militates against backgrounds of various kinds.
It is noteworthy that this setup (a composite of 330, 5"
photomultiplier tubes viewing several tons of liquid
scintillators) with its complexity born of necessity heralded an
early use in particle physics of very large detectors.

    We allowed that the first hint from the Hanford work was
fine, but our goal was to show beyond even an unreasonable doubt
that there's a neutrino. John Wheeler, hearing of our experiment
suggested that the Savannah River Plant with its well-shielded
700 megawatt (at that time) reactors would be a good place to
go. We located our detecting system about 11 meters from the
core, the whole setup was deep underground, about 12 meters in a
very massive building. "The great Kiva of the southeast," Wheeler

called it. By the way, these were production reactors, built to
provide material for the H-bomb. The several tests we devised
showed first that the reactor-associated delayed coincidence rate
is consistent with theoretical expectations. (This is more
carefully stated, as "not inconsistent" because the fission
neutrino spectra were not well known). Second, we had to prove
that the first pulse of that delayed coincidence was due to a
positron. Third, the second pulse had to be shown to be due to
the capture of a neutron. Then, to be redundant about it, we felt
it necessary to show that the signal was proportional to the
number of target protons. Finally a bulk absorption experiment,
in which one would shield against the neutrons and gamma rays from
the reactor must be shown not to affect this reactor- associated
signal.

Those were the criteria. If the signal has all of these
characteristics, what could it be but a neutrino? Well it could
conceivably be some other peculiar particle, but it seemed that
the most reasonable hypothesis, if it obeyed all those conditions,
called for it to be the Pauli-Fermi neutrino.

Summarizing the results in more detail: the reactor-
correlated signal rate equaled 3±0.2/hr. This signal was tested
in the ways indicated above. Various figures of merit were
achieved. The ratio of this signal to the total accidental
background was 4/1; the ratio to the reactor-independent
correlated background was 5/1; and the ratio of the signal to
reactor-associated accidental background, was 25/1. So there it
was, a husky signal well above background on all these counts.
What about the agreement between the measured signal and the
predicted value? Taking into account our best estimates of
detector efficiency we deduced the interaction cross-section to be
$12^{+7}_{-4}\text{x}10^{-44}$ cm^2. The theoretically predicted cross-section was
$5\pm1\text{x}10^{-44}$ cm^2. The ratio of observed to expected cross-sections
appeared to be off by a factor of 2 or so but the fission
antineutrino spectrum was not well known and were delighted with
such good agreement. We further tested the signal by placing a
thin lead sheet between the target water and the top detector.
The signal went down the way it should if you absorb one of the
positron annihilation gamma rays. In addition, we put some Cu^{64}
in the water target and checked that the magnitudes and shapes of
the pulses were as expected for e$^+$ annihilation. The second pulse
was checked as arising from neutron capture by the capture time
delay spectrum, which agreed with a Monte Carlo simulation.
Furthermore, when we removed the cadmium the correlated signal
disappeared. Then we diluted this light water with heavy water
(to the considerable chagrin of the Savannah River people who had
worked so hard to get pure heavy water) and the rate dropped by a
factor of 0.4±0.1, a number to be compared with the expected
0.5. We took each test seriously, insisting that the signal meet

each one. Finally, we decided to make a test which would
eliminate every other known particle except a neutrino, i.e. a
bulk shielding measurement. This shielding should have cut the
signal by a factor of 10 if it were due to gammas or neutrons.
The shielded configuration rate was 1.74±0.12 per hour, the
unshielded 1.69±0.17 per hour and so it was clear that we were not
looking at reactor neutrons or gamma rays. We felt pretty good
about these results as you might well imagine and we thought it
was time to tell the man who had started it all when, as a young
fellow, he wrote his famous letter in which he postulated the
neutrino saying something to the effect that he couldn't come to a
meeting and tell them about it in person because he had to go out
to a dance! We understand that on receiving our message (Fig. 6)
he interrupted Bernardini during a meeting they were attending to
tell about all this.

It took many years before anybody checked our findings. I've
never been able to figure out why because science is not made by
one person, or one small group doing a measurement. It has to be
done by separate groups so that they can either reinforce or
destroy each other. My guess is that the reason it wasn't
expeditiously checked by others was that almost everybody believed
the neutrino was there anyway and since it appeared to behave as
expected there was no need to check. That might be one reason,
and it has a measure of validity. Another reason is the enormous
effort required relative to the standards of the early to mid
50's--many tons of shielding, big trailers carrying tons of liquid
for the detectors, hundreds of photomultiplier tubes, etc. etc.
Besides the proof was so persuasive who needed to check it? It is
interesting that a careful check could have revealed the factor of
two due to non-conservation of parity!

In the ensuing years several reactions were studied by my
colleagues and me at the Case Institute of Technology and then and
currently at the University of California at Irvine

$$\bar{\nu}_e + p \rightarrow n + e^+ \qquad \text{(inverse beta decay)}$$

$$\bar{\nu}_e + d \rightarrow n + n + e^+ \qquad \text{(inverse beta decay)}$$

$$\bar{\nu}_e + d \rightarrow n + p + \bar{\nu}_e \qquad \text{(neutral current)}$$

$$\bar{\nu}_e + e^- \rightarrow \bar{\nu}_e + e^- \qquad \text{(purely leptonic)}$$

All these reactions were detected, answering several fundamental
questions regarding the nature of the weak interaction including
the parity factor. Incidentally, when this program started it
was, except for the initial free neutrino detection, without
persuasive theoretical motivation. It is just that it seemed to
us these were obvios things to try. As it turned out, though the
number of reactions was small, each one was rich in impli-

cations. Had the field of reactor neutrino physics been more
widely practiced I believe neutral currents would have been
discovered much earlier. Incidentally, the neutral current
reaction of $\bar{\nu}_e$ is obviously unique and should be checked and
measured more precisely.

As a post script I would like to remark on the question of
the identity of the "neutretto," as the neutrino associated with
$\pi \rightarrow \mu$ decay was sometimes referred to in the early literature.
Having just detected $\bar{\nu}_e$ it seemed to us appropriate to investigate
the matter experimentally. So in late 1956, early '57 we said to
the laboratory management, "Now we would like very much to take a
detector to Brookhaven and test the identity of
ν_e and ν_μ." The response we received was, "You fellows have had
enough fun playing around. Why don't you go back to work." We
were deeply disappointed.

In recent years two other groups in addition to ours at
Irvine have set up at reactors detectors designed to study the
intriguing question of neutrino oscillations. The situation is
made interesting by the results so far obtained: One experiment
gives negative results, one gives positive results and the third
is still gathering data. The consequences of a positive result,
i.e. massive neutrinos, would be profound for our understanding of
elementary particles and cosmology.

It is pleasing that such relatively simple experiments can
have an important bearing on such deep questions.

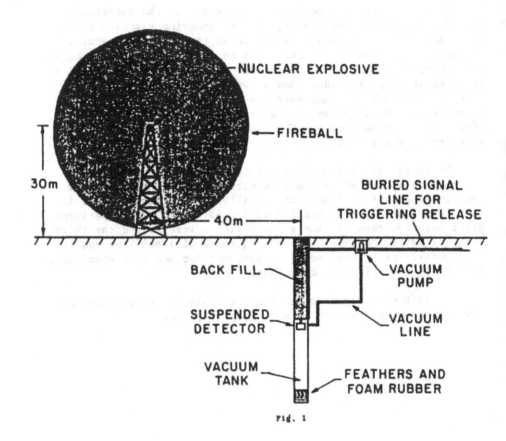

Fig. 1

THE UNIVERSITY OF CHICAGO
CHICAGO 37 · ILLINOIS
INSTITUTE FOR NUCLEAR STUDIES

October 8, 1952

Dr. Fred Reines
Los Alamos Scientific Laboratory
P.O. Box 1663
Los Alamos, New Mexico

Dear Fred:

Thank you for your letter of October 4th by Clyde Cowan and
yourself. I was very much interested in your new plan for
the detection of the neutrino. Certainly your new method
should be much simpler to carry out and have the great advantage that the measurement can be repeated any number of
times. I shall be very interested in seeing how your 10 cubic
foot scintillation counter is going to work, but I do not know
of any reason why it should not.

Good luck.

Sincerely yours,

Enrico

Enrico Fermi

EF:vr

Fig. 2. Letter from Fermi on hearing about our plan to use the
Hanford reactor to attempt to observe the neutrino.

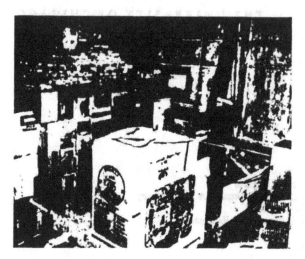

Fig. 3. Shield configuration.

Table 1. Listing of data from the Hanford experiment.

Run	Pile status	Length of run (sec)	Counts per minute[*]	
			Net delayed pair time	Accidental background rate
1	On	4000	2.56	0.84
2	On	2000	2.46	3.54
3	On	4000	2.58	3.11
4	Off	3000	2.20	0.45
5	Off	2000	2.02	0.15
6	Off	1000	2.19	0.13

[*]Delayed coincidence rates: reactor on (10,000 seconds),
2.55 ± 0.15 count/min.; reactor off (6,000 seconds),
2.14 ± 0.13 count/min. Reactor-associated delayed
coincidence rates, 0.41 ± 0.20 count/min.

Fig. 4

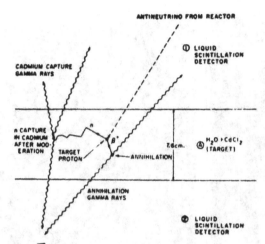

Fig. 5a Schematic of neutrino experiment.

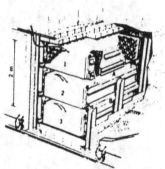

Fig. 5c. Sketch of detectors inside their lead shield. The tanks marked 1, 2, and 3 contained 1400 liters of triethyl-benzene (TEB) liquid scintillator solution, which was viewed in each tank by 110 5-inch photomultiplier tubes. The TEB was made to scintillate by the addition of p-terphenyl (3 grams per liter) and POPOP [1,4-bis-2-(5-phenyloxazolyl) benzene] wavelength shifter (0.2 g per liter). The tubes were immersed in pure nonscintillating TEB to make light collection more uniform. Tanks A and B were polystyrene and contained 200 liters of water, which provided the target protons and contained as much as 40 kilograms of dissolved $CdCl_2$ to capture the product neutrons.

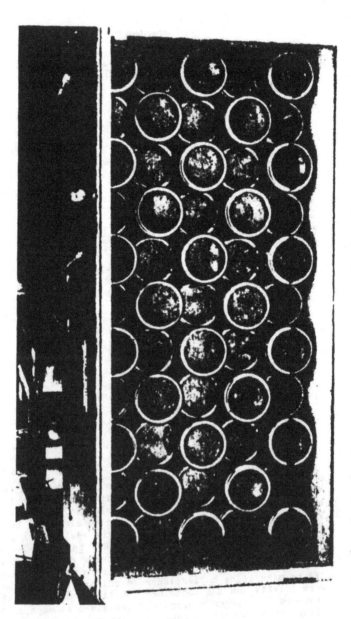

Fig. 5b. Inside end view of detector tank showing 55, 5" photomultiplier tubes.

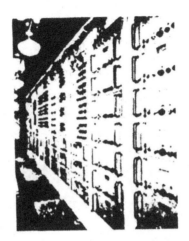

Fig. 5d. Inside view of electronics van showing equipment required to select and record neutrino signals.

WESTERN UNION

To Professor W. Pauli 14 June 1956
Zurich University
Switzerland

We are happy to inform you that we have definitely detected neutrinos from fission fragments by observing inverse beta decay of protons. Observed cross section agrees well with expected six times ten to minus forty-four square centimeters.

Frederick Reines Clyde Cowan
Box 1663 Los Alamos, New Mex

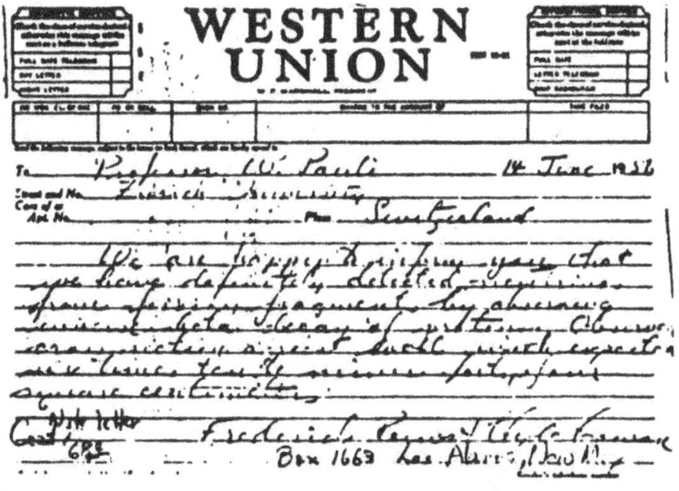

Fig. 6 Telegram to Pauli informing him of our results. The text read: "We are happy to inform you that we have definitely detected neutrinos from fission fragments by observing inverse beta decay of protons. Observed cross section agrees well with expected six times ten to minus forty-four square centimeters."

Probing the weak force with neutrinos

The study of high-energy neutrino scattering
is providing answers—some of them surprising—to some
basic questions concerning the weak interactions.

David B. Cline, Alfred K. Mann and Carlo Rubbia

Interest in neutrino physics has surged up recently, partly because the "little neutral ones" are being groomed for the job (for which they alone are qualified) of probing the interiors of stars. Among elementary particles the neutrino is unique. This is because—as far as we know—it alone interacts with other particles only through the Fermi, or weak, interaction. In this article we will discuss how recent experiments, with new accelerators and detectors, such as the ones at CERN and Fermi Lab, have shed new light on some of the fundamental questions regarding the weak interactions. We will review the recent discoveries of neutral weak currents that conserve strangeness, parity violation and point-like neutrino collisions, as well as some of the implications of these experiments to particle theory.

Many of us know the history of the neutrino, one of the great triumphs of scientific prediction. The charged products of nuclear beta decay were found to carry less energy than the rest energy of the parent nucleus, and Wolfgang Pauli proposed the existence of a new particle with no charge or rest mass. This particle, the neutrino, carried the missing energy and made possible linear and angular momentum conservation in the decay.[1] Shortly after-

ward, Enrico Fermi published his theory of nuclear beta decay,[2] which has since been generalized to an extensive theory of weak interactions. Until the crucial experimental detection of reactor-generated electron antineutrinos by Fred Reines and Clyde Cowan[3] in 1953, all information about neutrinos was obtained indirectly from nuclear decay processes.

The next step was taken about 1960, soon after the alternating-gradient synchrotron at the Brookhaven National Laboratory and the Proton Synchrotron at CERN were constructed. It was recognized by Bruno Pontecorvo in the Soviet Union[4] and independently by Melvin Schwartz,[5] that the high yield of pions and kaons that are produced in the collisions of very energetic protons with matter could be used to make neutrino and antineutrino beams as byproducts. Although the first neutrino beam at BNL was feeble by present standards, it was adequate to show that the neutrinos (of which there are two types) produced through the decays of pions and kaons were mainly those of the muon variety (ν_μ rather than ν_e, the electron neutrino), a result that had been suspected for some time because of the experimentally observed absence of certain decay modes of the muon.[6] The suggestion that electrons and muons differ by some quantum number, the conservation of which forbids some rare decays of muons, was independently made by Julian Schwinger and Kazuhiko Nishijima in 1957. This important discovery opened a new chapter in weak-interaction studies in which accel-

erator-produced high-intensity neutrino and antineutrino beams were used in conjection with massive targets that also served as detectors.

The neutrino is an ideal probe of weak interactions—and, somewhat surprisingly, also of strong interactions. Furthermore, if weak and electromagnetic interactions have a common origin, as the spontaneously broken gauge theories suggest,[7] the neutrino may also be able to probe the electromagnetic force at very small distances. The study of neutrino scattering by nucleons at the highest energies, where the cross section and the variety of final states are very large, has therefore become of considerable importance at the new proton synchrotron now in operation at Fermi Lab and at a similar accelerator soon to be completed at CERN.

Targets that detect

The first detector of neutrino interactions[3] was a relatively modest 10 tons of liquid scintillator and water-loaded $CdCl_2$. It was used to observe the collisions of few-MeV antineutrinos from beta decay with matter that resulted in the inverse beta decay reaction,

$$\bar{\nu}_e + p \longrightarrow n + e^+$$

The earliest detectors of neutrino scattering at BNL were massive, thick aluminum-plate optical spark chambers.[6] They were well suited to separating muons from electrons in the final states of the neutrino–nucleon interaction. Later, heavy-liquid (propane and heavy freon) bubble chambers were used extensively at CERN as neutrino detec-

David B. Cline is professor of physics at the University of Wisconsin, Madison; Alfred K. Mann is professor of physics at the University of Pennsylvania, Philadelphia, and Carlo Rubbia is professor of physics at Harvard University, Cambridge, Massachusetts.

And another particle?

Evidence accumulated in an experiment at Fermi Lab indicates that the high-energy interactions of neutrinos and antineutrinos with nucleons may be significantly different even after taking into account their opposite helicities. This appears to be a violation of a basic principle of semileptonic weak processes known as charge-symmetry invariance (A. Benvenuti et al, Phys. Rev. Lett. 33, 984, 1974). It has been conjectured by Alvaro De Rujula and Sheldon Glashow of Harvard University that the production of new particles, in particular those with a new quantum number called "charm," could cause such a violation. Some theories that attempt to unify the weak and electromagnetic interactions have suggested the existence of these particles.

Now, the same experimenters (Carlo Rubbia and Lawrence Sulak from Harvard; William Ford, Ta-Yung Ling, Alfred Mann and Frederick Messing from Pennsylvania; Alberto Benvenuti, David Cline, Richard Imlay, Robert Orr, Donald Reeder and

Peter Wanderer from Wisconsin, and Ray Stefanski from Fermi Lab) have recently reported (Phys. Rev. Lett. 34, 419, 1975) the substantial production, by neutrinos and antineutrinos, of events with two final-state muons (dimuon events). This follows their initial report of two such events at the London Conference on High Energy Physics last summer. It appears that these events require the production of one or more new particles, which could be either leptons or hadrons. Certain characteristics of these dimuon events lead to estimates of mass greater than 2 GeV/c^2 and less than about 5 GeV/c^2 and of lifetime less than 10^{-10} sec.

If the new particle is a hadron, one of the muons in the dimuon events would be a product of the decay of that hadron (along with a neutrino or antineutrino, and possibly other hadrons), while the other muon would be part of the initial neutrino interaction that gave rise to the new particle. These particles may be indirectly related to the narrow-width, neutral vector mesons recently discovered at Brookhaven and SLAC (see "Search and Discovery," page 17, this issue). In contrast to the vector mesons, however, the proposed new particle or particles would carry a new quantum number, and decay through the weak interaction. This is because neutrinos and antineutrinos can produce a single particle with any value of a quantum number that is not conserved in weak interactions.

One can not, however, rule out other possible origins of the dimuon events. For example, the production of a heavy neutral lepton with a partial decay mode into two muons and a neutrino is an alternative, though less probable, explanation of the observed dimuon events. The Harvard-Pennsylvania-Wisconsin-Fermi Lab group plans to repeat the experiment at Fermi Lab during the next few months, and expects to find additional answers to some of the vexing questions relating to these events.

---

tors. These are useful detectors of neutrino interactions, being relatively massive and allowing the vertex of the scattering event to be observed and reconstructed.

Neutrinos produced by 30-GeV proton accelerators often give rise to several hadrons in the final states of their collisions. The advent of higher-energy neutrino beams and the increased theoretical interest in inclusive lepton scattering, which concentrates on only a few gross properties of the final-state hadrons, such as total hadron energy, have suggested the use of a separated-function neutrino detector. This detector consists of two primary parts: an ionization calorimeter (target-detector) and a magnetic muon spectrometer. The ionization calorimeter serves as the target for the neutrino and also measures the total hadronic energy E_H released in the collision. The muon spectrometer measures the sign of the muon's charge and its momentum p_μ. The laboratory energy of the incident neutrino E_ν is then determined as the sum $E_\nu = E_H + E_\mu$, where to a good approximation, $E_\mu = p_\mu c$.

Target masses of about 100 metric tons have been used; an ionization calorimeter now being constructed at Fermi Lab will have a mass of about 300 metric tons, and will be capable of measuring the direction of the hadronic energy flow and of separating the hadronic component from the electromagnetic component in the final-state shower. Among other things, this should allow the detection of high-energy electron-neutrino interactions as well as muon-

neutrino interactions. At high neutrino energies, the shield employed to protect the target-detector from incident charged particles (particularly the muons from the decays of the initial pions and kaons) is necessarily long, which in turn means neutrino beams of relatively large cross-sectional area. The transverse dimensions of the detectors, especially those of the muon spectrometer, are correspondingly large if the full kinematic range of the outgoing muons is to be studied. For example, one detector at Fermi Lab uses four toroidal iron magnets, each 12 feet in diameter and 4 feet thick; a future detector will add toroids 24 feet in diameter.

The large-solid-angle detectors used in conjunction with the higher neutrino energies and intensities available at Fermi Lab, and the large total neutrino cross section, yield a substantial number of neutrino-induced events per hour of accelerator time. This encourages the performance of relatively brief experiments directed at specific points of enquiry. Occasionally they use neutrino beams of different properties, emphasizing various aspects of the detectors; this in itself is a new trend with significant implications for the future of neutrino physics.

Charged weak currents

A striking property of the weak-coupling constant $G = 1.01 \times 10^{-5}/m_p^2$ observed in low-energy weak interactions is that it has the dimensions of $1/$energy^2. By dimensional arguments, then, if no fundamental masses enter, the neutrino-nucleon total cross section σ^ν

should depend on the center-of-mass energy E as[8]

$$\sigma^\nu \propto G^2 E^2 \qquad (1)$$

For collisions of a massless particle with a nucleon at high energy,

$$E^2 \approx 2 M_p E_\nu c^2$$

and thus the cross section should rise linearly with neutrino laboratory energy,

$$\sigma^\nu \propto G^2 E_\nu \qquad (2)$$

Recent data from CERN in the few-GeV region, and from Fermi Lab in the region up to 200 GeV, are shown in figure 1, which shows that the cross section indeed depends linearly on neutrino energy within present experimental error.[9,10,11] It is interesting also to include in figure 1 the cross-section value obtained by Reines and Cowan[3] for the process

$$\bar\nu_e + p \longrightarrow n + e^+$$

and data[11] from Argonne at $E_\nu \lesssim 1$ GeV, where quasielastic scattering and nucleon-resonance production dominate. It might have been expected that above a few GeV some characteristic mass would influence the energy scale, but the observed rise of the total cross section from 0.1 GeV to 10 GeV in figure 1 indicates the absence of any such mass.

The neutrino cross section, in rising linearly with energy, fulfils a prediction of scale invariance that was proposed by James Bjorken[12] and suggested by the MIT-SLAC experiments on electron-nucleon scattering.[13] For the total cross section to continue to rise linearly with energy, however, the assumption that the weak interaction is local is required as well as scale invariance.[14] A deviation from linearity would arise, for example, if an intermediate vector boson should serve as the propagator of the weak interaction. Gauge theories that attempt to unify weak and electromagnetic interactions, with coupling constants G and α, respectively, suggest a fundamental mass Λ given by[7,15]

$$G \Lambda^2 \approx \alpha \text{ or } \Lambda \approx 37 \text{GeV} \quad (3)$$

Other estimates, following from attempts to calculate higher-order corrections to weak interactions, lead to much lower values of Λ, as indicated in the Table on page 26. An upper limit on Λ is presumably obtained directly from the weak-coupling constant itself, since $G^{-1/2} \approx 300$ GeV is the approximate CM energy at which unitarity is violated in the elastic scattering of neutrinos by electrons,[8] an interaction involving leptons only. The data of figure 1 rule out a propagator of mass less than about 10 GeV. In the next few years, the improved quality of the data should permit the search to be extended to energies as high as 30 GeV.

Maximal parity violation

Another striking feature of weak interactions at low energy is the fact that parity violation is almost at its theoretical maximum for the usual model. Before high-energy neutrino interactions were studied, there was no certainty that parity violation would continue to be observed in weak interactions at high energies. Recently, the cross section for muon-antineutrino scattering has been measured in the few-GeV region at CERN and at much higher energies at Fermi Lab.[9,10,11] This was done with an isoscalar target, which is one that (like He[4] or C[12]) has equal numbers of neutrons and protons, and hence zero isospin. These results are compared with the measured cross section for neutrino scattering in the energy range of 1 to 80 GeV in figure 2. From the lowest energy to the highest, the cross-section ratio plotted there stays quite close to 1/3, the lowest possible value and one that is expected for the scattering of neutrinos and antineutrinos from a relativistic point-like fermion, as figure 3 illustrates.

This result is surprising in light of the fact that high-energy neutrino and antineutrino collisions are probing the nucleon over very short time intervals (by the uncertainty principle) and virtual fermion–antifermion pairs of very large effective mass are expected to be abundant (as are e^+–e^- pairs in quantum electrodynamics). Thus *neutrino-hadron collisions at very high energy continue to show nearly maximal parity violation, which is consistent with the*

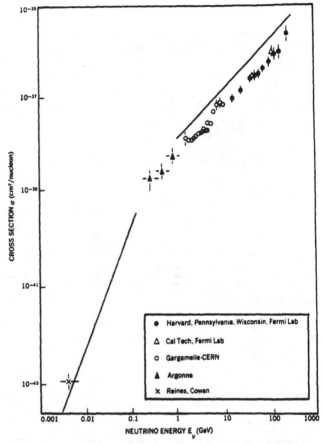

CROSS SECTION σ (cm²/nucleon)

NEUTRINO ENERGY E_ν (GeV)

- ● Harvard, Pennsylvania, Wisconsin, Fermi Lab
- △ Cal Tech, Fermi Lab
- ○ Gargamelle-CERN
- ▲ Argonne
- × Reines, Cowan

The neutrino cross sections that have been measured, at energies ranging from 8 MeV to 200 GeV, are compiled here. The point with lowest energy is for interactions of electron antineutrinos; all the others are for those of muon neutrinos. The lower line is the low-energy theoretical prediction and the upper line is that for point protons.   Figure 1

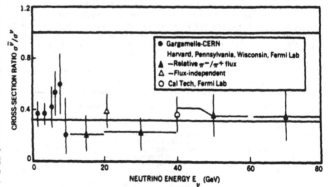

CROSS-SECTION RATIO $\sigma_{\bar{\nu}}/\sigma_\nu$

NEUTRINO ENERGY E_ν (GeV)

- ● Gargamelle-CERN
  Harvard, Pennsylvania, Wisconsin, Fermi Lab
- ▲ —Relative π^-/π^+ flux
- △ —Flux-independent
- ○ Cal Tech, Fermi Lab

The measured ratio of total cross section for antineutrinos to that for neutrinos. The black line at 1.0 represents the value expected when parity is conserved, such as when scattering from an equal mixture of fermions and antifermions. The colored line at 1/3 is that for maximal parity violation, expected with scattering from relativistic point-like fermions, as indicated in figure 3. It represents the ratio of $\bar{\nu}_e + e \rightarrow \bar{\nu}_e + e$ to $\nu_e + e \rightarrow \nu_e + e$ interactions.   Figure 2

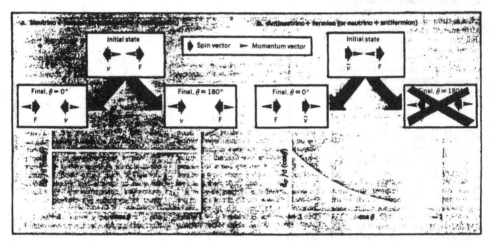

How the ratio of 1/3 arises. The three upper diagrams in (a) represent two extreme cases of neutrino-fermion collisions in the center-of-mass frame. As total angular momentum is zero, scattering angles of 0 or 180 deg are possible, as shown in the diagrams. The curve below generalizes this to all angles: The angular distribution is flat. For antineutrino–fermion collisions (b), a final state with $\theta = 180$ deg is impossible, because angular momentum would then not be conserved; the angular-distribution curve generalizes this. The ratio of the area under the right-hand curve to that under the left is 1/3: This is the total cross-section ratio of figure 2.   Figure 3

conjecture that nucleons are composed of low-effective-mass fermions and almost no antifermions. A model of the nucleon as made up of "partons" was invented by Richard Feynman. In such constituent models of the nucleon, the value of the effective mass is determined approximately by the slope of the linear rise of the neutrino total cross section.

These examples constitute two surprising results that have been obtained with high-energy neutrinos in the study of the weak interaction. They are, of course, surprises only in the sense that essentially the same behavior of the weak interaction is observed at high as at low energy. There is no evidence of space–time structure in the weak interaction up to a laboratory energy of about 200 GeV. These examples also raise the question: Why are these high-energy, high-momentum-transfer inter-

actions of nucleons, which are thought to be soft, complicated objects, so similar to the predicted weak interactions of primitive, point-like fermions, as suggested by figure 3?

It is amusing to note that, in 1964, the data from the CERN neutrino experiments, then being carried out in a small propane bubble chamber, frequently showed very wide-angle muons arising from neutrino–nucleon collisions, which suggested that the collisions were more point-like than predicted by any theory at that time. Now, ten years later and with laboratory energy about 100 times higher, we begin to recognize the deeper simplicity implied by these early results.

High-energy neutrino studies bear on a number of other questions in weak interactions, some of which are treated in question-and-answer form on page 27. The remainder of this article will be

concerned with three of those questions and the answers to them as far as they are available at present. These are the conserved-vector-current hypothesis, charge-symmetry invariance in deep-inelastic neutrino scattering and the existence of neutral weak currents. Although we will not discuss further the other subjects in the box on page 27, it is a tribute to the power of the neutrino as a probe that many of them will be addressed by neutrino experiments in the next few years.

A conservation, an invariance

Two theoretical predictions, conservation of vector current and charge-symmetry invariance, have been verified. Low-energy weak interactions involving one lepton are described in lowest order by the product of a leptonic current J^L and a hadronic current J^H. The latter is in turn composed of a vector part V^H and an axial-vector part A^H. For weak-interaction processes in which the strangeness of the initial and final states is the same, V^H and A^H currents in low-energy processes have some remarkably simple properties. One is the conservation of the vector current V^H, which was first postulated to explain the near equality of the vector-coupling constants observed in nuclear β decay and μ decay.[16] This situation is similar to that in electromagnetic processes, where the matrix elements of

$$e^+ + e^- \longrightarrow e^+ + e^-$$

and $e^- + p \longrightarrow e^- + p$ are identical at low momentum transfer. In spite of the presence of strong interactions, the matrix element is de-

Some mass scales that have been suggested

Mass	Theoretical conjecture
~2600 GeV	Unitarity violation for[a] $\nu + \bar{\nu} \to W + W$
~300 GeV	Unitarity violation for $\nu_\mu + e \to \mu_e + \mu$
~30 — 70 GeV	Gauge theory unification of weak[b] and electromagnetic interactions for leptons
~4 — 15 GeV	Perturbation calculation of the $K_L - K_S$ mass difference[c] and suppression of higher order weak interactions in $K_L \to \mu + \mu$
$\lesssim$ 10 GeV	New particles needed to incorporate[d] hadrons into gauge theories and remove $\Delta S = 1$ neutral currents.

a See reference 32.   c See reference 33.
b See references 3, 9, 10, 11 and 12.   d See reference 31.

termined by the charge of the targets, and electric charge is conserved.

A second important property of semileptonic interactions at low energy is charge-symmetry invariance, which is established by experiments on the beta decay of mirror nuclei (pairs of nuclei for which proton and neutron number are interchanged) and on the strangeness-conserving beta decay of the sigma particle[16,17]

$$\Sigma \longrightarrow \Lambda + e + \nu_e$$

Simply stated, this principle says that the weak hadronic current, when rotated through 180 deg in isotopic-spin space, becomes its hermitian adjoint. In a model in which the exchange of (possibly fictitious) charged intermediate vector bosons accounts for the weak interaction, charge-symmetry invariance requires the weak amplitude to be independent of the charge of the exchanged boson. However, the amplitude may still depend on the helicity of the exchanged boson.

The combination of the conserved vector current and charge-symmetry invariance results in the isotriplet current hypothesis, which directly links the isovector weak amplitude with the isovector electromagnetic amplitude. The classic low-energy tests of this hypothesis are measurements of the "weak-magnetism" terms in the decays

$$N^{12} \longrightarrow C^{12} + e^+ + \nu$$
$$\text{and } B^{12} \longrightarrow C^{12} + e^- + \bar{\nu}$$

and of the extremely small value of the branching ratio for the decay[20]

$$\pi^+ \longrightarrow \pi^0 + e^+ + \nu$$

The conservation of the vector current and charge-symmetry invariance reflect the conservation of isospin (the quantum number that distinguishes the proton from the neutron) by the strong interaction, and the isotriplet hypothesis follows from the conservation of hypercharge (related to baryon number and strangeness) by the electromagnetic interaction.[16,18]

Extension to high energies

The isotriplet hypothesis and charge-symmetry invariance are concepts derived from the study of semileptonic processes involving a limited number of hadrons at very low energy and momentum transfer. In contrast, at high energy and high momentum transfer, neutrino interactions proceed largely through processes that involve many hadrons in the final state; it is therefore by no means clear that these simple properties should continue to describe the semileptonic weak interaction.

The failure of these hypotheses for neutrino energies above some threshold energy would be a signal of an important change in either the weak interaction or the nature of the hadronic final states. For example, new hadronic de-

Some fundamental questions in high-energy weak interactions, circa 1960–1974.

● Question	● Answer
Is ν_μ identical with ν_e?	No, they are different, as observed in the first accelerator neutrino experiment at Brookhaven National Laboratory.
Do first-order neutral currents exist? Are they V-A interactions also?	No, for low-energy reactions that change strangeness. Yes, for neutrino collisions (strangeness non-changing). The V-A property is *not* tested yet.
What is the lepton conservation rule for ν_e and ν_μ?	Additive quantum numbers experimentally preferred in electron–neutrino interactions.
Does charge-symmetry invariance hold?	Probably yes, but it may appear to fail if new particles are made at high energy.
Is the vector current conserved at high energies?	Probably yes, but it has only been weakly tested.
Is the weak interaction pointlike at high energies?	Yes, as far as now tested, up to a mass of about 10 GeV.
Is there a univeral weak interaction that includes neutral and charged currents?	Not known, but it is conjectured that a new quantum number ("charm") can restore universality between strangeness-changing and strangeness-preserving neutral current processes.
What is the origin of CP violation?	Not known, but it is conjectured that a new superweak interaction may exist.
Is there a universality between ν_μ and ν_e neutral currents?	Not tested yet—can be tested when enriched electron-neutrino beams become available.
Do the weak interactions (charged currents and neutral currents) proceed by the exchange of massive vector bosons?	Not tested so far. The mass of the charged boson is greater than 10 GeV. No meaningful limit has been set on the neutral boson.
Why are higher-order weak interactions so "weak"?	Not known, but partially explained if gauge theories are correct and if low-mass "charmed" particles exist.
Are there heavier leptons in nature?	Not well tested, but for some varieties mass limits of 2-7 GeV have been set.

grees of freedom excited at high energy could cause a failure of both charge-symmetry invariance and the isotriplet hypothesis. The production of single strange particles also provides an example of how these two hypotheses can be broken. Because neither isospin nor hypercharge is conserved in such processes, there is no simple reason why the vector current should be conserved, or that charge-symmetry invariance should hold. However, the expected level of symmetry breaking in this case is only about 5%, due to the known small cross section for strange-particle production. Thus the test of the isotriplet hypothesis and of charge-symmetry invariance with high-energy neutrinos can be thought of as *a search for new particles* that have different isospin or hypercharge properties from nucleons.

A preliminary test of the isotriplet hypothesis can be made by comparing the integral of the structure function $F_2(x)$ of the nucleon that is obtained in electron–nucleon scattering with that in neutrino–nucleon scattering on isoscalar targets.[9] The electron–nucleon integral is taken directly from the MIT-

SLAC experiment.[13] To estimate the isovector part of the electromagnetic structure function, the electromagnetic current has been assumed to obey the SU(3) predictions for the relative coupling of the $I = 1$ and $I = 0$ components. This decomposition fits the experimental data on photoproduction.

We obtain the integrated vector part of the structure function for neutrino scattering from the slope of the energy dependence of the total cross section (figure 1), and a comparison of neutrino and antineutrino scattering (figures 2 and 3). The latter comparison permits the contribution of the structure function associated with A^H to be taken into account. Figure 4 shows a comparison of the two structure functions and a test of the isotriplet hypothesis.[9,10] Within experimental error there is good agreement between the two integrated structure functions.

The test of charge-symmetry invariance relies on the comparison of the details of the reactions

$$\nu_\mu + \text{nucleus} \longrightarrow \mu^- + \text{anything} \quad (4)$$
$$\bar{\nu}_\mu + \text{nucleus} \longrightarrow \mu^+ + \text{anything} \quad (5)$$

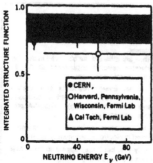

A test of the hypothesis of conserved vector current. The experimental points are integrated vector structure functions and the colored band is the estimated isovector electroproduction structure function.   **Figure 4**

and on the use of a hadronic target of zero isotopic spin, $I = 0$.

For these systems the prediction of charge-symmetry invariance is particularly simple: The three individual nucleon structure functions for neutrino and antineutrino processes with $I = 0$ are equal,

$$F_i' = F_i^{\,\bar{\imath}}, \; i = 1, 2, 3 \qquad (6)$$

Equation 6 is a consequence of the fact that the amplitudes for these processes are hermitian adjoints of each other and therefore charge-symmetry invariant. Under the assumption of charge-symmetry invariance these reactions can only differ if the helicities of the outgoing and incoming leptons are different (as, for example, in the model of collisions with relativistic fermions, figure 3).

The first test of the invariance of charge symmetry, in which the validity of scale invariance is assumed, has been carried out by comparing the ratio r of two structure functions for processes 4 and 5. If charge-symmetry invariance holds, the ratios r^r and $r^{\bar{r}}$ should be the same for neutrino- and antineutrino-initiated reactions. Comparison of these ratios has been made at various values of the dimensionless scaling variable $x = Q^2/2mE_H$, where Q^2 is the square of the four-momentum transfer and E_H is the hadronic energy measured in the neutrino collision. Figure 5 shows the results of the test.[19] We see that for the larger values of x (>0.1), the data are compatible with charge-symmetry invariance, while in the lower x region a discrepancy exists. This discrepancy at small x, if it survives future experimental tests, suggests a surprise may be in store for us: The violation of charge-symmetry invariance may be due to the production of particles with new quantum numbers in high-energy neutrino and antineutrino collisions. Recent devel-

opments in the search for these new particles are discussed in the box on page 24.

Observation of weak neutral currents

Historically, weak interactions were detected[24] through charge-current semileptonic nuclear decays,

$$A \longrightarrow B + e^- + \bar{\nu},$$

The observation of weak decays with two charge-conjugate leptons in the final state,

$$A \longrightarrow A' + e^+ + e^-$$
$$\text{or } A \longrightarrow A' + \bar{\nu} + \nu,$$

would have provided evidence for neutral weak currents. The experimental search for neutral currents with two charged leptons in the final state is effectively impossible in nuclear decays because the probability for a neutral-current transition is many orders of magnitude smaller than that for a corresponding electromagnetic transition. The search for weak nuclear decays with two neutrinos is equally difficult. Thus the question of weak neutral currents was never put to an experimental test in the study of nuclear decays.

The earliest definitive search for weak neutral currents focussed on the decays of strange particles. The semileptonic weak decays of K mesons by neutral currents,[20]

$$K^+ \longrightarrow \pi^+ + e^+ + e^- \qquad (7)$$
$$K^+ \longrightarrow \pi^+ + \nu + \bar{\nu} \qquad (8)$$

and

$$K^0 \longrightarrow \mu^+ + \mu^- \qquad (9)$$

are forbidden for electromagnetic interactions and, therefore, provide excellent tests for weak neutral currents that change strangeness. Over the past twelve years, intensive experimental searches for these and similar decays have been carried out. No example of a weak neutral-current decay of first order has been found—and the experimental limits are by now extremely low. Neutrino interactions provide another way to search for weak neutral currents without competition from electromagnetic interactions. In neutrino interactions it is possible to search for both strangeness-changing and strangeness-conserving weak neutral currents by searching for neutrino-induced events without charged leptons in the final state. In the neutrino experiments carried out in the 1960's, several experimental searches for weak neutral current interactions were reported, and limits were placed[22] on various neutral-current processes that conserve strangeness. Along with the null results obtained from strange-particle decays, these results from neutrino scattering were taken as further indication that first-order semileptonic neutral weak currents are absent in nature. There were even instances of rejections by

journals of theoretical papers that predicted large neutral-current rates because it was "well known that first-order neutral currents did not exist."

Is it a neutrino or a neutron?

The advent of a new generation of heavy-liquid bubblechambers, neutrino beams of higher energy and higher intensity and massive target–calorimeter detectors reopened the search for weak neutral currents.[23,24] Caution must be exercised in distinguishing neutrino events from the background, as there are two mechanisms that can stimulate neutrino-induced events without final-state charged leptons. The most important of these comes from neutral hadrons that are incident on the experimental detector. Proton accelerators produce copious quantities of neutrons and neutral kaons. The interactions of these particles in neutrino detectors are easy to distinguish from high-energy charged-current neutrino events because of the presence of a charged lepton in the final state. However, neutral-current events, those without the charged lepton in the final state, are very difficult to separate from neutron and K^0 interactions. Furthermore, neutrons and K^0 mesons are also produced in quantity by the neutrino beam itself in the shielding used to protect the neutrino detector from muons. A second source of background tracks for neutral-current events is provided by charged-current events in which the outgoing charged lepton either escapes from the detector before identification or is misidentified as a hadron.

Since muon neutrinos are the principal source of high-energy neutrino interactions, events with undetected

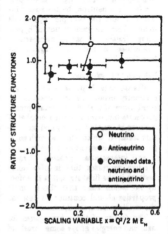

The ratio of structure functions for reactions initiated by neutrinos and antineutrinos versus scaling variable as a test of charge-symmetry invariance.   **Figure 5**

muons are the largest source of background from charged-current interactions. Thus, to search for neutral currents, good muon identification is required, and interactions due to incident neutral hadrons must be either suppressed or subtracted out.

One of the observations of events without muons in the final state has been carried out at CERN with the large bubble chamber, Gargamelle. In this experiment a semiempirical estimate of the neutron flux and subtraction of the background eliminated neutron and K^0 interactions.[23] The charged-current background is eliminated by selecting events in which all particles are identified as hadrons through scattering or decay.[23]

The electronic experiments with large-mass target-detectors cope with the background in another way. It is experimentally observed that the neutral hadron interactions are eliminated by choosing a target volume that requires the hadrons to traverse many protective interaction lengths of material, in which the neutrons and K^0 mesons are attenuated.[24] Identification and rejection of charged-current events is carried out by placing a hadron filter near the detector through which the muons pass.[24,25]

The primary uncertainty is related to the semiempirical estimate of the efficiency for muon detection. In the experiments at Fermi Lab that were carried out by physicists from Harvard, Pennsylvania and Wisconsin as well as from Fermi Lab, the detection efficiency was obtained directly from the muon angular distribution that was measured out to 500 mrad, and a muonless event signal was observed after correcting for detection efficiency.[24,25] The existence of this signal does not depend on the muon's angular distribution outside of the measured angular region, as can be shown by the following simple argument:

The kinematics of high-energy neutrino collisions require that, in events in which muons were produced at angles greater than 500 mrad, the hadronic energy carried away in the collision is nearly equal to the incident neutrino energy; this is a direct consequence of the conservation of transverse momentum in the collision. Thus, if the observed muonless signal were actually due to misidentified charged-current events at wide angles, the spectrum of visible energy for the muonless events should approximate the incident neutrino spectrum.

The comparison shown in figure 6 of the spectra for the muonless events with that for the charged-current events indicates a substantial difference. The muonless events also have the uniform spatial distribution in the detector that is expected from neutrino interactions.

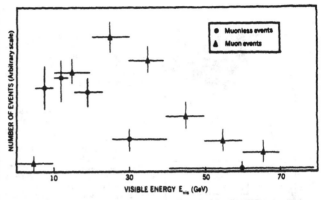

Comparison of the visible energy spectra for events with and without muons led to the conclusion that muonless events induced by neutrinos have been observed. If the muonless events were actually misidentified muon events, the two distributions should be the same, but the average visible energy is 15 GeV for muonless and 35 GeV for muon events.   Figure 6

This evidence led to the conclusion that neutrino-induced muonless events had been observed.[24,25]

Visible energy

The primary conclusion from the CERN-Gargamelle and the Harvard-Pennsylvania-Wisconsin-Fermi Lab experiments is that, unlike earlier neutrino-scattering experiments, they show a significant number of neutrino- and antineutrino-induced events without muons in the final state. It is important to note that this positive signal is obtained in both experiments even though the experimental methods are so different. It is also of interest that the average energy of the events observed in the two experiments is quite different; in the CERN experiment it was about 3 GeV, as compared to about 40 GeV in the HPWF experiment. Other evidence is now available to support these observations.[26,27]

The energy spectra in figure 6 also indicate that, on the average, approximately one-half of the incident neutrino energy does not appear as "visible" energy in the final state; energy, that is, that was carried by particles detected in the bubble or spark chambers. Since the events are born in the middle of a massive detector, the escaping energy must be carried by a long lived, weakly interacting particle. The simplest hypothesis concerning the nature of these events is that they are examples of the neutral-current processes

$$\nu_\mu + nucleon \longrightarrow \nu_\mu + anything \quad (10)$$

$$\bar{\nu}_\mu + nucleon \longrightarrow \bar{\nu}_\mu + anything \quad (11)$$

which have been long sought after. However, we cannot rule out the possibility that they are manifestations of other, as yet unrecognized, processes. For now we shall call them weak neu-

tral-current events, although keeping in mind that this interesting possibility is not conclusively proven.

The rate for the observed weak neutral currents in neutrino and antineutrino collisions is not greatly different from the corresponding charged-current processes, quite in contrast to the very low limits on strangeness-changing neutral currents obtained for K decays. A convenient comparison is provided by the ratio of the neutral-current rate to the charged-current rates, R^ν for neutrino and $R^{\bar{\nu}}$ for antineutrino processes. Figure 7 shows the published values of these ratios.[23,25] While the actual experimental values need to be further refined by better data, this figure shows the latitude of physical possibilities for this new phenomenon. For example, if neutral currents conserved parity, a ratio of 3 would be expected between R^ν and $R^{\bar{\nu}}$, which would in turn reflect the maximal parity violation in the charged-current interaction discussed previously.

Alternatively, the Weinberg-Salam gauge model[7] of the weak and electromagnetic interactions, in which weak neutral currents as well as weak charged currents make a natural appearance, suggests a different relation between R^ν and $R^{\bar{\nu}}$. It may be that weak neutral currents are a part of the weak interaction that has been missing for 77 years, which forms a direct connection between the weak and electromagnetic interactions. Whatever their origin, the neutral weak currents promise to be a rich field of study in high-energy neutrino collisions.

Future directions

Major advances in weak-interaction studies have occurred in every one of the past four decades. They started

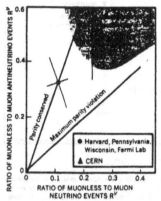

RATIO OF MUONLESS TO MUON NEUTRINO EVENTS R^ν

Measured values of the ratio of neutral-current to charged-current rates for neutrino and antineutrino processes are here compared with the predictions of three different theoretical models. Figure 7

with the neutrino hypothesis and the Fermi theory in the 1930's, continued with the universality of the weak interaction for hadrons and leptons in the 1940's, parity violation and V–A coupling in the 1950's, CP violation and SU(3) universality in the 1960's, and so far include strangeness-conserving neutral weak currents and point-like high-energy neutrino collisions in the 1970's.

The preliminary results of the study of high-energy charged-current reactions with neutrino and antineutrino beams have already provided some interesting surprises. These results indicate that the deep inelastic scattering cross sections from protons and neutrons are similar to the scattering expected from elementary or primitive fermions with approximately one-half of the nucleon mass. Tests of charge-symmetry invariance and the isotriplet hypothesis at high energies show qualitative agreement with low-energy data, with the possible exception of a deviation at higher energies and at low x values.

Models that have been invented to explain these and other results on deep inelastic scattering of charged leptons, such as the quark–parton model, suffer from the difficulty that the constituent quarks have never been dislodged from the nucleon and experimentally observed. Also the results of recent experiments on high-energy e^+–e^- annihilation to hadrons appear to disagree with these models.[28]

Thus we are left with experimental results that suggest a profound simplicity in the interaction of leptons with hadrons at high energy, but without a complete picture of the origin of that simplicity. Perhaps there is a deeper reason that underlies these results, hav-

ing to do with a common origin of the weak, electromagnetic and strong forces, as suggested in attempts to formulate unified field theories. Steven Weinberg has written an excellent introduction to this subject.[29]

New low-mass hadrons?

The charged-current experiments have also set a new lower limit of about 10 GeV on the effective mass of any charged particle that might serve as the propagator of the weak interaction. This can be interpreted as a lower limit on the mass of the charged intermediate vector boson. Thus, charged-current weak interactions appear to be point-like at least down to distances of the order of 10^{-15} cm.

The existence of muonless events—presumably neutral-current weak interactions—is now established in high-energy neutrino collisions. The next step is to delineate the matrix element of this interaction. Let us hope that the study will take less than the twenty years needed to go from the first Fermi theory to the V–A theory!

Finally, we may note that the ratio of the strangeness-changing to the strangeness-conserving neutral current amplitude is less than about 10^{-4}, if we compare results from K decay and neutrino interactions. For charged currents this ratio is approximately $\frac{1}{5}$, as observed from low-energy decay processes. Thus universality of the weak interaction, insofar as it incorporates all known weak processes, is no longer apparent. The origin of this non-universality is unknown, but it is conjectured that universality would be restored if a new class of massive hadrons exist and carry a new quantum number called "charm."[30,31] The possible violation of charge-symmetry invariance discussed earlier might be related to the production of these new hadrons. Direct searches for such particles through their leptonic or semileptonic decays are in progress. The recently discovered narrow vector bosons (called ψ on the West coast and J in the East) have also been conjectured to be bound states of charm–anticharm. The remarkable possibility exists that a new family of hadrons lies roughly 1 GeV from the well established hadron states that were discovered more than twenty years ago.

• • •

This work has been supported in part by the US Atomic Energy Commission.

References

1. W. Pauli, *Septième Conseil Solvay 1933*, Gauthier-Villars, Paris (1934).
2. E. Fermi, Z. Physik, 88, 161 (1934).
3. F. Reines, C. L. Cowan, Phys. Rev. 92, 830 (1953).
4. B. Pontecorvo, JETP 37, 175 (1959).
5. M. Schwartz, Phys. Rev. Lett. 4, 306 (1960).
6. G. Danby, J. M. Gaillard, K. Goulianos, L. M. Lederman, N. Mistry, M. Schwartz, J. Steinberger, Phys. Rev. Lett. 9, 36 (1962).
7. S. Weinberg, Phys. Rev. Lett. 19, 1264 (1967); A. Salam in *Elementary Particle Physics* (N. Svartholm, ed.), Almquist and Wiksells, Stockholm (1968), page 367.
8. T. D. Lee, PHYSICS TODAY, April 1972, page 23.
9. T. Eichten et al., Phys. Lett. B46, 274 (1973); A. Benvenuti et al., Phys. Rev. Lett. 32, 800 (1974); B. Aubert et al., report submitted to the London Conference (1974).
10. B. Barish et al., report submitted to the London Conference (1974).
11. M. Derrick, ANL/HEP 7350 (1973).
12. J. D. Bjorken, Phys. Rev. 179, 1547 (1969).
13. G. Miller et al., Phys. Rev. D5, 528 (1972); A. Bodek et al., Phys. Rev. Lett. 30, 1087 (1973).
14. J. D. Bjorken, E. A. Paschos, Phys. Rev. 185, 1975 (1969).
15. J. Schechter, Y. Ueda, Phys. Rev. D2, 736 (1970).
16. S. S. Gerstein, Ia. B. Zel'dovich, JETP 1, 576 (1956); R. P. Feynman, M. Gell-Mann, Phys. Rev. 109, 193 (1958).
17. M. Gell-Mann, Phys. Rev. 111, 362 (1958).
18. J. J. Sakurai, Annals of Physics 11, 1 (1960).
19. B. Aubert et al., Phys. Rev. Lett. 33, 984 (1974).
20. U. Camerini, D. Cline, W. Fry, W. M. Powell, Phys. Rev. Lett. 13, 318 (1964); M. Bott-Bodenhausen et al., Phys. Lett. 24B, 194 (1967).
21. A Clark et al., Phys. Rev. Lett. 26, 1667 (1971); D. Cable et al., Phys. Rev. D8, 3807 (1973); D. Ljung, D. Cline, Phys. Rev. D8, 1307 (1973); R. J. Cence et al., Phys. Rev. D10, 776 (1974).
22. M. M. Block et al., Phys. Lett. 12, 281 (1964); D. Cundy et al., Phys. Lett. 31B, 478 (1970).
23. F. J. Hasert et al., Phys. Lett. 46B, 138 (1973) and Nucl. Phys. B73, 1 (1974).
24. A. Benvenuti et al., paper 288, Sixth International Symposium, Bonn (1973) and Phys. Rev. Lett. 32, 800 (1974).
25. B. Aubert et al., Phys. Rev. Lett. 32, 1454 and 1457 (1974).
26. S. Barish et al., Phys. Rev. Lett. 33, 448 (1974).
27. B. C. Barish et al., CIT-FNAL report given at the London Conference, July 1974; W. Lee et al., report at the London Conference.
28. A. Litke et al., Phys. Rev. Lett. 30, 1189 (1973); B. Richter, invited talk at the University of California, Irvine, Conference, December 1973.
29. S. Weinberg, Rev. Mod. Phys. 46, 255 (1974).
30. J. D. Bjorken, S. L. Glashow, Phys. Lett. 11, 255 (1964).
31. S. L. Glashow, J. Iliopoulos, L. Maiani, Phys. Rev. D2, 1285 (1970).
32. M. Gell-Mann, M. Goldberger, N. Kroll, F. E. Low, Phys. Rev. 179, 1518 (1969).
33. B. L. Ioffe, JETP 11, 1158 (1960).

OBSERVATION OF HIGH-ENERGY NEUTRINO REACTIONS AND THE EXISTENCE OF TWO KINDS OF NEUTRINOS*

G. Danby, J-M. Gaillard, K. Goulianos, L. M. Lederman, N. Mistry,
M. Schwartz,[†] and J. Steinberger[†]

Columbia University, New York, New York and Brookhaven National Laboratory, Upton, New York
(Received June 15, 1962)

In the course of an experiment at the Brookhaven AGS, we have observed the interaction of high-energy neutrinos with matter. These neutrinos were produced primarily as the result of the decay of the pion:

$$\pi^{\pm} \to \mu^{\pm} + (\nu, \overline{\nu}). \qquad (1)$$

It is the purpose of this Letter to report some of the results of this experiment including (1) demonstration that the neutrinos we have used produce μ mesons but do not produce electrons, and hence are very likely different from the neutrinos involved in β decay and (2) approximate cross sections.

Behavior of cross section as a function of energy. The Fermi theory of weak interactions which works well at low energies implies a cross section for weak interactions which increases as phase space. Calculation indicates that weak interacting cross sections should be in the neighborhood of 10^{-38} cm^2 at about 1 BeV. Lee and Yang[1] first calculated the detailed cross sections for

$$\nu + n \rightarrow p + e^-,$$
$$\bar{\nu} + p \rightarrow n + e^+, \qquad (2)$$
$$\nu + n \rightarrow p + \mu^-,$$
$$\bar{\nu} + p \rightarrow n + \mu^+, \qquad (3)$$

using the vector form factor deduced from electron scattering results and assuming the axial vector form factor to be the same as the vector form factor. Subsequent work has been done by Yamaguchi[2] and Cabbibo and Gatto.[3] These calculations have been used as standards for comparison with experiments.

Unitarity and the absence of the decay $\mu \rightarrow e + \gamma$. A major difficulty of the Fermi theory at high energies is the necessity that it break down before the cross section reaches $\pi \lambda^2$, violating unitarity. This breakdown must occur below 300 BeV in the center of mass. This difficulty may be avoided if an intermediate boson mediates the weak interactions. Feinberg[4] pointed out, however, that such a boson implies a branching ratio $(\mu \rightarrow e + \gamma)/(\mu \rightarrow e + \nu + \bar{\nu})$ of the order of 10^{-4}, unless the neutrinos associated with muons are different from those associated with electrons.[5] Lee and Yang[6] have subsequently noted that any general mechanism which would preserve unitarity should lead to a $\mu \rightarrow e + \gamma$ branching ratio not too different from the above. Inasmuch as the branching ratio is measured to be $\leq 10^{-8}$,[7] the hypothesis that the two neutrinos may

be different has found some favor. It is expected that if there is only one type of neutrino, then neutrino interactions should produce muons and electrons in equal abundance. In the event that there are two neutrinos, there is no reason to expect any electrons at all.

The feasibility of doing neutrino experiments at accelerators was proposed independently by Pontecorvo[8] and Schwartz.[9] It was shown that the fluxes of neutrinos available from accelerators should produce of the order of several events per day per 10 tons of detector.

The essential scheme of the experiment is as follows: A neutrino "beam" is generated by decay in flight of pions according to reaction (1). The pions are produced by 15-BeV protons striking a beryllium target at one end of a 10-ft long straight section. The resulting entire flux of particles moving in the general direction of the detector strikes a 13.5-m thick iron shield wall at a distance of 21 m from the target. Neutrino interactions are observed in a 10-ton aluminum spark chamber located behind this shield.

The line of flight of the beam from target to detector makes an angle of 7.5° with respect to the internal proton direction (see Fig. 1). The operating energy of 15 BeV is chosen to keep the muons penetrating the shield to a tolerable level.

The number and energy spectrum of neutrinos from reaction (1) can be rather well calculated, on the basis of measured pion-production rates[10] and the geometry. The expected neutrino flux from π decay is shown in Fig. 2. Also shown is

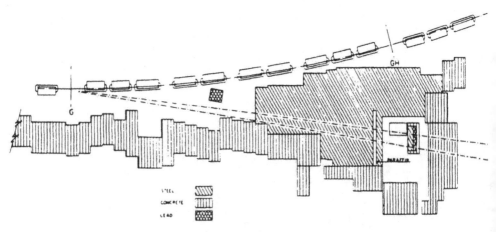

FIG. 1. Plan view of AGS neutrino experiment.

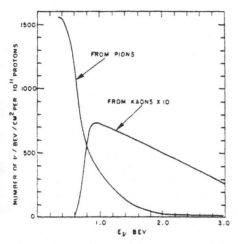

FIG. 2. Energy spectrum of neutrinos expected in the arrangement of Fig. 1 for 15-BeV protons on Be.

an estimate of neutrinos from the decay $K^{\pm} \rightarrow \mu^{\pm} + \nu(\bar{\nu})$. Various checks were performed to compare the targeting efficiency (fraction of circulating beam that interacts in the target) during the neutrino run with the efficiency during the beam survey run. (We believe this efficiency to be close to 70%.) The pion-neutrino flux is considered reliable to approximately 30% down to 300 MeV/c, but the flux below this momentum does not contribute to the results we wish to present.

The main shielding wall thickness, 13.5 m for most of the run, absorbs strongly interacting particles by nuclear interaction and muons up to 17 BeV by ionization loss. The absorption mean free path in iron for pions of 3, 6, and 9 BeV has been measured to be less than 0.24 m.[11] Thus the shield provides an attenuation of the order of 10^{-24} for strongly interacting particles. This attenuation is more than sufficient to reduce these particles to a level compatible with this experiment. The background of strongly interacting particles within the detector shield probably enters through the concrete floor and roof of the 5.5-m thick side wall. Indications of such leaks were, in fact, obtained during the early phases of the experiment and the shielding subsequently improved. The argument that our observations are not induced by strongly interacting particles will also be made on the basis of the detailed structure of the data.

The spark chamber detector consists of an array of 10 one-ton modules. Each unit has 9 aluminum plates 44 in. × 44 in. × 1 in. thick, separated by $\frac{3}{4}$-in. Lucite spacers. Each module is driven by a specially designed high-pressure spark gap and the entire assembly triggered as described below. The chamber will be more fully described elsewhere. Figure 3 illustrates the arrangement of coincidence and anticoincidence counters. Top, back, and front anticoincidence sheets (a total of 50 counters, each 48 in. × 11 in. × $\frac{1}{2}$ in.) are provided to reduce the effect of cosmic rays and AGS-produced muons which penetrate the shield. The top slab is shielded against neutrino events by 6 in. of steel and the back slab by 3 ft of steel and lead.

Triggering counters were inserted between adjacent chambers and at the end (see Fig. 3). These consist of pairs of counters, 48 in. × 11 in. × $\frac{1}{2}$ in., separated by $\frac{3}{4}$ in. of aluminum, and in fast coincidence. Four such pairs cover a chamber; 40 are employed in all.

The AGS at 15 BeV operates with a repetition period of 1.2 sec. A rapid beam deflector drives the protons onto the 3-in. thick Be target over a period of 20-30 μsec. The radiation during this interval has rf structure, the individual bursts being 20 nsec wide, the separation 220 nsec. This structure is employed to reduce the total "on" time and thus minimize cosmic-ray background. A Čerenkov counter exposed

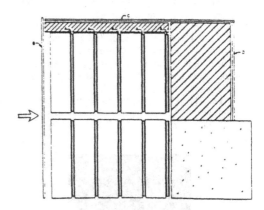

FIG. 3. Spark chamber and counter arrangement. A are the triggering slabs. B, C, and D are anticoincidence slabs. This is the front view seen by the four-camera stereo system.

to the pions in the neutrino "beam" provides a train of 30-nsec gates, which is placed in coincidence with the triggering events. The correct phasing is verified by raising the machine energy to 25 BeV and counting the high-energy muons which now penetrate the shield. The tight timing also serves the useful function of reducing sensitivity to low-energy neutrons which diffuse into the detector room. The trigger consists of a fast twofold coincidence in any of the 40 coincidence pairs in anticoincidence with the anticoincidence shield. Typical operation yields about 10 triggers per hour. Half the photographs are blank, the remainder consist of AGS muons entering unprotected faces of the chamber, cosmic rays, and "events." In order to verify the operation of circuits and the gap efficiency of the chamber. cosmic-ray test runs are conducted every four hours. These consist of triggering on almost horizontal cosmic-ray muons and recording the results both on film and on Land prints for rapid inspection (see Fig. 4).

A convenient monitor for this experiment is the number of circulating protons in the AGS machine. Typically, the AGS operates at a level of $2-4 \times 10^{11}$ protons per pulse, and 3000 pulses per hour. In an exposure of 3.48×10^{17} protons, we have counted 113 events satisfying the following geometric criteria: The event originates within a fiducial volume whose boundaries lie 4 in. from the front and back walls of the chamber and 2 in. from the top and bottom walls. The first two gaps must not fire, in order to exclude events whose origins lie outside the chambers. In addition, in the case of events consisting of a single track, an extrapolation of the track backwards (towards the neutrino source) for two gaps must also remain within the fiducial volume. The production angle of these single tracks relative to the neutrino line of flight must be less than 60°.

These 113 events may be classified further as follows:

(a) 49 short single tracks. These are single tracks whose <u>visible</u> momentum. if interpreted as muons. is less than 300 MeV, c. These presumably include some energetic muons which leave the chamber. They also include low-energy neutrino events and the bulk of the neutron produced background. Of these, 19 have 4 spark or less. The second half of the run (1.7×10^{17} protons) with improved shielding yielded only three tracks in this category. We will not consider these as acceptable "events."

(b) 34 "single muons" of more than 300 MeV/c. These include tracks which, if interpreted as muons, have a visible range in the chambers such that their momentum is at least 300 MeV/c. The origin of these events must not be accompanied by more than two extraneous sparks. The latter requirement means that we include among "single tracks" events showing a small recoil. The 34 events are tabulated as a function of momentum in Table I. Figure 5 illustrates 3 "single muon" events.

(c) 22 "vertex" events. A vertex event is one whose origin is characterized by more than one track. All of these events show a substantial energy release. Figure 6 illustrates some of these.

(d) 8 "showers." These are all the remaining events. They are in general single tracks, too irregular in structure to be typical of μ mesons and more typical of electron or photon showers. From these 8 "showers," for purposes of comparison with (b), we may select a group of 6 which are so located that their potential range within the chamber corresponds to μ mesons in excess of 300 MeV/c.

In the following, only the 56 energetic events of type (b) (long μ's) and type (c) (vertex events) will be referred to as "events."

Arguments on the neutrino origin of the ob-

FIG. 4. Land print of Cosmic-ray muons integrated over many incoming tracks.

Table I. Classification of "events."

Single tracks			
$p_\mu < 300$ MeV/c [a]	49	$p_\mu > 500$	8
$p_\mu > 300$	34	$p_\mu > 600$	3
$p_\mu > 400$	19	$p_\mu > 700$	2
	Total "events" 34		
Vertex events			
Visible energy released < 1 BeV	15		
Visible energy released > 1 BeV	7		

[a] These are not included in the "event" count (see text)

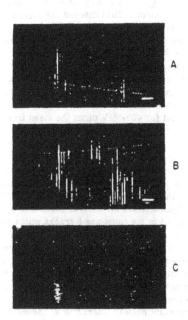

FIG. 5. Single muon events. (A) $p_\mu > 540$ MeV and δ ray indicating direction of motion (neutrino beam incident from left); (B) $p_\mu > 700$ MeV/c; (C) $p_\mu > 440$ with δ ray.

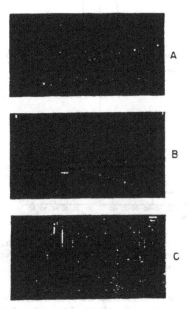

FIG. 6. Vertex events. (A) Single muon of $p_\mu > 500$ MeV and electron-type track; (B) possible example of two muons, both leave chamber; (C) four prong star with one long track of $p_\mu > 600$ MeV/c.

served "events."

1. The "events" are not produced by cosmic rays. Muons from cosmic rays which stop in the chamber can and do simulate neutrino events. This background is measured experimentally by running with the AGS machine off on the same triggering arrangement except for the Čerenkov gating requirement. The actual triggering rate then rises from 10 per hour to 80 per second (a dead-time circuit prevents jamming of the spark chamber). In 1800 cosmic-ray photographs thus obtained, 21 would be accepted as neutrino events. Thus 1 in 90 cosmic-ray events is neutrino-like. Čerenkov gating and the short AGS pulse effect a reduction by a factor of $\sim 10^{-6}$ since the circuits are "on" for only 3.5 μsec per pulse. In fact, for the body of data represented by Table I, a total of 1.6×10^6 pulses were counted. The equipment was therefore sensitive for a total time of 5.5 sec. This should lead to $5.5 \times 80 = 440$ cosmic-ray tracks which is consistent with observation. Among these, there should be 5 ± 1 cosmic-ray induced "events." These are almost evident in the small asym-

metry seen in the angular distributions of Fig. 7. The remaining 51 events cannot be the result of cosmic rays.

2. The "events" are not neutron produced. Several observations contribute to this conclusion.

(a) The origins of all the observed events are uniformly distributed over the fiduciary volume, with the obvious bias against the last chamber induced by the $p_\mu > 300$ MeV/c requirement. Thus there is no evidence for attenuation, although the mean free path for nuclear interaction in aluminum is 40 cm and for electromagnetic interaction 9 cm.

(b) The front iron shield is so thick that we can expect less than 10^{-4} neutron induced reactions in the entire run from neutrons which have penetrated this shield. This was checked by removing 4 ft of iron from the front of the thick shield. If our events were due to neutrons in line with the target, the event rate would have increased by a factor of one hundred. No such effect was observed (see Table II). If neutrons penetrate the shield, it must be from other di-

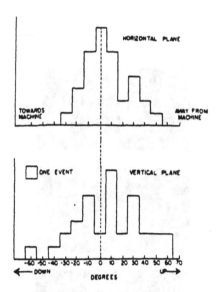

FIG. 7. Projected angular distributions of single track events. Zero degree is defined as the neutrino direction.

rections. The secondaries would reflect this directionality. The observed angular distribution of single track events is shown in Fig. 7. Except for the small cosmic-ray contribution to the vertical plane projection, both projections are peaked about the line of flight to the target.
(c) If our 29 single track events (excluding cosmic-ray background) were pions produced by neutrons, we would have expected, on the basis of known production cross sections, of the order of 15 single π^0's to have been produced. No cases of unaccompanied π^0's have been observed.

Table II. Event rates for normal and background conditions.

	Circulating protons × 10^{16}	No. of Events	Calculated cosmic-ray[c] contribution	Net rate per 10^{16}
Normal run	34.8	56	5	1.46
Background I[a]	3.0	2	0.5	0.5
Background II[b]	8.6	4	1.5	0.3

[a] 4 ft of Fe removed from main shielding wall.
[b] As above, but 4 ft of Pb placed within 6 ft of Be target and subtending a horizontal angular interval from 4° to 11° with respect to the internal proton beam.
[c] These should be subtracted from the "single muon" category.

3. **The single particles produced show little or no nuclear interaction and are therefore presumed to be muons.** For the purpose of this argument, it is convenient to first discuss the second half of our data, obtained after some shielding improvements were effected. A total traversal of 820 cm of aluminum by single tracks was observed, but no "clear" case of nuclear interaction such as large angle or charge exchange scattering was seen. In a spark chamber calibration experiment at the Cosmotron, it was found that for 400-MeV pions the mean free path for "clear" nuclear interactions in the chamber (as distinguished from stoppings) is no more than 100 cm of aluminum. We should, therefore, have observed of the order of 8 "clear" interactions; instead we observed none. The mean free path for the observed single tracks is then more than 8 times the nuclear mean free path.

Included in the count are 5 tracks which stop in the chamber. Certainly a fraction of the neutrino secondaries must be expected to be produced with such small momentum that they would stop in the chamber. Thus, none of these stoppings may, in fact, be nuclear interactions. But even if all stopping tracks are considered to represent nuclear interactions, the mean free path of the observed single tracks must be 4 nuclear mean free paths.

The situation in the case of the earlier data is more complicated. We suspect that a fair fraction of the short single tracks then observed are, in fact, protons produced in neutron collisions. However, similar arguments can be made also for these data which convince us that the energetic single track events observed then are also non-interacting.[12]

It is concluded that the observed single track events are muons, as expected from neutrino interactions.

4. **The observed reactions are due to the decay products of pions and K mesons.** In a second background run, 4 ft of iron were removed from the main shield and replaced by a similar quantity of lead placed as close to the target as feasible. Thus, the detector views the target through the same number of mean free paths of shielding material. However, the path available for pions to decay is reduced by a factor of 8. This is the closest we could come to "turning off" the neutrinos. The results of this run are given in terms of the number of events per 10^{16} circulating protons in Table II. The rate of "events" is reduced from 1.46 ± 0.2 to 0.3 ± 0.2 per 10^{16} in-

cident protons. This reduction is consistent with that which is expected for neutrinos which are the decay products of pions and K mesons.

Are there two kinds of neutrinos? The earlier discussion leads us to ask if the reactions (2) and (3) occur with the same rate. This would be expected if ν_μ, the neutrino coupled to the muon and produced in pion decay, is the same as ν_e, the neutrino coupled to the electron and produced in nuclear beta decay. We discuss only the single track events where the distinction between single muon tracks of $p_\mu > 300$ MeV/c and showers produced by high-energy single electrons is clear. See Figs. 8 and 4 which illustrate this difference.

We have observed 34 single muon events of which 5 are considered to be cosmic-ray background. If $\nu_\mu = \nu_e$, there should be of the order of 29 electron showers with a mean energy greater than 400 MeV. Instead, the only candidates which we have for such events are six "showers" of qualitatively different appearance from those of Fig. 8. To argue more precisely, we have exposed two of our one-ton spark chamber modules to electron beams at the Cosmotron. Runs were taken at various electron energies. From these we establish that the triggering efficiency for 400-MeV electrons is 67%. As a quantity characteristic of the calibration showers, we have taken the total number of observed sparks. The mean number is roughly linear with electron energy up to 400 MeV/c. Larger showers saturate the two chambers

which were available. The spark distribution for 400 MeV/c showers is plotted in Fig. 9, normalized to the $\frac{2}{3} \times 29$ expected showers. The six "shower" events are also plotted. It is evident that these are not consistent with the prediction based on a universal theory with $\nu_\mu = \nu_e$. It can perhaps be argued that the absence of electron events could be understood in terms of the coupling of a single neutrino to the electron which is much weaker than that to the muon at higher momentum transfers, although at lower momentum transfers the results of β decay, μ capture, μ decay, and the ratio of $\pi \to \mu + \nu$ to $\pi \to e + \nu$ decay show that these couplings are equal.[13] However, the most plausible explanation for the absence of the electron showers, and the only one which preserves universality, is then that $\nu_\mu \neq \nu_e$; i.e., that there are at least two types of neutrinos. This also resolves the problem raised by the forbiddenness of the $\mu^+ \to e^+ + \gamma$ decay.

It remains to understand the nature of the 6 "shower" events. All of these events were obtained in the first part of the run during conditions in which there was certainly some neutron background. It is not unlikely that some of the events are small neutron produced stars. One or two could, in fact, be μ mesons. It should also be remarked that of the order of one or two electron events are expected from the neutrinos produced in the decays $K^+ \to e^- + \nu_e + \pi^0$ and

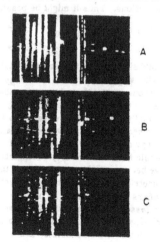

FIG. 8. 400-MeV electrons from the Cosmotron.

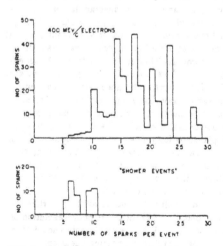

FIG. 9. Spark distribution for 400-MeV/c electrons normalized to expected number of showers. Also shown are the "shower" events.

$K_2^0 \to e^{\pm} + \nu_e + \pi^{\mp}$.

The intermediate boson. It has been pointed out[1] that high-energy neutrinos should serve as a reasonable method of investigating the existence of an intermediate boson in the weak interactions. In recent years many of the objections to such a particle have been removed by the advent of $V-A$ theory[14] and the remeasurement of the ρ value in μ decay.[15] The remaining difficulty pointed out by Feinberg,[4] namely the absence of the decay $\mu \to e + \gamma$, is removed by the results of this experiment. Consequently it is of interest to explore the extent to which our experiment has been sensitive to the production of these bosons.

Our neutrino intensity, in particular that part contributed by the K-meson decays, is sufficient to have produced intermediate bosons if the boson had a mass m_W less than that of the mass of the proton (m_p). In particular, if the boson had a mass equal to $0.6 m_p$, we should have produced ~20 bosons by the process $\nu + p \to w^+ + \mu^- + p$. If $m_W = m_p$, then we should have observed 2 such events.[16]

Indeed, of our vertex events, 5 are consistent with the production of a boson. Two events, with two outgoing prongs, one of which is shown in Fig. 6(B), are consistent with both prongs being muons. This could correspond to the decay mode $w^- \to \mu^+ + \nu$. One event shows four outgoing tracks, each of which leaves the chamber after traveling through 9 in. of aluminum. This might in principle be an example of $w^+ \to \pi^+ + \pi^- + \pi^+$. Another event, by far our most spectacular one, can be interpreted as having a muon, a charged pion, and two gamma rays presumably from a neutral pion. Over 2 BeV of energy release is seen in the chamber. This could in principle be an example of $w^+ \to \pi^+ + \pi^0$. Finally, we have one event, Fig. 6(A), in which both a muon and an electron appear to leave the same vertex. If this were a boson production, it would correspond to the boson decay mode $w^+ \to e^+ + \nu$. The alternative explanation for this event would require (i) that a neutral pion be produced with the muon; and (ii) that one of its gamma rays convert in the plate of the interaction while the other not convert visibly in the chamber.

The difficulty of demonstrating the existence of a boson is inherent in the poor resolution of the chamber. Future experiments should shed more light on this interesting question.

Neutrino cross sections. We have attempted to compare our observations with the predicted cross sections for reactions (2) using the theory.[1-3] To include the fact that the nucleons in (2) are, in fact, part of an aluminum nucleus, a Monte Carlo calculation was performed using a simple Fermi model for the nucleus in order to evaluate the effect of the Pauli principle and nucleon motion. This was then used to predict the number of "elastic" neutrino events to be expected under our conditions. The results agree with simpler calculations based on Fig. 2 to give, in terms of number of circulating protons,

from $\pi \to \mu + \nu$,   0.60 events/10^{16} protons,

from $K \to \mu + \nu$,   0.15 events/10^{16} protons,

Total   0.75 events/$10^{16} \pm$ ~30%.

The observed rates, assuming all single muons are "elastic" and all vertex events "inelastic" (i.e., produced with pions) are

"Elastic": 0.84 ± 0.16 events/10^{16} (29 events),

"Inelastic": 0.63 ± 0.14 events/10^{16} (22 events).

The agreement of our elastic yield with theory indicates that no large modification to the Fermi interaction is required at our mean momentum transfer of 350 MeV/c. The inelastic cross section in this region is of the same order as the elastic cross section.

Neutrino flip hypothesis. Feinberg, Gursey, and Pais[17] have pointed out that if there were two different types of neutrinos, their assignment to muon and electron, respectively, could in principle be interchanged for strangeness-violating weak interactions. Thus it might be possible that

$$\pi^+ \to \mu^+ + \nu_1 \qquad \text{while} \qquad K^+ \to \mu^+ + \nu_2$$
$$\pi^+ \to e^+ + \nu_2 \qquad\qquad\qquad K^+ \to e^+ + \nu_1.$$

This hypothesis is subject to experimental check by observing whether neutrinos from $K_{\mu 2}$ decay produce muons or electrons in our chamber. Our calculation of the neutrino flux from $K_{\mu 2}$ decay indicates that we should have observed 5 events from these neutrinos. They would have an average energy of 1.5 BeV. An electron of this energy would have been clearly recognizable. None have been seen. It seems unlikely therefore that the neutrino flip hypothesis is correct.

The authors are indebted to Professor G. Feinberg, Professor T. D. Lee, and Professor C. N. Yang for many fruitful discussions. In particular, we note here that the emphasis by Lee and Yang on the importance of the high-energy behavior of weak interactions and the likelihood of the existence of two neutrinos played an important part in stimulating this research.

We would like to thank Mr. Warner Hayes for technical assistance throughout the experiment. In the construction of the spark chamber, R. Hodor and R. Lundgren of BNL, and Joseph Shill and Yin Au of Nevis did the engineering. The construction of the electronics was largely the work of the Instrumentation Division of BNL under W. Higinbotham. Other technical assistance was rendered by M. Katz and D. Balzarini. Robert Erlich was responsible for the machine calculations of neutrino rates, M. Tannenbaum assisted in the Cosmotron runs.

The experiment could not have succeeded without the tremendous efforts of the Brookhaven Accelerator Division. We owe much to the cooperation of Dr. K. Green, Dr. E. Courant, Dr. J. Blewett, Dr. M. H. Blewett, and the AGS staff including J. Spiro, W. Walker, D. Sisson, and L. Chimienti. The Cosmotron Department is acknowledged for its help in the initial assembly and later calibration runs.

The work was generously supported by the U. S. Atomic Energy Commission. The work at Nevis was considerably facilitated by Dr. W. F. Goodell, Jr., and the Nevis Cyclotron staff under Office of Naval Research support.

*This research was supported by the U. S. Atomic Energy Commission.

†Alfred P. Sloan Research Fellow.

[1]T. D. Lee and C. N. Yang, Phys. Rev. Letters 4, 307 (1960).

[2]Y. Yamaguchi, Progr. Theoret. Phys. (Kyoto) 6, 1117 (1960).

[3]N. Cabbibo and R. Gatto, Nuovo cimento 15, 304 (1960).

[4]G. Feinberg, Phys. Rev. 110, 1482 (1958).

[5]Several authors have discussed this possibility. Some of the earlier viewpoints are given by: E. Koropinski and H. Mahmoud, Phys. Rev. 92, 1045 (1955; J. Schwinger, Ann. Phys. (New York) 2, 407 (1957) I. Kawakami, Progr. Theoret. Phys. (Kyoto) 19, 45 (1957); M. Konuma, Nuclear Phys. 5, 504 (1958); S. Bludman, Bull. Am. Phys. Soc. 4, 80 (1959); S. On and J. C. Pati, Phys. Rev. Letters 2, 125 (1959); K. Nishijima, Phys. Rev. 108, 907 (1957).

[6]T. D. Lee and C. N. Yang (private communicatio: See also Proceedings of the 1960 Annual Internationa Conference on High-Energy Physics at Rochester (Interscience Publishers, Inc., New York, 1960), p. 567.

[7]D. Bartlett, S. Devons, and A. Sachs, Phys. Rev Letters 8, 120 (1962); S. Frankel, J. Halpern, L. H way, W. Wales, M. Yearian, O. Chamberlain, A. L onick, and F. M. Pipkin, Phys. Rev. Letters 8, 123 (1962).

[8]B. Pontecorvo, J. Exptl. Theoret. Phys. (U.S.S. 37, 1751 (1959) [translation: Soviet Phys. – JETP 10, 1236 (1960)].

[9]M. Schwartz, Phys. Rev. Letters 4, 306 (1960).

[10]W. F. Baker et al., Phys. Rev. Letters 7, 101 (19

[11]R. L. Cool, L. Lederman, L. Marshall, A. C. Melissinos, M. Tannenbaum, J. H. Tinlot, and T. Yamanouchi, Brookhaven National Laboratory Internal Report UP-18 (unpublished).

[12]These will be published in a more complete report.

[13]H. L. Anderson, T. Fujii, R. H. Miller, and L. T Phys. Rev. 119, 2050 (1960); G. Culligan, J. F. Latl rop, V. L. Telegdi, R. Winston, and R. A. Lundy, Ph Rev. Letters 7, 458 (1961); R. Hildebrand, Phys. Re Letters 8, 34 (1962); E. Bleser, L. Lederman, J. Re en, J. Rothberg, and E. Zavattini, Phys. Rev. Lette 8, 288 (1962).

[14]R. Feynman and M. Gell-Mann, Phys. Rev. 109, 193 (1958); R. Marshak and E. Sudershan, Phys. Rev 109, 1860 (1958).

[15]R. Plano, Phys. Rev. 119, 1400 (1960).

[16]T. D. Lee, P. Markstein, and C. N. Yang, Phys. Rev. Letters 7, 429 (1961).

[17]G. Feinberg, F. Gursey, and A. Pais, Phys. Rev. Letters 7, 208 (1961).

CHAPTER 3
THE SEARCH FOR OTHER FORMS
OF THE WEAK INTERACTION

Changing Flavors (1963 – 1970)

3.1. INTRODUCTION

The history of weak interactions in the 1960s was largely that of attempting to understand the concept of flavor-changing weak decays (flavor was discovered at the same time as K mesons in the late 1940s – early '50s). The real importance and difference of flavor-changing charged weak decays (*e.g.*, $K^+ \to \pi e^+ v_e$) and flavor-conserving ones (*e.g.*, $n \to p + e^- + \bar{v}_e$) was not initially appreciated; in fact, it was not even certain that the same weak interaction mediated both processes. The assumption of a universal weak interaction was a strong unifying concept, and later it was shown that all weak interactions with charged currents follow the V-A interaction, at least at the quark level. The search for neutral current interactions, which changed flavor and the limits that were derived, was not really appreciated until the advent of the GIM mechanism in 1970. Nevertheless, these searches set the stage for one of the key ingredients of the standard model of elementary particle physics that is still being tested today.

3.2. THE EARLY SEARCH FOR WEAK NEUTRAL CURRENTS: The Search for Flavor-Changing Neutral Current during 1963 – 1970

To trace the history of the weak interaction, it is necessary to place particular emphasis on the discovery of the WNC and the underlying theory of the standard model. The different types of weak interaction processes are best illustrated by the types of reactions for neutrino quark scattering as shown in Table 3.1. In this table, we illustrate the different types of weak interactions through a hypothetical scattering of neutrinos from quarks. Reactions (A) and (B) have been observed to occur in Nature; so far, FCNC [reaction (C)] has not been observed. We also date the first observation (or the first dedicated search). In the lower part of the table, we illustrate the quark families in Nature that have been observed so far.

Let us trace the early history of the WNC concept. The earliest ideas concerning the unity of weak interactions and WNC seems to have been described by the Swedish physicist, Oscar Klein, in 1938 in a very obscure paper presented to the League of Nations meeting in France and in pre-wartime Warsaw, Poland [3.1]. The concept of the IVBs evolved over the early period and, in many cases, both charged and neutral bosons were postulated, even in the absence of any evidence for the existence of WNCs, which will be discussed further in Chapter 7.

The search for FCNCs was carried out with strange particles, mainly in the period 1963 – 1970. The first definitive search for WNCs that change the flavor was undertaken by the author and his colleagues using the LBL kaon beam. [See articles (A) and (B) by, respectively, D. Cline and U. Camerini *et al.*] In the mid-1970s, the absence of FCNC prompted the concept of "natural flavor conservation" for the WNCs [3.2]. The GIM model was also partially invented to explain these results

Table 3.1. Fundamental processes – neutrino quark scattering.

Current	Equation	First Observation or Dedicated Search
(A) Charged	$\nu + q \rightarrow q' + e$ $q = u, c, t$ $q' = d, s, b$	1896
(B) WNC	$\nu + q \rightarrow q + \nu$ $q = u, c, t, d, s, b$	1973
(C) FCNC,	$\nu + q \not\rightarrow q' + \nu$	1963
e.g.,	$g = \begin{pmatrix} u \\ d \\ b \\ \vdots \end{pmatrix}$, $g' = \begin{pmatrix} c \\ s \\ s \\ \vdots \end{pmatrix}$	

$$\begin{pmatrix} +2/3 \\ -1/3 \end{pmatrix} \qquad \begin{pmatrix} u \\ d \end{pmatrix} \qquad \begin{pmatrix} c \\ s \end{pmatrix} \qquad \begin{pmatrix} t \\ b \end{pmatrix}$$

Quark Families	(1)	(2)	(3)

[3.3]. This has come to be a key development in the rise of the standard model and strongly limits the types of quarks that exist in Nature. This is an example of how null experiments can strongly influence the direction of a field of science. In Table 3.2, we trace some of the early experimental searches for FCNC processes.

3.3. THE ROLE OF THE SEARCH FOR FLAVOR-CHANGING NEUTRAL CURRENT IN THE RISE OF THE STANDARD MODEL

After the initial hypothesis of charged IVBs in the late 1940s, many people believed it would be natural to introduce neutral IVBs and, hence, to expect neutral current processes. Unfortunately it was impossible to detect current processes in nuclear reactions at the time since, in competition with electromagnetic interactions, these processes were essentially invisible. For example, it was impossible to detect a $\nu\bar{\nu}$

Table 3.2. Best experimental limits on neutral current K decays (1967).

Process	Branching Ratio*	CC** Ratio	Upper Limit on CC Ratio*	Reference
$K^+ \to \pi^+ e^+ e^-$	$< 8.8 \times 10^{-7}$	$g^V(e^+e^-)/g^V(e^+\nu_e)$	7.0×10^{-4}	Cline et al. [3.4] and Camerini et al. [3.5]
$K^+ \to \pi^+ \mu^+ \mu^-$	$< 3.0 \times 10^{-6}$	$g^V(\mu^+\mu^-)/g^V(\mu^+\nu_\mu)$	1.4×10^{-2}	Camerini et al. [3.5], [3.6]
$K^+ \to \pi^+ \nu \bar{\nu}$	$< 1.1 \times 10^{-4}$	$g^V(\nu\nu)/g^V(e^+\nu_e)$	6.0×10^{-2}	Cline [3.7], $55 < T_\pi < 80$ MeV
$K_2^0 \to \mu^+ \mu^-$	$< 1.6 \times 10^{-6}$	$g^A(\mu^+\mu^-)/g^A(\mu^+\nu_\mu)$	7.4×10^{-4}	Bott-Bodenhauser et al. [3.8]
$K_1^0 \to \mu^+ \mu^-$	$< 7.3 \times 10^{-5}$	$g^A(\mu^+\mu^-)/g^A(\mu^+\nu_\mu)$	1.6×10^{-3}	Bott-Bodenhauser et al. [3.8]
$K_2^0 \to \pi^0 \mu^+ \mu^-$	No estimate	$g^V(\mu^+\mu^-)/g^V(\mu^+\nu_\mu)$	No estimate	
$K_2^0 \to \pi^+ e^+ e^-$	No estimate	$g^V(e^+e^-)/g^V(e^+\nu_e)$	No estimate	
$K^+ \to \pi^0 \pi^+ e^+ e^-$	$< 8.0 \times 10^{-6}$		No estimate	Cline [3.7]
$K^+ \to \pi^+ \mu^+ e^-$	$< 3.0 \times 10^{-5}$		Violates lepton conserv.	Cline [3.7]
$K_2^0 \to \mu^\pm e^\mp$	$< 9.0 \times 10^{-6}$		Violates lepton conserv.	Bott-Bodenhauser et al. [3.8]

*All estimates are 90% confidence limits.

**Coupling constant.

pair in the final state of the reaction,

$$\text{Nucleus} \rightarrow \text{Nucleus'} + e^+e^- \ ,$$

since it would be dominated by the electromagnetic interactions. Because it was impossible to search for neutral currents in nuclear processes, the search was started in K^+, K^0 decays. Dedicated searches for neutral current processes started in the early 1960s [3.4],[3.5], however these processes necessarily required flavor-changing reactions. We will presently review these early results. By 1965, it was clear that the level of FCNCs was much smaller than the standard charged-current reactions. We can compare the limits on

$$K^+ \rightarrow \pi^+ + \nu + \overline{\nu} \ , \qquad
\begin{array}{ll}
(\leq 10^{-4}, \ 1965) & \text{[3.7], [3.9]} \\
(\leq 5 \times 10^{-5}, \ 1969) & \text{[3.10]} \\
(\leq 10^{-6}, \ 1970) & \text{[3.11]}
\end{array} \qquad (3.1)$$

and

$$K^+ \rightarrow \pi^0 + e^- + \nu_e \ , \qquad (5 \times 10^{-2}) \ , \qquad\qquad (3.2)$$

to illustrate this point of view. At that time, there were arguments that explained the suppression of processes like

$$K_L^0 \rightarrow \mu^+\mu^- \qquad\qquad (3.3)$$

and

$$K^+ \rightarrow \pi^+ e^+ e^- \qquad\qquad (3.4)$$

through a combination of weak and electromagnetic interactions. [See article (A) by D. Cline for an illustration of these arguments at the time.] However, there was no way to suppress reaction (3.1) compared to (3.2) except by the absence of FCNC processes in Nature. The bulk of the data on weak interactions (charged current) fit well with the simple Fermi-type theory.

In the very early days of the study of K decays, little attention was paid to reactions (3.1), (3.3), and (3.4). Only in the early 1960s were sufficient K decays available to enable a sensitive search for these decays. The first explicit searches for FCNCs occurred in 1963 with the search for the $K^+ \rightarrow \pi^+ e^+ e^-$ process (3.4). By 1967, it was possible to review the limits on FCNC and to compare them to the theories of the time ([3.9], article (A), reproduced here so the reader can gauge the picture at that time). In the past 20 years, the limits on these decay modes have been reduced another 1 to 2 orders of magnitude, and no explicit example of an FCNC process has been detected. However, reactions (3.3) and (3.4) have been detected and are presumed to be due to a combined charged-current reaction and electromagnetic effect.

The present observation on charmed and beauty particles has failed to observe effects due to FCNCs as well. However, the limits are much less restrictive than those for strangeness-changing reactions even today. (This will be discussed further in Chapter 8.)

A crucial process to study in the search for FCNC is

$$K^+ \to \pi^+ \nu \bar{\nu}$$

which, unlike final states with all charged particles (*i.e.*, $K \to \mu^+\mu^-$ or $K \to \pi^+e^+e^-$), would be a direct indication of the existence or non-existence of FCNC. A chart of the progress in the search for this decay mode is shown in Fig. 3.1. (Table 3.2 shows the limits on FCNC amplitudes in 1967.)

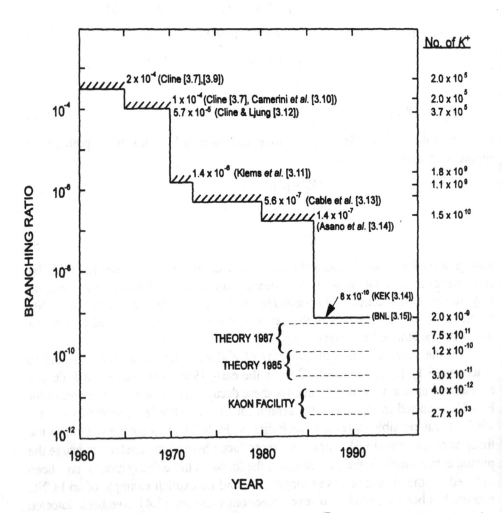

Figure 3.1. The history of the search for the decay mode $K^+ \to \pi^+ + \nu + \bar{\nu}$. This decay was crucial to indicate the absence of FCNC in the standard weak interaction.

After the very early search for FCNC, there were two important developments in the late 1960s:

1. A clever experiment was mounted to search for $K^+ \rightarrow \pi^+ + \nu + \bar{\nu}$ to the $\leq 10^{-7}$ branching ratio level by Klems *et al.* [3.11], reproduced here [article (C)].
2. An LBL group failed to find the decay $K_L^0 \rightarrow \mu^+\mu^-$ at the level predicted from purely electromagnetic processes. (This was later shown to be an experimental problem.)

The result of the first development provided additional confidence in the fundamental nature of the absence of FCNC, whereas the result of the second generated a great many theoretical papers on higher order corrections that attempted to explain the second result. In retrospect, the outpouring of theory helped formulate a better understanding of the electromagnetic corrections in K decays and was likely the forerunner of the more extensive calculations on the higher order corrections in the SU(2) × U(1) standard model. (See [3.16] for extensive references.)

It was very clear by the start of the 1970s when the $K_L^0 \rightarrow \mu^+\mu^-$ puzzle was resolved, that the absence of FCNC was a fundamental aspect of weak interaction physics. In many ways the absence of FCNC at the very sensitive level being probed at the end of the 1960s was one of the few negative results in particle physics that cried out for an explanation. It was extremely hard to understand how the GWS [1.1]–[1.3] model predicted WNC in the leptonic (and most likely hadronic) section, while FCNC were not observed at a very low level in the amplitude. Ultimately the explanation came in the form of the GIM model, but this was far from clear to experimentalists working at the time [3.2], [3.3].

The process

$$K_L^0 \rightarrow \nu\bar{\nu} \qquad\qquad (3.5)$$

is forbidden or highly suppressed for massless (or low-mass) left-handed neutrinos, thus the only processes that could be used to make a compelling search for FCNC was to look for $K^+ \rightarrow \pi^+ + \nu + \bar{\nu}$ or $K^+ \rightarrow \pi^0 + \nu + \bar{\nu}$. By 1970, the limit on $K^+ \rightarrow \pi^+\nu\bar{\nu}$ reached $\leq 10^{-6}$, and it was clear that there were no first-order FCNC processes in Nature. (See Fig. 3.1 for the recent history of the search for this reaction.) All subsequent experiments have reinforced this conclusion. (However, only light quark systems have been studied to great sensitivity.)

It is clear that the absence of FCNC is a unique feature of the weak interaction that had to be explained by the theoretical models. We have attempted to illustrate the impact of the absence of FCNC on various aspects of the quark model in Table 3.3. The absence of FCNC also played a role in the difficult observation of WNCs in 1973, as we shall see in Chapter 5.

Table 3.3. The impact of FCNC search on the rise of the standard model.

Process	Period	Implication
$s \not\to d$	~ 1963	FCNC absent at first level of weak interaction (GIM mechanism)
$K \to \pi^+ \nu \bar{\nu}$	~ 1970	
$s \not\to$ loop $\not\to d$	~ 1960–1974	New quark in loop, $m_q \sim 2$ GeV (charm)
$K^0 - \bar{K}^0, K_s - K_L$ mixing		
$s \not\to d$	~ 1970s	Natural flavor conservation for neutral current implies only $Q = -1/3$; $Q = 2/3$ quarks exist in doublets in Nature
$c \not\to u$		
etc.		
$b \not\to d$	~ 1980s	Necessary existence of (massive) t quark
$b \not\to s, B^0 - \bar{B}^0$ mixing		
Strong Limits on		
$s \not\to d$	~ 1990s	Limits on supersymmetric interactions and other exotics
$b \not\to d$		
$\nu_e \not\to \nu_\mu$		
$K \to \mu e$		

REFERENCES

3.1. O. Klein, in *Les Nouvelles Théories de la Physique, Proceedings of a Symposium held in Warsaw, May 30–June 3, 1938* (Institut International de Coopération Intellectuelle, Paris, 1939), p. 6.

3.2. S. Glashow and S. Weinberg., *Phys. Rev.*, **D15**, 1958 (1977).

3.3. S. J. Glashow, J. Iliopoulos, and L. Maiani, *Phys. Rev.*, **D2**, 1285 (1970).

3.4. D. Cline *et al.*, "Further Search for Neutral Leptonic Currents in K^+ Decay," submitted to the Heidelberg Int. Conf. on High Energy Physics (1972).

3.5. U. Camerini, D. Cline, W. F. Fry, and W. M. Powell, *Phys. Rev. Lett.*, **13**(9), 318 (1964) [article (B)].

3.6. U. Camerini *et al.*, *Nuovo Cimento*, **37**, 1795 (1965).

3.7. D. Cline, Ph.D. thesis, Univ. of Wisconsin (1965).

3.8. M. Bott-Bodenhausen *et al.*, *Phys. Lett.*, **24B**, 194 (1967).

3.9. D. Cline, "Experimental Search for Weak Neutral Currents," in *Methods in Subnuclear Physics*, v. III (Proc., Int. School of Elementary Particle Physics, Herceg-Novi, Yugoslavia, 1967), M. Nikolić, ed. (Gordon Breach, New York, 1969), p. 355 [article (A)].

3.10. U. Camerini *et al.*, *Phys. Rev. Lett.*, **23**, 326 (1969).

3.11. J. Klems *et al.*, *Phys. Rev. Lett.*, **24**, 1086 (1970) [article (C)].

3.12. D. Cline and D. Ljung, *Phys. Rev. Lett.*, **28**, 1287 (1972).

3.13. G. D. Cable *et al.*, *Phys. Rev.*, **D8**, 3807 (1973).

3.14. Y. Asano *et al.*, *Phys. Lett.*, **107B**, 159 (1981).

3.15. M. S. Atiya *et al.*, *Phys. Rev. Lett.*, **64**, 21 (1990).

3.16. M. K. Gaillard, B. Lee, and J. Rosner, *Rev. Mod. Phys.*, **47**, 227 (1975).

Experimental Search for Weak Neutral Currents*

D. CLINE

University of Wisconsin, Physics Department, Madison, Wisconsin, USA.

CONTENTS

* Supported in part by United States Atomic Energy Commission under Contract AT (11-1)-881, COO-881-128.

I. INTRODUCTION

Weak interaction dynamics remains one of the fascinating puzzles of elementary particle physics. There is little doubt that we presently have only a very vague understanding of these interactions. That is to say there remain as many mysteries about the interactions as there do successful explanations. The purpose of my lecture is to discuss the results of recent experimental searches for weak processes requiring neutral current interactions. In addition some of the large variety of theories which have been put forward to explain the possible absence of such currents, will be discussed.

For a variety of reactions the interaction responsible for the reaction is assumed to be phenomologically describable in terms of the coupling of two currents.[1] That is to say the weak interaction Lagrangian is of the current-current type

$$\mathscr{L}(x) = \frac{G}{2} (l_\lambda l_\lambda^* + J_\lambda l_\lambda^* + J_\lambda J_\lambda^* + \cdots) \tag{1}$$

where G is the Fermi constant, l_λ is a current involving only leptons and J_λ is a current involving hadrons. The $l_\lambda l_\lambda^*$ interaction describes purely leptonic processes. Only one such process has so far been observed, namely μ decay. In order to describe μ decay the lepton current is of the form

$$l_\lambda = [l\gamma_\lambda(1 + \gamma_5) \, \nu_l] \tag{2}$$

is required and in addition at least two components of the current should exist:

$$l_\lambda(\bar{e}\nu_e) + l_\lambda(\bar{\mu}\nu_\mu).$$

Unfortunately since only one leptonic process has been so far observed we have no knowledge about other possible components of this current or whether other leptonic processes will in fact be described by such a current. Our knowledge about the form of the hadronic current is even more vague. There are presently three kinds of hadronic processes that have been observed [(F, B) represent Fermions and bosons, respectively.]

(a) Semi-leptonic
$$B_1 \rightarrow B_2 + l + \nu$$
$$F_1 \rightarrow F_2 + l + \nu$$

(b) Non-leptonic involving bosons in the initial and final state.

(c) Non-leptonic involving fermions in the initial and final states $F_1 + F_2 \rightarrow F_3 + F_4$ or $F_1 \rightarrow F_2 + B$.

Present experimental evidence indicates that processes of type (a) are describable by the current-current interaction. The current-current description of processes (b) and (c) is so far an open question because of the perturbing effects of the strong interaction for these processes. Nevertheless many attempts have been made to relate these processes to a current-current type of interaction with varying degrees of success.

Thus there is a class of reactions for which the form of the interaction appears to be of the current X current type. These reactions will therefore be useful in deciding which currents nature has chosen. The most general form of lepton current which satisfies lepton conservation is

$$l_\lambda = l_\lambda(e\nu_e) + l_\lambda(\mu\nu_\mu) + l_\lambda(\bar{\mu}\mu) + l_\lambda(\bar{\nu}_e\nu_e) \tag{3}$$

$$+ \; l_\lambda(\bar{\nu}_\mu\nu_\mu) + l\lambda(\bar{e}e)$$

whereas the most general form of hadron current involving Fermions is

$$J_{\lambda F} = J_\lambda(\bar{p}n) + J_\lambda(\bar{p}\Lambda) + J_\lambda(\bar{p}p) + J_\lambda(\bar{n}n) + J_\lambda(\bar{n}\Lambda) \tag{4}$$

$$+ \; J_\lambda(\bar{n}\bar{\Sigma}) + J_\lambda(\Sigma\Lambda) + \cdots \text{ etc.}$$

and the current involving bosons in the initial and final state is given by

$$J_{\lambda B} = J_\lambda(\bar{K}^+\pi^0) + J_\lambda(\bar{K}^+\pi^-) + J_\lambda(\bar{K}^0\pi^0) + J_\lambda(\bar{\pi}^+, \pi^0) \tag{5}$$

$$+ \; J_\lambda(\bar{K}^+, \pi^+\pi^-) + J_\lambda(\bar{K}^+, \pi^-\pi^0)$$

$$+ \; J_\lambda(K, \text{vacuum}) + \cdots$$

If all the current components corresponding to (3), (4) and (5) were operative a very large number of decay and scattering process would be allowed. Table 1 indicated some of the pure leptonic processes and Table 2 indicates a representative sample of allowed semileptonic processes. (We are at present ignoring currents with $\Delta S > 1$ and the $\Delta S = -\Delta Q$ currents.)

Table 1

Leptonic Processes

$\Delta Q = 1$	$\mu^+ \to e^+\nu_\mu\nu_e$	$(\mu \to e\gamma, \mu \to eee$
	$e^+ - e^- \to \nu_e - \bar{\nu}_e$	are forbidden by
	$\nu_e + e \to \nu_e - e$	μonic lepton conservation)
$\Delta Q = 0$	$\nu_\mu + e^- \to e^- - \nu_\mu$	
	$e^+e^- \xrightarrow{\nu} \mu^+\mu^-$	
	$\nu_\mu + Z \to \nu_\mu - Z - e^+ - e^-$	

In the spirit of the $(V - A)$ leptonic theory the current components in (4) are assumed to have at least axial vector and vector parts. For the axial vector transition the current can be thought of as a 1^+ object and for the vector the current is like a 1^- object. For example the axial vector $(\bar{n}p)$ current is like the $(\bar{n}, p)$ system in a $^3P_1$ state whereas the vector part is like a $(\bar{n}, p)$ state in the $^3S_1$.

Leptonic decays with more than two particles in the final state usually proceed through a variety of possible transitions depending on the number of invariants that can be formed out of the 4 vectors available in the problem. A simple way to decide on the type and number of form factors is given in the following section. (Note that $(V - A)$ theory favors transition with a di-lepton system in the spin 1 state over those with singlet di-lepton systems.)

(a) $F_1 \rightarrow F_2 l_1 l_2$

look at

$$F_1 + \bar{F}_2 \rightarrow l_1 + l_2$$

because of locality only the $'S_0(0^-)$ and $^3S_1(1^-)$ states will be considered. In the Dirac theory this implies that for relativistic leptons there will also be states

Table 2

Mesic Transitions

	$\Delta S = 0$	$\Delta S = 1$
$\Delta Q = 1$	$\pi^+ \rightarrow \pi^0 e^+ \nu$	$K^+ \rightarrow \pi^0 e^+ \nu$
	$K^0 \rightarrow K^+ e^- \nu$	$K^0_{1,2} \rightarrow \pi^\pm e^\mp \nu$
		$K^+ \rightarrow \pi^+ \pi^- e^+ \nu$
$\Delta Q = 0$	$\pi^0 \xrightarrow{\nu} e^+ e^-$	$K^+ \rightarrow \pi^+ e^+ e^-$
	$\eta \xrightarrow{\nu} \pi^0 e^+ e^-$	$\rightarrow \pi^+ \mu^+ \mu^-$
		$\rightarrow \pi^+ \nu \bar{\nu} \pi$
		$\rightarrow \pi^+ \nu_\mu \bar{\nu}_\mu$
		$\rightarrow \pi^+ \nu_\mu \bar{\nu}_\mu$
		$K^0_2 \rightarrow \mu^+ \mu^-$
		$K^0_2 \rightarrow \pi^0 \mu^+ \mu^-$
		$K^0_1 \rightarrow \mu^+ \mu^-$
		$\rightarrow \pi^0 \mu^+ \mu^-$
		$K^+ \rightarrow \pi^+ \pi^0 e^+ e^-$
		$K^0_{1,2} \rightarrow \pi^+ \pi^- e^+ e^-$

$^3P_0(0^+)$ and $^3P_1(1^+)$. The vector current describes transitions from $l_1 + l_2$ states with an odd $(l + S)$ ($^3S_1$ etc.) whereas the axial vector current describes states with $(l + S)$ even. ($^1S_0, {}^3P_1 \ldots$)

(a) G parity selection rules

$$\bar{p}n \rightarrow l^- \nu_l \quad \text{or} \quad \bar{p}p \rightarrow ll$$

Since $G = (-1)^{l+S+T}$ for $\bar{N}N$ annihilation and for $\bar{p}n$, $T = 1$ and $(l = 0$ because of locality) therefore

$G = -1$ $^1S_0 \rightarrow$ axial vector transition

$G = +1$ $^3S_1 \rightarrow$ vector transition

(b) $\bar{p}p \rightarrow \bar{l}l$; both $T = 1$ and $T = 0$ transitions are allowed. We have the following transitions

$$T = 1 \begin{cases} G = -1 \ ^1S_0 \rightarrow \text{axial vector} \\ G = +1 \ ^3S_1 \rightarrow \text{vector transition} \end{cases} \quad T = 0 \begin{cases} G = +1 \ ^1S_0 \rightarrow \text{axial} \\ G = -1 \ ^3S_1 \rightarrow \text{vector} \end{cases}$$

A simple way to describe meson decays is to think of the process $m \rightarrow F_1 + F_2 \rightarrow l_1 \bar{l}_2$.

Three examples are given (we assume nonrelativistic leptons for simplicity).

(c) $\begin{pmatrix} K \\ \pi^- \end{pmatrix} \rightarrow \begin{pmatrix} \Lambda\ p \\ \bar{p}\ n \end{pmatrix} \xrightarrow{w} l_1 \bar{l}_2$ since $\begin{pmatrix} \Lambda n \\ \bar{p}\ n \end{pmatrix}$ must be in an $^1S_0$ state, this is an axial vector transition. Note that the lepton system is in the $^1S_0$ state also which is very unfavorable for $e^+\nu$ and forbidden for $\bar{\nu}\nu$.

(d) $\begin{pmatrix} \pi K \\ \pi\pi \end{pmatrix} \rightarrow \begin{pmatrix} \Lambda N \\ N N \end{pmatrix} \xrightarrow{w} l_1 l_2$ for a $\pi^+\pi^-$ system since $CP = +1$ only the $^3S_1$ inter-

mediate state is allowed. Thus this transition is vector like requiring the $\pi\pi$ or πK system to be in a p wave in the initial state. The 0^+ initial state goes through a (ΛN) $^3P_0$ state and is sometimes called the induced scalar interaction. The latter interaction results in the two leptons being in a spin zero state which is unfavorable and, therefore, these transitions are expected to be suppressed in the case of $e\nu$ final states and completely forbidden for $\nu\bar{\nu}$ final states.

(e) $[K(\pi\pi)] \rightarrow l_1 l_2$. This system is more complicated because of the three particles in the initial state. For simplicity we consider cases (i) $(\pi\pi)$ system in 0^+ state and (ii) $(\pi\pi)$ system in 1^- state.

(i) $K(0^+) \rightarrow (\Lambda N) \xrightarrow{w} l_1 l_2$; the intermediate state is $^1S_0$ with s wave relative momentum in the initial state and $^3P_1$ with p wave relative momentum. Thus this is an axial vector transition with the $^1S_0$ state resulting in an induced pseudo-scalar transition. The vector transition ($^3S_1$ or $^1P_0$) are forbidden.

(ii) $K(1^-) \rightarrow (\Lambda N) \xrightarrow{w} l_1 l_2$.

The $^1S_0$ and $^3P_1$ intermediate states come from the p wave initial states. There is also the possibility of the $^1P_1$ intermediate state for relative p wave in the initial state and, therefore, there is a vector transition allowed for this process.

II. VECTOR BOSON THEORY OF WEAK INTERACTIONS

It has been pointed out by many people that the phenomological Lagrangian of equation (1) might be justified in a field theoretic sense if there exists a vector boson field to "transmit" the weak interaction.[2] Such an interaction is visualized as the exchange of a boson between pairs of particles as shown below.

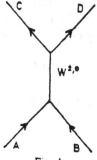

Fig. 1

The number of bosons needed to describe weak interactions seems to be a function of the year in which the boson theory is reformulated. Figure 2 shows the history of the intermediate boson theory.

If the intermediate boson theory is correct then the electric charge of the currents required to describe the experimental data gives a direct indication of the

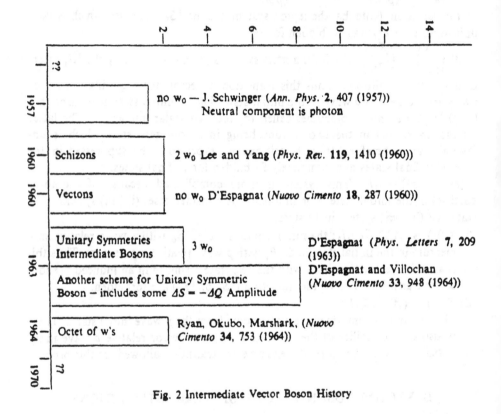

Fig. 2 Intermediate Vector Boson History

charges of the vector bosons, provided the coupling of the bosons to all currents is the same. The nonexistence of neutral leptonic currents either implies that W^0's do not exist or that for some reason the coupling of leptons to W^0's is very weak.

Thus a clear cut way to observe neutral currents is to observe neutral intermediate bosons. Unfortunately if these objects are not coupled to leptons, unambiguous detection will be very difficult. However, processes like

$$\bar{p}p \to W^0 \to \pi K$$

might disclose the existence of such objects as resonances in the direct channel.

III. EXPERIMENTAL EVIDENCE CONCERNING NEUTRAL CURRENTS

A. Leptonic Processes

There is at present no experimental evidence concerning neutral currents in these interactions. The absence of decays of the type

$$\mu \rightarrow e + \gamma$$

$$\rightarrow eee$$

was at one time thought to indicate the absence of neutral currents but is now considered to be due to the separate conservation of electronic and muonic lepton numbers.

Experimentally the most accessible leptonic process in which to observe leptonic neutral currents is[3]

$$\nu_\mu + Z \rightarrow \nu_\mu + Z + e^+ + e^- \tag{6}$$

which is simply related to $\nu_\mu + e^- \rightarrow \nu_\mu + e^-$.

Unfortunately this process is very difficult to distinguish from pair production by photons. In order to measure the neutral current coupling constant it is necessary to compare process (6) with the charged current process

$$\nu_\mu + Z \rightarrow \mu^- + Z + e^+ + \nu_e. \tag{7}$$

Neither process 6 nor 7 have been observed so far in neutrino experiments.

B. Semi-Leptonic Processes Involving Bosons

There is a large amount of experimental evidence concerning these processes. Table 2 shows a representative list of the processes that are allowed by lepton conservation (μonic and electronic).

At present there is no evidence concerning neutral currents for $\Delta S = 0$ processes because these processes all compete unfavorably with the electromagnetic interaction.

Since the electromagnetic interaction conserves strangeness the $\Delta S \neq 0$ processes may be a more sensitive place to look for such currents unless the couplings are extremely weak. If the coupling is extremely weak then the combination of weak interaction, electromagnetic and strong interactions can give rise to processes that simulate neutral current decays. These "induced neutral current" processes will be discussed in some detail later.

If we assume that leptonic and hadronic currents are only of the vector and axial vector type then there are 10 coupling constants characterizing these currents,

$$g^A(\mu\mu),\ g^A(e\bar{e}),\ g^A(\nu\bar{\nu}),\ g^A(\mu\nu_\mu)\ g^A(e,\nu_e)$$

$$G^V(\mu\bar{\mu}),\ g^V(e\bar{e}),\ g^V(\nu\bar{\nu}),\ g^V(\mu\nu_\mu)\ g^V(e,\nu_e)$$

where $g^A(l_1 l_2)$ stands for the axial vector neutral hadronic current and $l_1 l_2$ stands for the neutral lepton current involving l_1 and l_2 leptons. Here all types of lepton currents that conserve leptonic charge are included.

These coupling constants are directly proportional to the rates for the following processes

$$K^+ \rightarrow \pi^0 \mu \nu_\mu \qquad\qquad\qquad g^V(\mu^+ \nu_\mu) \qquad\qquad (8a)$$

$$\rightarrow \pi^0 e^+ \nu_e \qquad\qquad\qquad g^V(e^+ \nu_e) \qquad\qquad (8b)$$

$$\rightarrow \pi e^+ e^- \qquad\qquad\qquad g^V(e^+ e^-) \qquad\qquad (8c)$$

$$\rightarrow \pi^+ \mu^+ \mu^- \qquad\qquad\qquad g^V(\mu^+ \mu^-) \qquad\qquad (8d)$$

$$\rightarrow \pi^+ (\nu_\mu \bar\nu_\mu) \qquad\qquad\qquad g^V(\nu\bar\nu) \qquad\qquad (8e)$$
$$\nu_e \nu_e$$

$$\rightarrow \mu^+ \nu_\mu \qquad\qquad\qquad g^A(\mu^+ \nu_\mu) \qquad\qquad (8f)$$

$$\rightarrow e^+ \nu_e \qquad\qquad\qquad g^A(e^+ \nu_e) \qquad\qquad (8g)$$

$$K_2^0 \rightarrow \mu^+ \mu^- \qquad\qquad\qquad g^A(\mu^+ \mu^-) \qquad\qquad (8h)$$

$$\rightarrow e^- e^- \qquad\qquad\qquad g^A(e^+ e^-)$$

$$\rightarrow \nu_\mu \nu_\mu \qquad\qquad\qquad g^A(\nu_{\mu,e}, \bar\nu_{\mu,e})$$

In addition the processes[4]

$$K_1^0 \rightarrow \mu^+ \mu^- \qquad\qquad (9a)$$

$$K_2^0 \rightarrow \pi^0 e^+ e^- \qquad\qquad (9b)$$

are allowed by lepton conservation but are possibly CP violating modes provided the weak interaction is very local. These modes will be discussed later in connection with CP violation.

The process $K_2^0 \rightarrow \nu_\mu \bar\nu_\mu$ is absolutely forbidden for a massless ν because of helicity conservation (provided the ν and $\bar\nu$ are not identical particles). Likewise because of the small electron mass $K_2^0 \rightarrow e^+ e^-$ would be extremely unlikely even if the coupling constant $g^A(e^+ e^-)$ were very large. In addition it is not likely that the processes $K^+ \rightarrow \pi^+ \nu_e \bar\nu_e$ and $K^+ \rightarrow \pi^+ \nu_\mu \bar\nu_\mu$ can be separately identified, at least in the near future. Thus we are left with (8) possible leptonic processes. At present only processes (8a), (8b), (8f) and (8g) have been observed.

A variety of experimental searches for neutral current decays of the K mesons have been carried out in the past few years. No unambiguous events have been observed but several suggestive candidates have been recorded. The limits on K^+ decays come from experiments using heavy liquid bubble chambers. As a by-product of the studies of $K_2^0 \rightarrow \pi^+ \pi^-$ good limits on the neutral current decays of K_2^0 mesons have been obtained. Table 4 shows the present experimental limits on the various branching ratios as well as the limit on the ratio of the neutral current coupling constant to charged current couplings.

de Rafael has estimated the rates of the neutral current decays of the K^+ and $K^0_{L,S}$ in terms of the ratio of coupling constants.[10] He obtains for K^+ decays

$$\Gamma(K^+ \to \pi^+\mu^+\mu^-) = 5.5[g^V(\mu^+\mu^-)/g^V(\mu^+\nu)]^2 \, 10^7 \, (\text{sec}^{-1})$$

$$\Gamma(K^+ \to \pi^+ e^+ e^-) = 1.2 \frac{[g^V(e^+e^-)]^2}{g^V(e^+\nu)} \, 10^8 \, (\text{sec}^{-1}).$$

The ratio of rates for K^0_L decay is given by Lee and Yang[2] as

$$\frac{\Gamma(K^0_2 \to \mu^+\mu^-)}{\Gamma(K^+ \to \mu^+\mu^-)} = 4[g^A(\mu^+\mu^-)/g^A(\mu^+\mu\nu)]^2 \frac{m^3_K(m^2_K - 4m^2_\mu)^{1/2}}{(m^2_K - m^2_\mu)^2}.$$

The coupling constants limits in table 4 were obtained using these formulas.

Because the neutral current processes have been searched for in a variety of reactions including both axial vector and vector strongly interacting current and all possible combinations of leptons, the absence of these processes is very suggestive of a fundamental property of the weak interaction or of the lepton system. It is thus apparent that if primary neutral currents exist their coupling strength is at least three orders of magnitude less than that of charged currents.

The 4 body final states such as

$$K^0 \to \pi^+\pi^- e^+ e^-$$

$$\pi^0\pi^0 e^+ e^-$$

deserve special attention. The two and three body mesic transitions involve either axial vector or vector transitions. The four body processes are reactions where interference terms between axial and vector transitions can occur. In addition for the $\pi\pi$ system in the 0^+ state, these processes are strongly forbidden as Dalitz electromagnetic decays because of the $0 \to 0$ transition. Thus such states may be a good place to look for primative neutral currents.

C. Semi-Leptonic Processes Involving Hadronic Fermions

Table 3 lists a representative sample of processes which should be observed if neutral leptonic currents couple to B_1B_2 currents. Unfortunately the experimental evidence concerning such processes is much more restricted than for the meson case. In addition a new class of decays can be studied with fermion processes, namely ($\Delta S = 2$ and $\Delta Q = 0$). There have been no reported examples of processes like

$$\Sigma^+ \to p e^+ e^- \quad \text{or} \quad \Lambda \to n e^+ e^-$$

as of yet, however, it is doubtful whether the branching ratio limits on these decays are as small as the branching ratios for $\Sigma^- \to n e^- \nu$ or $\Lambda \to p e^- \nu$.

No evidence concerning the ($\Delta S = 2$, $\Delta Q = 0$) processes

$$\Xi^0 \to n e^+ e^-$$

exists or is likely to exist for sometime because of the difficulty of obtaining a large sample of Ξ^0 events.

The process

$$\nu_\mu + p \rightarrow \nu_\mu + p$$

Table 3

(Fermion Transitions)

	$\Delta S = 0$	$\Delta S = 1$	$\Delta S = 2$
$\Delta Q = 1$	$n \rightarrow pe^-\nu$	$\Sigma^- \rightarrow ne^-\nu_e$	$\Xi^- \rightarrow ne^-\nu$
	$\Sigma^\pm \rightarrow \Lambda e^\pm \nu$	$\Lambda \rightarrow pe^-\nu_e$	$\Omega^- \rightarrow \Lambda e^-\nu$
		$\Xi^- \rightarrow \Lambda e^-\nu$	
$\Delta Q = 0$	$(Z, A) \rightarrow (Z, A-1) +$	$\Sigma^+ \rightarrow pe^-e^-$	$\Xi^0 \rightarrow ne^+e^-$
	$e^+e^- + n$	$\Lambda \rightarrow ne^+e^-$	$\Omega^- \rightarrow \Sigma^-e^+e^-$
	$\nu_\mu + p \rightarrow \nu_\mu + p$	$\Lambda \rightarrow n\nu\bar{\nu}$	
	$\Sigma^0 \xrightarrow{w} \Lambda e^+e^-$		
	$\nu_\mu + p \rightarrow \nu_\mu + \pi^+ + n$		

has been searched for in recent ν experiments. The best limit on the cross section for this process is given in Ref. 11 as

$$\sigma_{\nu_\mu \nu_\mu} \langle 10^{-40}\ \text{cm}^2 \quad \text{for} \quad q^2 \rangle \cdot 5\ (\text{BeV}/C)^2$$

as compared to the cross section for $\nu_\mu + n \rightarrow \mu^- + p$ of

$$\sigma_{\nu_\mu \mu}^- \sim 5 \times 10^{-39}\ \text{cm}^2$$

Thus, although the limit is not as good as the corresponding limit for mesic decays this limit gives a strong indication that the $\Delta S = 0, \Delta Q = 0$ transition is suppressed relative to the $\Delta S = 0, \Delta Q = 1$.

Unfortunately all other $\Delta S = 0, \Delta Q = 0$ transitions are likely to be swamped by the electromagnetic interaction. There is possibly one exception, if a decay of the kind $0^- \rightarrow 0^+ + e^+ + e^-$ could be found, (in nuclear decays for example); then the weak interaction might compete favorably with the electromagnetic interaction because of the $0 \rightarrow 0$ transition.

IV. HADRONIC VS. LEPTONIC NEUTRAL CURRENTS

From the previous discussion it is clear that the coupling constants associated with mesonic and Fermion decays that have zero charged leptons in the final state are very small compared to the coupling constants associated with charged leptons. The evidence for reduced coupling constants for such a variety of possible currents leads to the expectation that we are here dealing with a universal phenomenon. How are we to interpret these reduced couplings? If these decays are describable within the context of the current-current interaction they would involve the product of a neutral hadronic current with a neutral leptonic current. Therefore there are at least three alternatives:

(a) There are no primary neutral leptonic currents
(b) There are no neutral hadronic currents
(c) The neutral leptonic currents are somehow decoupled from the neutral hadronic currents.

Table 4

Best Experimental Limits on Neutral Current K Decays

Process	Branching Ratio	Coupling Constant Ratio	Upper Limit on the C. C. Ratio	Reference
$K^+ \to \pi^+ e^+ e^-$	$< 8.8 \times 10^{-7}$*	$g^V(e^+e^-)/g^V(e^+\nu_e)$	7×10^{-4}	Cline et al.[5,6]
$K^+ \to \pi^+ \mu^+ \mu^-$	$< 3 \times 10^{-6}$	$g^V(\mu^+\mu^-)\, g^V(\mu^-\nu_\mu)$	1.4×10^{-2}	Camerini et al.[6,7]
$K^+ \to \pi^+ \nu\bar\nu$	$< 1.1 \times 10^{-4}$	$g^V(\nu\bar\nu)\,g^V(e^-\nu_e)$	6×10^{-2}	Cline[8] $55 < T_\pi < 80$ MeV
$K_2^0 \to \mu^+\mu^-$	$< 1.6 \times 10^{-6}$	$g^A(\mu^+\mu^-)/g^A(\mu^+\nu_\mu)$	7.4×10^{-4}	Bott-Bodenhausen et al.[9]
$K_1^0 \to \mu^+\mu^-$	$< 7.3 \times 10^{-5}$	$g^A(\mu^-\mu^-)/g^A(\nu^+\mu_\mu)$	1.6×10^{-3}	Bott-Bodenhausen et al.[9]
$K_2^0 \to \pi^0\mu^+\mu^-$	No estimate	$g^V(\mu^+\mu^-)/g^V(\mu^+\nu)$	No estimate	
$K_2^0 \to \pi^0 e^+ e^-$	No estimate	$g^V(ee)\,g^V(e^-\nu_e)$	No estimate	
$K^+ \to \pi^+\pi^0 e^- e^-$	$< 8 \times 10^{-6}$		No estimate	Cline[8]
$K^+ \to \pi^+\mu^+ e^-$	$< 3 \times 10^{-5}$		violates lepton conservation	Cline[8]
$K_2^0 \to \mu^\pm e^\mp$	$< 9 \times 10^{-6}$		violates lepton conservation	Bott-Bodenhausen et al.[9]

* All estimates are 90% confidence limits.

As previously discussed there is no experimental evidence concerning possibility (a). Possibilities (b) and (c) have been widely discussed in the literature and it is, therefore, appropriate to spend some time considering theoretical and experimental implications.

A) Experimental Evidence Concerning Hadronic Neutral Currents
If the strong interactions were turned off the existence of the weak reactions

$$\Lambda + n \xrightarrow{w} n + n \tag{10}$$

$$n + n \xrightarrow{w} n + n \tag{11}$$

would be unambiguous evidence for hadronic neutral currents. The corresponding charged current processes would be

$$\Lambda + p \xrightarrow{w} n + p \tag{12}$$

$$n + p \xrightarrow{w} n + p. \tag{13}$$

With the strong interaction turned on Λ's can be bound in nuclei for long times and thus the weak interaction transitions 10 and 12 occur. Experimentally process 10 is distinguished from process 12 by the emission of two fast neutrons in the final state as compared to a fast (np) pair. Experimentally nonmesic hypernuclear decays do sometimes involve two fast neutrons. For example for He^4 decays the following ratio of process 10 to 12 has been estimated as $1.1 \pm . 4$.[12]

Although process 10 is well established at present it is not possible to separate the part of weak interaction and strong interaction contributions to this process. For example even if neutral hadronic currents do not exist, process 10 would be expected to occur with an appreciable probability via the process[13,14]

$$\varLambda + n \rightarrow (\pi^- + p) + n \rightarrow n + n$$

$$\rightarrow (\pi^0 + n) + n \rightarrow n + n$$

In addition there would be a contribution to process 12 from

$$\varLambda + p \rightarrow (\pi^- + p) + p \rightarrow p + n$$

$$\rightarrow (\pi^0 + n) + p \rightarrow n + p$$

Also the $\varSigma$ intermediate state processes can be important

$$\varLambda + n \rightarrow \varSigma^- + p \rightarrow (\pi^- + n) + p \rightarrow n + n$$

Since the $\varLambda \rightarrow \varSigma$ transition is known to be very strong this latter process may be very important. There should also be a close connection between these transitions and the occurance of π^+ mesic decays which probably come from the process $\varLambda + p \rightarrow \varSigma^+ + n \rightarrow \pi^+ + n + n$.

In the past there have been a number of calculations of the ratio of process 10 to 12 assuming a variety of intermediate states.[15] Unfortunately an added difficulty for such calculations is that processes 10 and 12 can only be studied in the presence of several nucleons and in many cases the neutron to proton ratio in the nucleus is not equal. Because of the messy nuclear physics involved and the presence of strong radiative corrections for process 10 and 12 it is not likely that an unambiguous statement concerning the existence of neutral hadron currents can be obtained in the near future.

Indirect evidence concerning the existence of neutral hadronic currents is provided by the success of the leptonic and nonleptonic $\Delta T = 1/2$ rule. As pointed out by Dalitz[14] it is possible to construct a fourfermion interaction that leads to the $\Delta T = 1/2$ selection rule for the interaction involving the $\varLambda$ hyperon in the form

$$(\bar{\varLambda}n)\,(\bar{n}n) - (\bar{\varLambda}p)\,(\bar{p}n) \rightarrow \left(\sqrt{1/2} + \sqrt{1/6} + \sqrt{2/3}\right) A(\Delta T = 1/2).$$

The $(\bar{\varLambda}p)\,(\bar{p}n)$ term alone would allow both $\Delta T = 1/2$ and $\Delta T = 3/2$ processes in the combination $\sqrt{1/3}\,A(\Delta T = 3/2) - \sqrt{2/3}\,A(\Delta T = 1/2)$. Thus, to the extent that the current-current form of the hadronic current is valid the existence of the $|\Delta T| = 1/2$ rule seems to support the neutral hadronic current hypothesis. However, there is by no means universal agreement that the $|\Delta T| = 1/2$ rule actually gives strong support to the existence of neutral hadronic currents. In particular one apparent recent success of current algebra techniques is the successful description of the $|\Delta T| = 1/2$ rule *without* the need for neutral hadronic currents.

V. INDUCED NEUTRAL CURRENTS AND CP VIOLATION IN NEUTRAL CURRENT PROCESSES

It has been suggested by many people[16,17,18] that even in the absence of fundamental neutral lepton currents the combination of the weak interaction and electromagnetic interaction can give rise to processes which have two charged leptons in the final state (Induced neutral current processes). Since such processes always involve $\Delta Q = 0$ for the hadrons it is possible that the existence of hadronic neutral currents is coupled to the existence of such processes. In particular as emphasized by Beg[19] if there are neutral hadronic currents, induced processes are to be expected. The converse probably need not be true, however, because of strong interaction effects although this is perhaps an open question worthy of detailed theoretical analysis.

Table 5

Predicted Rates for Induced
Neutral Current Processes

Process	Branching Ratio Estimate	Reference and Remarks
$K^+ \to \pi^+ e^+ e^-$	$\sim 10^{-7}$	Cabibbo and Ferrari, *N. C.* **18**, 928 (1960)
	$\sim 10^{-6}$	Baker and Glashow, *N. C.* **25**, 857 (1962)
	$\sim 10^{-6}$	M. A. Beg, *Phys. Rev.* **132**, 426 (1963).
	$\sim 10^{-6}$	K. Tanaka, Ohio State University Preprint
	$(1.8 - 4) \times 10^{-7}$	Ignatovich and Strummsky, *Phys. Letters* 24B, 69 (1967).
$K^+ \to \pi^+ \mu^+ \mu^-$	Reduced by a factor of ~ 4 by phase space over the $\pi^+ e^+ e^-$ estimates	
$K_1^0 \to \pi^0 e^+ e^-$	$\sim 10^8$	Baker and Glashow, *N. C.* **25**, 857 (1962).
$K_2^0 \to \mu^+ \mu^-$	$\sim 4 \times 10^{-8}$	M. A. Beg, *Phys. Rev.* **132**, 426 (1963).
	$\sim 10^{-8} - 10^{-9}$	L. Seghal, Carnegie Institute Preprint
$\Sigma^+ \to p e^+ e^-$	1.4×10^{-5}	Obtained from $\frac{1}{130} \cdot R(\Sigma^+ \to p\gamma)$
		Lygin and E. Ginzburg, *JETP* **14**, 653 (1962).

A number of estimates of the rate for various induced processes have been given in the literature. Table 5 lists these rate estimates. At present all experimental results are consistent with these rate estimates, at least to within an order of magnitude.

There have been suggestions that the observed CP violation in the decay mode

$$K_2 \to \pi^+ \pi^-$$

may be related to the existence of neutral leptonic currents, the smallness of the CP violation effect being coupled to the reduced coupling of the neutral currents. If such a theory were to be correct the decay modes

$$K^+ \to \pi^+ \mu^+ \mu^-$$

have ample measurable correlation functions that violate T; for example the correlations $\langle \sigma_{\mu^+} \cdot P_\mu \times P_\pi \rangle$ or $\langle \sigma_{\mu^-} \cdot \sigma_{\mu^+} \times P_\pi \rangle$.

Good, Michel and de Rafael[4] have pointed out that because of the locality of the coupling of the leptons the decay mode

$$K_1^0 \to \mu^+ \mu^-,$$

if it occurs in first order weak interaction and without the intervention of the electromagnetic field (i.e. the $\mu^+ \mu^-$ being in the $^1S_0$ state) would be CP violating. If such decays are detected it should be possible to distinguish the $^1S_0(\mu^-\mu^-)$ state from the $'P_0$ state through the correlation between the μ^+ and μ^- polarization vectors. Similarly the decay modes

$$K_1^0 \to \pi^0 \mu^- \mu^-$$

$$\to \pi^0 e^- e^-$$

$$\to \pi^0 \nu \bar{\nu}$$

are vector transitions (although the induced scalar component might be dominant for $\pi^0 \mu^- \mu^-$) and therefore, the $l^- l^+$ state will be predominately $^3S_1$. In this case the CP of the system will be (-1) $[CP = (-1)^{s+1}$ for a $l\bar{l}$ system] thus violating CP invariance. Because of the possibility of CP violation in these modes they are of great interest.

VI. SOME THEORETICAL PROPOSALS CONCERNING THE ABSENCE OF NEUTRAL LEPTON CURRENTS

A large number of explanations have been offered for the absence of neutral current processes. Here we shall only sketch the outline of a few of these theories.
(1) Bludman has proposed a scheme incorporating only symmetric neutral currents of the form $\mu\mu$, $\nu\nu$, and ee.[21] He proposes to classify the leptons into isotopic spin doublets. In this scheme a process like

$$\nu_e + e \to \nu_e + e$$

does not exist because the neutral currents compensate for the charged currents causing a cancellation in this amplitude. Pontecorvo[22] has pointed out that such a scheme does not allow processes like

$$K^+ \to \pi^+ e^+ e^-$$

because the $(K^+ \pi^-)$ neutral current is not symmetric. An interesting prediction of this model is the occurrence of elastic ν_μ scattering. F. Michel[23] has suggested a variant of this model which requires the $\Delta S = 0$ interaction to be symmetric in isotopic spin space and the leptons arranged in isotopic spin doublets. In this model the small $\Delta Q = 0$, $\Delta S = 1$ coupling is to be regarded as a fluke.
(2) Salam and Ward[24] have proposed to couple electromagnetic and weak interactions. Such a scheme is attractive because superficially the absence of

neutral currents (or existence of only charged currents) seems to imply a connection between the weak interaction and the *electromagnetic* charge carried by the particles. In this model the neutral vector boson and the photon are connected in such a way that only $\Delta S = 0$ weak processes are followed for the $\Delta Q = 0$ weak interaction. The vector boson mass is assumed to be ~ 137 BeV which makes the primary weak interaction and the electromagnetic interaction of comparable strength. Since the effective weak interaction is related to the inverse of the vector boson mass the electromagnetic interaction will always overwhelm the $\Delta Q = 0$ weak interaction. The primary test of the model will therefore come from future searches for elastic ν_μ scattering.

(3) Recently Good, Michel, and de Rafael[4] have proposed an elegant model based on d'Espagnat's vector boson scheme. In this model the neutral lepton-neutral hadron coupling is reduced to a small value because of cancellations between the amplitudes of the neutral vectors bosons. In addition the following consequences follow from the model:

(1) the separate conservation of μ and e leptons.
(2) A neutral self energy loop for the μ but not for the e thus possibly giving rise to a mass difference between the μ and e.
(3) Possibility of *CP* violation if the decoupling of the neutral lepton currents and neutral hadron currents is upset.

Thus far there is little experimental evidence to support the above conjectures. In fact the crucial test of such models will probably come from experimental studies of lepton-lepton scattering which presently seem virtually impossible. Nevertheless the successful explanation of the absence of neutral lepton processes (and possibly of primative neutral hadron couplings) will undoubtedly be a very significant factor in the ultimate theory of the weak interaction.

REFERENCES

1. R. P. Feynman and M. Gell-Mann, *Phys. Rev.* **109**, 193 (1958).
2. See, for example, T. D. Lee, and C. N. Yang, *Phys. Rev.* **119**, 1410 (1960).
3. T. T. Wu, *Phys. Rev.* **147**, 1033 (1966).
4. M. L. Good, L. Michel, and E. de Rafael, *Phys. Rev.* **151**, 1195 (1966).
5. D. Cline, H. Haggerty, W. Singleton, W. Fry, and N. Sehgal, "Further Search for Neutral Leptonic Currents in K^+ Decay", submitted to the Heidelberg International Conference on High Energy Physics.
6. U. Camerini, D. Cline, W. Fry, and W. M. Powell, *Phys. Rev. Letters* **13**, 318 (1964).
7. U. Camerini, D. Cline, G. Gidal, G. Kalmus, and A. Kernan, *Nuovo Cimento* **37**, 1795 (1965).
8. D. Cline, thesis University of Wisconsin, unpublished 1965.
9. M. Bott-Bodenhausen, X De Bouard, D. G. Cassel, Dl Dekkers, R. Felst, R. Mermod, I. Savin, P. Scharff, M. Vivargent, T. R. Willits, and K. Winter, *Phys. Letters* **24B**, 194 (1967).
10. E. de Rafael, BNL Preprint (1967).
11. M. M. Block *et al.*, *Phys. Letters* **12**, 281 (1964).
12. M. M. Block, R. Gessaroli *et al.*, Proceedings of the International Conference on Hyperfragments at St. Cergue, Switzerland, 1963. This method was suggested by M. Baldo-Ceolin, C. Dilworth, W. F. Fry *et al.*, *Nuovo Cimento* **10**, 328 (1958).

13. M. Ruderman and R. Karplus, *Phys. Rev.* **26**, 1458 (1949); also F. Cerulus, *Nuovo Cimento* **5**, 1685 (1957).
14. R. H. Dalitz, *Rev. Mod. Phys.* **31**, 823 (1959).
15. See for references, R. Kreman and G. Dass, *Phys. Rev.* **151**, 1244 (1966).
16. L. B. Okun and A. Rudik, *Soviet Physics JETP* **12**, 422 (1961).
17. N. Cabibbo and E. Ferrari, *Nuovo Cimento* **18**, 928 (1960).
18. M. Baker and S. Glashow, *Nuovo Cimento* **25**, 857 (1962).
19. M. A. B. Beg, *Phys. Rev.* **132**, 426 (1963).
20. E. M. Lipmann, *Soviet Physics JETP* **48**, 750 (1965).
21. S. Bludman, *Nuovo Cimento* **9**, 433 (1958).
22. B. Pontecorvo, *Soviet Physics JETP* **16**, 1073 (1963).
23. F. C. Michel, *Phys. Rev.* **138**, B408 (1965).
24. A. Salam and J. Ward, *Physics Letters* **13**, 168 (1964).

SEARCH FOR NEUTRAL LEPTONIC CURRENTS IN K⁺ DECAY*

U. Camerini, D. Cline, and W. F. Fry

Physics Department, University of Wisconsin, Madison, Wisconsin

and

W. M. Powell

Lawrence Radiation Laboratory, University of California, Berkeley, California

(Received 4 August 1964)

A basic assumption of most present models of the weak interaction is that primitive neutral leptonic currents, to first order in the weak coupling constant, do not exist.[1] However, some models propose the existence of neutral nonleptonic currents in order to explain the $|\Delta T| = \frac{1}{2}$ rule.[2] Recently it has been suggested that primitive neutral leptonic currents of strength comparable to that of charged currents might exist, but some reactions where they would appear could be inhibited by selection rules among the strongly interacting particles.[3] Even if primitive neutral currents do not exist, the combined effects of weak and electromagnetic interactions can cause induced neutral currents which may be observable.[4,5]

In order to look for evidence of neutral currents in strangeness-changing interactions, the possible decay mode

$$K^+ \to \pi^+ + e^+ + e^- \tag{1}$$

has been searched for in a sample of 1.7×10^6 stopped-K^+ decays. The K^+ mesons were stopped in the Lawrence Radiation Laboratory 30-inch heavy-liquid chamber filled with C_3F_8. No unambiguous events have been found corresponding to decay mode (1).

The detection procedure consisted of initially scanning for three-track decays that were not examples of the ordinary τ decay of the K^+. About two thirds of the film was scanned twice. Each event was then carefully looked at again on the scanning table and was classified in one of the following three categories: (a) ordinary Dalitz pair with obvious missing momentum; (b) apparent momentum-conserving event; (c) electron pairs which converted very near the K^+ decay.

The events in categories (a) and (b) were used to compute the absolute scanning efficiency from the number of Dalitz decays expected. About

6000 ordinary Dalitz-pair events were found, giving a scanning efficiency of 84%. In category (b) only events with an angle between the electron and positron of greater than 10° were accepted as candidates for mode (1). This reduces the background considerably and does not significantly reduce the detection efficiency. The remaining events were measured and constrained to the hypothesis of decay mode (1). The electron energies were corrected for bremsstrahlung energy loss using the Behr-Mittner method. Because of the inability to measure any momentum to a precision of better than 20%, the events were tested for decay mode (1) using mainly a one-constraint fit, which does not include the measured momenta. Since this fit depends only on angle measurements, it also contains a good coplanarity test ant it is expected to be quite reliable. The $\chi^2(1C)$ distribution for all events with $\chi^2 < 50$ is shown in Fig. 1.

The most serious background for decay mode (1) comes from

$$K^+ \to \pi^+ + \pi^0 \to \pi^+ + e^+ + e^- + \gamma, \qquad (2)$$

where the γ ray does not materialize and comes off at the right center-of-momentum angles to make the charged particles nearly coplanar. The configurations of mode (2) that fit the hypothesis for mode (1) always have missing γ-ray momentum in the same direction as the positive pion momentum; this leads to a fitted pion momentum for this hypothesis that is greater than or equal to the unique momentum of decay mode (2): namely, 205 MeV/c. Modes (1) and (2) can be separated by selecting events with fitted momentum of less than 205 MeV/c as examples of (1).

Unfortunately there are also other three-body decay K^+ modes with Dalitz pairs that can fake mode (1) and give a fitted momentum below 205 Mev/c. In general, there is no way to separate these events from mode (1) unless the assumed pion stops in the chamber, thus allowing an accurate momentum measurement by range. The number of such background events is expected to be at least an order of magnitude below that of Reaction (2).

Figure 2 shows a histogram of all events with $\chi^2(1C) \leqslant 10$ plotted as a function of the pion momentum obtained from the 1C fit. For the events with $P_\pi > 205$ MeV/c there are five that have stopping pions in the chamber with the range expected for decay mode (2); using geometrical loss for this range we expect 21 examples of decay mode (2) compared to 26 events above 205 MeV/c. Thus all events above 205 MeV/c are consistent with mode (2). The important characteristics of the three events with pion momentum below 205 MeV/c are summarized in Table I. The first two events cannot be examples of (1) and are most likely examples of

$$K^+ \to \pi^0 + \binom{\mu^+}{e^+} + \nu \to e^+ + e^- + \gamma + \binom{\mu^+}{e^+} + \nu. \qquad (3)$$

The third event has a large invariant mass and is unlikely to be an example of (2) or (3). We expect 0.2 events of type (2) with an e^+-e^- invariant mass between 115 and 125 MeV and a negligible number of type (3).[6] However, because of the large uncertainty in invariant mass of this event we are unable to conclude that it is an unambiguous event and shall consider it as an upper limit. Assuming that decay mode (1) comes about

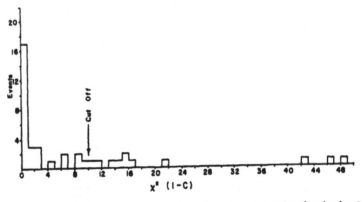

FIG. 1. $\chi^2(1C)$ distribution for events that have the configuration expected for $K^+ \to \pi^+ + e^+ + e^-$.

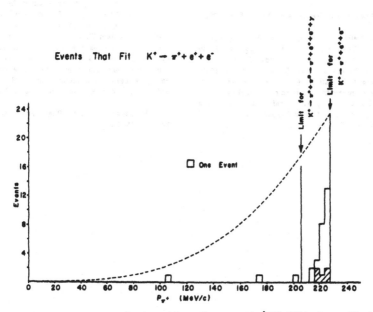

FIG. 2. The pion-momentum spectrum (as deduced from all events with $\chi^2(1C) < 10$) is shown. The dashed line represents "the expected spectrum" for induced or primitive neutral currents. The cross-hatched events are pions that stopped in the chamber indicating that they are examples of decay mode (2).

through primitive neutral currents, it is reasonable to expect that the pion spectrum should be that expected for the ordinary K_{e3}. The same spectrum is theoretically predicted for the induced neutral current contribution to decay (1).[4] On the basis of this spectrum for $P_\pi < 205$ MeV/c, the total detection efficiency is calculated to be 55%, which gives an effective sample size of 9.4×10^8 stopped-K^+ decays. An invariant phase-space pion spectrum leads to a larger detection efficiency. Since one possible event was found, the branching ratio is

$$\frac{\Gamma(\pi ee)}{\Gamma(\text{all})} \leq \frac{1}{9.4 \times 10^8} = 1.1 \times 10^{-9};$$

the 90% confidence level is 2.45×10^{-9}. The 90% confidence level for the upper limit[7] of the ratio of primitive neutral leptonic current to charged leptonic current coupling constants is

$$|g_{e\bar{e}}|^2 \leq 2.5 \times 10^{-5} |g_{e\bar{\nu}}|^2.$$

It has been shown in a model for decay mode (1) through induced neutral currents that the branching ratio is proportional to $f_{K\pi}$, the weak K-π coupling constant.[5] To the extent that this model describes the actual decay mode (1), we can put

a conservative limit on $f_{K\pi}$:

$$f_{K\pi}(-m_K^2) \leq 7 \times 10^{-8} m_K^2.$$

Biswas and Bose[8] pointed out that the K_1-K_2 mass difference demands a rate for decay mode (1) larger than the limit for this rate found in this experiment, if the dominant contribution to K_1-K_2 mass difference comes from the π^0 and η^0 pole. The value of $f_{K\pi}^2$ required to give a reasonable value of the K_1-K_2 mass difference is 40 times larger than the 90% confidence level found in this experiment.

Table I. Characteristics of events with fitted pion momentum of less than 205 MeV/c. χ^2 is for the 1C fit; $P\pi^+$ is the momentum from the same fit; $P\pi_\gamma^+$ is the lower limit on the momentum from the observed range of the π^+ in the chamber.

Event	P_π^+	$P_{\pi_\gamma}^+$	$M_{e^+ - e^-}$ (MeV)	$\chi^2(1C)$
146389	106 ± 3	115.8	40 ± 20	0.92
116859	174 ± 3	180	69 ± 33	0.60
187088	>201 ± 3.7	189.2	120 ± 55	0.77

U. Camerini *et al.*

*Work supported in part by the U. S. Atomic Energy
Commission and in part by the Graduate School from
funds supplied by the Wisconsin Alumni Research
Foundation.

[1]R. Feynman and M. Gell-Mann, Phys. Rev. <u>109</u>,
13 (1958).

[2]T. D. Lee and C. N. Yang, Phys. Rev. <u>119</u>, 1410
(1960); S. B. Treiman, Nuovo Cimento <u>15</u>, 916 (1960).

[3]S. Okubo and R. E. Marshak, Nuovo Cimento <u>28</u>,
56 (1963); Y. Ne'eman, Nuovo Cimento <u>27</u>, 922 (1963).

[4]Estimates of the rate for $K^+ \to \pi^+ + e^+ + e^-$ due to induced neutral currents have been calculated by several
authors. For a list of previous references see Mirza A.
Baqi Bég, Phys. Rev. <u>132</u>, 426 (1963).

[5]M. Baker and S. Glashow, Nuovo Cimento <u>25</u>, 857
(1962). They predict a branching ratio for decay mode
(1) of $\sim 10^{-4}$.

[6]N. P. Samios, Phys. Rev. <u>121</u>, 275 (1961).

[7]The best previously reported estimate comes from
the limit on $K_2^0 \to \mu^+ + \mu^-$. The 90% confidence level is
$|g_{\mu\bar{\mu}}|^2 < 10^{-3}|g_{\mu\bar{\nu}}|^2$: M. Barton, K. Lande, L. M. Lederman, and William Chinowsky, Ann. Phys. (N.Y.) <u>5</u>,
156 (1958). The absence of the decay mode $\mu^+ \to e^+ + e^+
+ e^-$ is not a good test for the existence of neutral cur-
rents since this decay mode may be absolutely forbidden by conservation of muon number: G. Feinberg
and L. M. Lederman, Ann. Rev. Nucl. Sci. <u>13</u>, 465
(1963).

[8]S. N. Biswas and S. K. Bose, Phys. Rev. Letters
<u>12</u>, 176 (1964).

LIMIT ON THE $K^+ \to \pi^+ + \nu + \bar{\nu}$ DECAY RATE*

J. H. Klems and R. H. Hildebrand

Lawrence Radiation Laboratory and University of Chicago, Chicago, Illinois 60637

and

R. Stiening

Lawrence Radiation Laboratory, University of California, Berkeley, California 94707

(Received 27 March 1970)

The branching ratio for the process $K^- \to \pi^- + \nu + \bar{\nu}$ is shown by a counter spark chamber experiment to be less than 1.2×10^{-6} of all decay modes, assuming a pion energy spectrum like that of $K^- \to \pi^0 + e^- + \nu$. Our apparatus was sensitive to pions in the kinetic energy range 117-127 MeV.

In 1964 Camerini, Cline, Fry, and Powell[1] reported the results of a search for the reaction $K^+ \to \pi^+ + e^+ + e^-$. They set an upper limit of 2.5×10^{-6} on the branching ratio for this decay mode. Other experiments[2] have been made to search for $K_L^0 \to e^+ + e^-$, $K_{L,S}^0 \to \mu^+ + \mu^-$, and $K^+ \to \pi^+ \mu^+ \mu^-$.

These decays have not been observed. In the experiment described here, we have searched for the decay

$$K \to \pi^+ + \nu +, \bar{\nu}. \qquad (1)$$

We have observed no examples of this decay. If

we assume that the energy spectrum of the π^+ is the same as that of the π^0 in the observed reaction[3] $K^+ \to \pi^0 + e^+ + \nu$, we can set an upper limit on the branching ratio[4] for the K^+ to decay in this manner of 1.2×10^{-6} (90% confidence level).

The significance of our result depends upon the manner in which we account for the absence of the reactions discussed above. We may suppose that $K^+ \to \pi^+ + \nu + \bar{\nu}$ should result from the same interaction that gives rise to $K^+ \to \pi^0 + e^+ + \nu$. The matrix element for this latter decay is known to be of the form[5]

$$2^{-1/2}G[\bar{U}_\nu \gamma_\lambda (1 + \gamma_5) U_e] \langle \pi^0 | J_\lambda | K^+ \rangle. \qquad (2)$$

If we substitute $\langle \pi^+ | J_\lambda | K^+ \rangle$ for $\langle \pi^0 | J_\lambda | K^+ \rangle$ and $\bar{U}_\nu \gamma_\lambda (1 + \gamma_5) U_\nu$ for $\bar{U}_\nu \gamma_\lambda (1 + \gamma_5) U_e$, and if $\langle \pi^0 | J_\lambda | K^+ \rangle = \langle \pi^+ | J_\lambda | K^+ \rangle$, the above expression for the matrix element is practically unchanged. The energy release in the decay is so high that the electron mass is negligible. That our upper limit on the branching ratio for $K^+ \to \pi^+ + \nu + \bar{\nu}$ is at most very small in comparison with the branching ratio for $K^+ \to \pi^0 + e^+ + \nu$ (which is 0.05) can be accounted for by assuming either that $\langle \pi^+ | J_\lambda | K^+ \rangle$ vanishes or that some lepton selection rule is violated by a current $\bar{U}_\nu \gamma_\lambda (1 + \gamma_5) U_\nu$.

The current J_λ is known empirically to obey the $\Delta I = \frac{1}{2}$ rule. If we assume that the $\langle \pi^+ | J_\lambda | K^+ \rangle$ component of this current vanishes, it is impossible to account for the $\Delta I = \frac{1}{2}$ selection rule of nonleptonic strange particle decays in the usual fashion as the result of a current-current interaction where one current carries $\Delta I = \frac{1}{2}$, $\Delta S = 1$, and the other carries $\Delta I = 1$, $\Delta S = 0$. Thus it is necessary to abandon the hypothesis that all weak interactions occur as the self-interaction of a current made up of many parts. Our experiment is consistent with the assumption that the matrix element $\langle \pi^+ | J_\lambda | K^+ \rangle$ vanishes since both $K^+ \to \pi^+ + e^+ + e^-$ and $K^+ \to \pi^+ + \nu + \bar{\nu}$ would then vanish. On the other hand, if $\langle \pi^+ | J_\lambda | K^+ \rangle \neq 0$ there must be a selection rule among leptons which prohibits currents of the form $\bar{U}_e \gamma_\lambda (1 + \gamma_5) U_e$. Our experiment then shows that the combination $\bar{U}_\nu \gamma_\lambda (1 + \gamma_5) U_\nu$ is also forbidden. It is impossible to decide at present whether it is this leptonic current or the hadronic current matrix element $\langle \pi^+ | J_\lambda | K^+ \rangle$ that vanishes.

Oakes[6] has suggested that although $K^+ \to \pi^+ + \nu + \bar{\nu}$ may occur in the framework of conventional weak-interaction theory for one of the reasons discussed above, there may be an additional type of weak-interaction current which does not conserve CP and gives rise to $K_2^0 \to 2\pi$ decay. The branching ratio[7] for $K^+ \to \pi^+ + \nu + \bar{\nu}$ in the theory of Oakes is 1.8×10^{-5}. Our result is inconsistent with this prediction. Other authors[8] have calculated $K^+ \to \pi^+ + \nu + \bar{\nu}$ on the basis of higher order weak-interaction theories. A simple second-order application of weak-interaction theory as it is now known leads to a divergent result for the $K^+ \to \pi^+ + \nu + \bar{\nu}$ decay rate. Various models have been made to ameliorate the difficulties caused by this divergence.[8] The interpretation of our result hinges then on details of the model employed.

The experiment depends on the fact that no observed K^+ decay at rest produces a π^+ with an energy greater than that from $K^+ \to \pi^+ + \pi^0$ ($T_\pi = 109$ MeV; branching ratio = 0.21). In order to produce a π^+ of higher energy the K^+ must decay into a π^+ and a neutral system with rest mass less than that of the π^0. If we neglect decays into four or more particles, the only possibilities are $K^+ \to \pi^+ + e^+ + e^-$ (branching ratio $< 2.5 \times 10^{-6}$),[1] $K^+ \to \pi^+ + \gamma + \gamma$ (branching ratio $< 1.1 \times 10^{-4}$),[9] and Reaction (1). The last two reactions may give pions with energies up to 127 MeV. Hence that we observe no π^+ emitted with energy between 117 and 127 MeV unaccompanied by high-energy γ's or charged particles in the opposite hemisphere is sufficient to exclude the $\pi^+ \nu \bar{\nu}$ decay.

The experimental arrangement is shown in Fig. 1. Kaons in the incoming beam from the Beva-

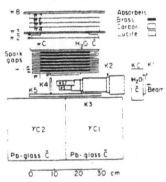

FIG. 1. Apparatus. Kaons stopping in the target scintillators $KS1$ and $KS2$ are selected from the incoming beam by signals $K1$, $\overline{KC}$, $K2$, $K3$, $KS1$ and/or $KS2$, $\overline{K4}$, $\overline{K5}$, where $K2$ and $K3$ have pulse heights $\geq 1.5 \times$ (pion pulse height). Low-velocity decay particles are selected by signals $\pi1$, $\overline{\pi C}$, $\pi2$, $\pi3$, $\overline{\pi8}$, $\overline{KC}$, $\overline{K4}$, $\overline{K5}$, where $\pi1$ is delayed ≥ 6 nsec after $K3$. Events which emit gammas into the opposite hemisphere are eliminated by signals $\overline{\gamma C1}$ and/or $\overline{\gamma C2}$. Pions are distinguished from stopping muons by scope displays of the π-μ-e decay pulses in counters $\pi4$-$\pi7$.

tron are brought to rest in the "K-stop" counters $KS1$ and $KS2$. Scattered and transmitted particles are suppressed by the anticoincidence counters $K4$ and $K5$. Those scattered toward the π counters are suppressed by the requirement that the pulses from counters $\pi1$ and $\pi2$ must be delayed ≥ 6 nsec after the pulse from the stopping K. Pions in the beam are excluded by (i) a water Cherenkov counter KC [actually consisting of two counters connected in parallel: $\beta(K) < \beta(\text{threshold}) < \beta(\pi)$], (ii) two dE/dX counters [$(dE/dX_K > 1.5(dE/dX)_\pi$], and (iii) range ($R_\pi \gg R_K$ for the same initial momentum). In summary the K^+ signal is [$K1$, $K2$, $K3$, $KS1$ and/or $KS2$, $\overline{KC}$, $\overline{K4}$, $\overline{K5}$]. We require that a subsequent π signal occur between 6 and 54 nsec after the K^+ signal, and we denote the K^+ signal together with this additional timing requirement as the "K-decay" signal.

The triggering system of the π^+ detector ($\pi1$, $\overline{\pi C}$, $\pi2$, $\pi3$, $\overline{\pi8}$, $\overline{KC}$, $\overline{K4}$, $\overline{K5}$) does not distinguish between stopping π's and stopping μ's, but high-velocity μ's from $K_{\mu2}$ are vetoed by the water Cherenkov counter πC and by the maximum-range counter $\pi8$. A large counter, KO, which completely covers the incoming beam (not shown in Fig. 1) is used to detect events in which more than one beam particle enters the apparatus during the $K - \pi - \mu - e$ decay sequence. These events are excluded if a beam particle enters in coincidence with the π, μ, or e. The requirement $\overline{KC}$ in the π^+ triggering system further insures against detecting scattered beam pions.

Whenever the whole triggering system ("K decay," π, and γ) indicates that a K^+ has stopped in the target, that later a slow charged particle has passed through the π telescope and stopped in one of the decay counters $\pi4$ to $\pi7$, and that no high-energy γ has entered the lead-glass Cherenkov counters $\gamma C1$ or $\gamma C2$, then the spark chambers are pulsed and the signals from the decay counters are displayed on each of two four-beam oscilloscopes. One of the oscilloscopes has a sweep range of 200 nsec. The four traces on this oscilloscope are examined for the stopping π^+ and the $\pi^+ - \mu^+$ decay. The μ^+ energy loss is determined by measuring the μ^+ pulse height. Since the μ^+ in $\pi^+ - \mu^+ + \nu$ decay has an energy of 4.4 MeV, this measurement is helpful in eliminating accidental backgrounds. The other oscilloscope has a sweep range of 3 μsec. The traces on this oscilloscope are examined for the $\mu^+ - e^+$ decay. We require that an e^+ pulse occur in the counter in which the π^+ stopped, and that

either the e^+ have an energy loss of >4.5 MeV in that counter or that it make a pulse in at least one adjacent counter.

The pion range is computed using the absorber thickness (which is varied according to the portion of the spectrum to be examined), the pion trajectory as seen in the spark gaps, and the positions of the counters showing the K stop and the π decay.

The $K^+ - \pi^+ + \pi^0$ decays are used to calibrate the apparatus. The measured pion mean life using pions from $K^+ - \pi^+ + \pi^0$ is 26.4 $\pm$ 1.0 nsec in good agreement with the accepted value. The inefficiency of the γC anticoincidence counters (determined by comparing the $K^+ - \pi^+ + \pi^0$ event rates with γ's vetoed versus the rate with γ's required) is 6×10^{-4}. The branching ratio $(K^+ - \pi^+ + \pi^0)/(K^+ - \mu^+ + \nu)$ is found to be 0.36 ± 0.03 in satisfactory agreement with the accepted value 0.33. This agreement checks the assumed value for π^+ absorption.

Events with an apparent $\pi^+ - \mu^+ - e^+$ decay sequence and with a π^+ range of at least 50 g cm^{-2} were considered "$K^+ - \pi^+ + \nu + \bar{\nu}$" events. The most important source of background was $K^+ - \mu^+ + \nu + \gamma$ events where an accidental particle struck the decay counters causing the $\mu - e$ decay to be mistaken for a $\pi - \mu - e$ decay sequence. The probability of this was determined by examining a sample of 30 000 stopping μ^+ from $K^+ - \mu^+ + \nu$ for apparent $\pi - \mu - e$ decays. The range of the stopping particles ($R > 70$ g cm^{-2}) guaranteed that they could not be pions. Most of the spurious "π"-"μ"$-e$ events were found to have low "μ" pulse heights. The same probability for spurious events was assumed for the 32 000 stopping μ^+ (from $K^+ - \mu^+ + \nu + \gamma$) seen during the search for $K^+ - \pi^+ + \nu + \bar{\nu}$ (50 $\leqslant R \leqslant$ 59 g cm^{-2}). Another source of background was due to pions in the K beam which the KC counter failed to veto.

The numbers of "$K^+ - \pi^+ + \nu + \bar{\nu}$" events and the expected background events were considered for various cutoff values of the K lifetime, the π lifetime, and the "μ" pulse height. The numbers of "$K^+ - \pi^+ + \nu + \bar{\nu}$" were consistent with the expected background for a wide variety of cutoff values. In the final sample of "$K^+ - \pi^+ + \nu + \bar{\nu}$" events "$\mu$" pulses were required to have between 0.5 and 1.5 times the mean pulse height of muons from $\pi - \mu$ decays. The π and K lifetimes were required to be within two mean lives after our detection thresholds. Measurement with $K^+ - \pi^+ + \pi^0$ events showed that these cuts excluded 33 % of the pions. After the cuts the expected background is 0.8

events. There were no "$K^+ \to \pi^+ + \nu + \bar{\nu}$" events in the final sample.

Our detector efficiency for $\pi^+ \nu \bar{\nu}$ events is shown in curve II of Fig. 2(a). The meaning of observing one event in our experiment depends on the convolution of this efficiency and an assumed pion spectrum. We denote this convolution by $\epsilon_{\pi^+ \nu \bar{\nu}}$. We have considered the possibilities shown in Table I.

Assuming a vector interaction if one event had been found, the branching ratio would have been

$$\frac{\Gamma(K^+ \to \pi^+ + \nu + \bar{\nu})}{\Gamma(\text{all modes})} = \frac{(\pi^+ \nu \bar{\nu})/K^+}{(\pi^+ \pi^0)/K^+} \frac{\epsilon_{\pi^+ \pi^0}}{\epsilon_{\pi^+ \nu \bar{\nu}}} \frac{T_I}{T_{II}}$$

$$\times \frac{\Gamma(\pi^+ \pi^0)}{\Gamma(\text{all modes})} = 5.3 \times 10^{-7}. \quad (3)$$

Here $\pi^+ \nu \bar{\nu}/K^+$ stands for the ratio of the number of "$K^+ \to \pi^+ + \nu + \bar{\nu}$" events found to the number of K^+ signals examined by the triggering system (7.2×10^{-10} assuming 1 event); $\pi^+ \pi^0/K^+$ is the ratio of the number of $\pi^+ \pi^0$ events found to the number of K^+ signals examined when the apparatus was used to detect $K^+ \to \pi^+ \pi^0$ (1.5×10^{-3}); $\epsilon_{\pi^+ \pi^0}$

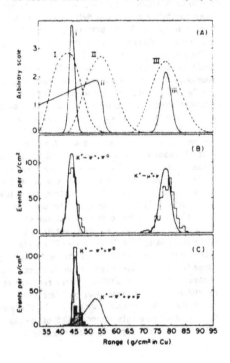

Table I. Effective detection efficiency $\epsilon_{\pi^+ \nu \bar{\nu}}$ (117 MeV $\le T_\pi \le$ 127 MeV) and resultant branching ratio (90% confidence level) for several assumed π^+ spectra. p_π and T_π are the momentum and kinetic energy, respectively, of the π^+. $T_{max} = 127$ MeV is the kinematic upper limit for $K^+ \to \pi^+ \nu \bar{\nu}$. The spectra were normalized to have unit area between 0 and T_{max}.

$\frac{dN}{dT_\pi}$ assumed	Type of first-order interaction	$\epsilon_{\pi^+ \nu \bar{\nu}}$	Branching ratio
p_π^3	vector	0.0105	1.2×10^{-6}
$p_\pi^3 (T_{max} - T_\pi)$	tensor	0.0034	3.8×10^{-6}
$p_\pi (T_{max} - T_\pi)$	scalar	0.00072	1.9×10^{-5}

(0.054) is the detection efficiency for $\pi^+ \pi^0$ [the convolution of curve I with curve (i) in Fig. 2(a)]; $\epsilon_{\pi^+ \nu \bar{\nu}}$ (0.0105) is the detection efficiency for $\pi^+ \nu \bar{\nu}$ assuming a vector spectrum [the convolution of curve II with curve (ii) in Fig. 2(a)]; and T_I/T_{II} (1.05) is the ratio of π^+ transmissions for the absorbers corresponding to curves I and II in Fig. 2(a) (44 and 52 g cm^{-2} Cu equivalent, respectively).

Table I contains the branching ratios which we infer from (3). By the 90% confidence level we mean the rate which we would compute had we found 2.3 events.

We are grateful to Professor E. Segrè for his encouragement and support and to Dr. C. Wiegand and D. Brandshaft for valuable assistance, especially in the early stages of this experiment. We also wish to thank W. Davis, J. Gallup, N. Green, E. Hahn, D. Hildebrand, P. Newman, and J. Wild for their help with scanning, analysis, and operation. One of us (RHH) wishes to acknowledge with thanks his support by the John Simon Guggenheim Foundation during the course of this experiment.

FIG. 2. Range distributions. (a) Calculated distributions for K^+ decays into $\pi^+ \pi^0$, $\pi^+ \nu \bar{\nu}$ (vector), and $\mu^+ \nu$ with straggling and small-angle multiple scattering taken into account. Dashed curves I, II, and III show detector efficiencies for different absorber thicknesses (curve II is weighted sum of curves corresponding to two nearly equal thicknesses). (b) Expected event distributions for $\pi^+ \pi^0$ and $\mu^+ \nu$ (curves I and III folded into i and iii) and corresponding observed distributions (histograms). (c) Expected and observed $\pi^+ \pi^0$ and "$\pi^+ \nu \bar{\nu}$" distributions for absorber corresponding to curve II, (a). [γC1 and/or γC2 required in coincidence for $\pi^+ \pi^0$ (open histogram) and in anticoincidence for "$\pi^+ \nu \bar{\nu}$" (shaded histogram).]

*Research supported by the U. S. Atomic Energy Commission and by the National Science Foundation Grant No. Gp 14521.

[1] U. Camerini, D. Cline, W. F. Fry, and W. M. Powell, Phys. Rev. Letters 13, 318 (1964).

[2] Branching ratio limits (90% confidence level): $K_L^0 \to e^+ + e^- < 1.5 \times 10^{-7}$, $K_L^0 \to \mu^+ + \mu^- < 2.1 \times 10^{-7}$, H. Foeth et al., Phys. Letters 30B, 282 (1969); $K_S^0 \to \mu^+ + \mu^- < 7.3 \times 10^{-6}$, B. D. Hyams et al., Phys. Letters 29B, 521 (1969); $K^+ \to \pi^+ + \mu^+ + \mu^- < 2.4 \times 10^{-6}$, V. Bisi, R. Cester, A. Marzari Chiesa, and M. Vigone, Phys. Letters 25B, 572 (1967). See also D. W. Carpenter et al., Phys. Rev. 142, 871 (1966); C. Alff-Steinberger et al., Phys. Letters 21, 595 (1966); Camerini, Cline, Fry, and Powell, Ref. 1; and U. Camerini, D. Ljung, M. Sheaff, and D. Cline, Phys. Rev. Letters 23, 326 (1969).

[3] For a compilation of the experimental results on this reaction see Particle Data Group, Rev. Mod. Phys. 42, 87 (1970).

[4] Previous upper limit: $K^+ \to \pi^+ + \nu + \bar{\nu} < 1.0 \times 10^{-4}$ per K^+ decay (90% confidence level), see Camerini, Ljung, Sheaff, and Cline, Ref. 2.

[5] P. Dennery and H. Primakoff, Phys. Rev. 131, 1334 (1963).

[6] R. J. Oakes, Phys. Rev. Letters 20, 1539 (1968).

[7] R. J. Oakes, Phys. Rev. 183, 1520 (1969).

[8] In a model containing a massive neutral muon-type lepton G. Segrè [Phys. Rev. 181, 1996 (1969)] estimates the branching ratio $(K^+ \to \pi^+ + \nu + \bar{\nu}) \approx [10^{-4} \times$ branching ratio $(K^+ \to \pi^0 + e^+ + \nu)] \approx 5 \times 10^{-6}$. See also N. Christ, Phys. Rev. 176, 2086 (1968), and M. Gell-Mann, M. L. Goldberger, N. M. Kroll, and F. E. Low, Phys. Rev. 179, 1518 (1969).

[9] M. Chen, D. Cutts, P. Kijewski, R. Stiening, C. Wiegand, and M. Deutsch, Phys. Rev. Letters 20, 73 (1968).

CHAPTER 4
THE ELECTROWEAK INTERACTION PICTURE EMERGES

Everyone Missed It (1962 – 1973)

4.1. OBSERVATION OF POINT-LIKE SCATTERING FROM NUCLEONS USING HIGH-ENERGY NEUTRINO BEAMS

One of the key developments in the early 1970s was the start-up of the 400-GeV machine at Fermilab. Two experiments were approved to study high-energy neutrino interactions: the E1A Harvard–Penn–Wisconsin–FNAL (HPWF) experiment, for which the author was the spokesperson, and the E21–Caltech experiment [4.1]–[4.3]. At the same time, the GGM at CERN was taking neutrino beam data [4.4]. When these experiments started taking data in late 1972, no one could imagine the revolution that lay ahead.

During the latter part of 1972, the FNAL neutrino beam sputtered into existence. The HPWF detector started to collect some small amount of data and, to the surprise of everyone, nearly every event was a "hard" scatter, high transverse-momentum event. This was exactly what was expected for point-like neutrino collisions with the nucleon (the quark–parton substructure). This result was published [4.2].

Another significant result from these experiments and the GGM [4.4] was to observe almost trivially a total cross section that increased directly with neutrino energy. (This was to be a key development in the later discovery of WNCs.) Figure 4.1 shows the cross sections measured in the GGM experiment [4.4]. We reprint here [article (A)] the very first observation of the HPWF experiment, indicating the point-like substructure in the proton, and the paper for the E21 experiment [article (B)], where a detailed measurement was made of the way the total neutrino cross section behaves with energy, namely $\sigma \propto E_v$ [4.5] – again, a key expectation of a point-like behavior in the proton. Perhaps even more amazing was that the GGM, using a much lower-energy neutrino beam, also observed a linear rise of the cross section with energy [4.4], as shown in Fig. 4.1.

In retrospect, there should have been no real surprise, since the SLAC–MIT deep-inelastic electron-scattering experiment at SLAC had shown similar behavior [4.6]. However, at the time it was not really clear that electron and neutrino interactions would observe the same point-like behavior. [See the *Physics Today* article of CMR [Chapter 2 (D)] for further clarification of these ideas.]

We can show these reactions with the exchange of the relevant particle (W or γ) by the diagrams (A) and (C) shown in Fig. 4.2. [The difficult experimental observation of the process shown in Fig. 4.2 (B) will be the subject of Chapter 5.] In one sense, the observation of deep-inelastic neutrino and antineutrino scattering and comparison with deep-inelastic electron scattering is a sort of unification of the weak and electromagnetic interaction. However, while there had been many low-energy electron-scattering experiments, the small weak-interaction cross section did not allow detailed studies of low-energy neutrino interaction with the same precision. Therefore, it was only at the very high energies that these tests could have been made.

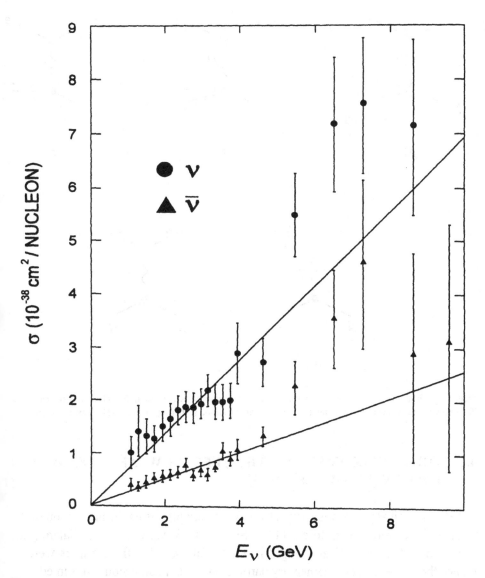

Figure 4.1. Total neutrino and antineutrino cross sections as functions of energy as measured by the GGM group in 1973.

One way to understand the concept of an electroweak force is to observe the diagrams (D) and (E) shown in Fig. 4.2. In the first case, we have a purely electromagnetic interaction that occurs, for example, in the atom; in the second case, a WNC process. However, except for the different strengths of these interactions, (D) and (E) have the same properties – the amplitudes can interfere with each other – thus the Z^0 was sometimes called "weak light" – a weak photon!

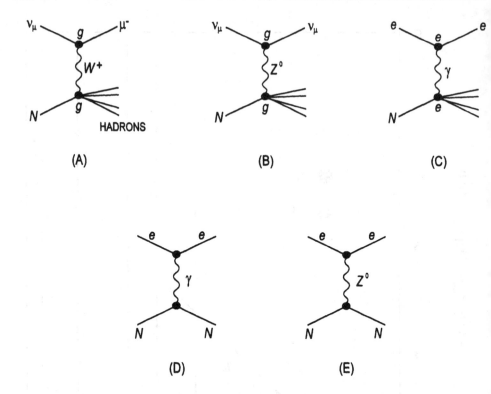

Figure 4.2. Diagrams that describe (A) neutrino charged-current inelastic scattering, (B) WNC inelastic scattering, (C) electromagnetic inelastic scattering, (D) electromagnetic elastic scattering, and (E) WNC elastic scattering.

4.2. THE DEVELOPMENT OF THE ELECTROWEAK INTERACTION THEORY (FROM 1968 - 1973)

In 1968, a seminal paper was written by S. Weinberg proposing the unification of the weak and electromagnetic forces [4.7]. Similar work was done by A. Salam [4.8] and even earlier by S. Glashow [4.9]. One could say that these papers were to initiate the modern age of elementary particle physics. In an account reprinted here [article (C)], S. Bludman, who also worked on these problems at the time, recounts this hectic period and makes his own important contributions to the so-called "electroweak" revolution. These theoretical models predicted the existence of flavor-conserving WNCs - a truly prophetic prediction!

The major problem of the weak interaction at the time was the disastrous behavior at high energy. While the $W^{\pm}$ bosons were proposed to partially cure this problem, the weak production of $W^{+}W^{-}$ pairs was shown to have a divergent behavior at high energy. The GWS theory helped cure these problems, but in turn implied new physics - most notably the existence of WNC processes! A key idea was the concept of spontaneous symmetry breaking advocated by Peter Higgs and many others at the time [4.10] (to be discussed in Chapter 9). It was also not clear in the GWS theory if the divergences were really gone. This issue was resolved by G. 't Hooft and M. Veltman in the early 1970s [4.11]. Thus by 1973, when the FNAL and CERN experiments were taking data, the theoretical model was already fairly mature! [The Bludman paper (C) gives many more details of this period.]

From an experimental standpoint, the expectations of the theory were mainly the existence of WNCs, but there was no way to know at what level these would be observed, since the theory had a free parameter, $\sin^2\theta_W$ – the Weinberg mixing angle [4.7]. In principle, this could have been very small, and the level of WNC would also have been very small. In this theory, there are four massless particles: W^+, W^-, W^0, and a neutral B (the Higgs mechanism gives the W's mass). The W^0 and B mix to give two physical particles: the photon γ and the Z^0 (with $\sin^2\theta_W$ the mixing angle). Because of gauge invariance, the photon γ remains massless, while the Z^0 acquires a mass (by the Higgs mechanism). There are two powerful predictions:

1. The mass of the W and Z particles are related by $m_Z^2 = m_W^2/\cos\theta_W$, and
2. The "strength" of the WNC processes compared to electromagnetic processes go like $\sin^2\theta_W$. Thus, if $\sin^2\theta_W$ is of a moderate value of 0.1 to 0.5, the WNC should be observable!

There was a great deal of phenomenology at the time to help guide the experimental search [4.12]-[4.14].

4.3. RECONCILING THE EXPERIMENTAL ABSENCE OF FCNC AND THE PREDICTED EXISTENCE OF WNCS – THE GIM MECHANISM

A major problem with the unified electroweak theory [4.7]-[4.9] was the expectation of WNC interactions when it was known that FCNCs were extremely suppressed (see Chapter 3). To understand the magnitude of this problem, we note that the FCNC had been shown to be suppressed by a factor of almost 10^{-3} in amplitude compared to other weak processes [4.15].

In 1970, a fourth seminal paper was written by S. Glashow, J. Iliopoulos, and L. Maiani [4.16], showing for the first time how one could understand the experimental absence of the FCNC (Chapter 3) and the possible existence of WNC – with a bonus: the prediction of the existence of a charm quark [4.17]. This has come to be known simply as the GIM effect in honor of this seminal work [4.16]. The basic concept was that the quarks should come in doublets, with the structure

$$\begin{pmatrix} u \\ d \end{pmatrix} \begin{pmatrix} c \\ s \end{pmatrix} \text{ ,}$$

where the fourth quark would have a charge of $Q = +2/3$ and is in the same doublet as the strange quark, which has $Q = -1/3$ [4.16]. Since the early 1960s, it was known that the quarks must have weak mixing angles first identified by N. Cabibbo [4.18]. In order to "cancel" the FCNC, the same angle had to apply to the charm quark mixing, and the mass of the quark should be about 1.5 to 2 GeV [4.19]. Thus, the absence of FCNC could be explained if a new quark existed! An example of the power of a null effect in particle physics!

REFERENCES

4.1. Proposal for the EIA Experiment at Fermilab, Harvard, Univ. of Penn., Wisconsin, and Fermilab, D. Cline, spokesperson, unpublished (1970).

4.2. A. Benvenuti *et al.*, *Phys. Rev. Lett.*, **30**, 1084 (1973) [article (A)].

4.3. Proposal for E21 at FNAL–California Institute of Technology, B. Barish and F. Scuilli, spokespersons, unpublished (1970).

4.4. A. Eichten *et al.*, *Phys. Rev. Lett.*, **46B**, 274 (1973).

4.5. B. C. Barish *et al.*, *Phys. Rev. Lett.*, **35**, 1316 (1975) [article (B)].

4.6. M. Breidenbach *et al.*, *Phys. Rev Lett.*, **23**, 935 (1969).

4.7. S. Weinberg, *Phys. Rev. Lett.*, **19**, 1264 (1967).

4.8. A. Salam, in *Proceedings of the VIII Nobel Symposium*, N. Svartholm, ed. (Almquist and Wiksell, Stockholm, 1968), p. 367.

4.9. S. L. Glashow, *Nucl. Phys.*, **22**, 579 (1961).

4.10. P. W. Higgs, *Phys. Lett.*, **12**, 132 (1964); *Phys. Rev.*, **145**, 1156 (1966); F. Englert and R. Brout, *Phys. Rev. Lett.*, **13**, 321 (1964); T. W. Kibble, *Phys. Lett.*, **155**, 1554 (1967); G. S. Guralnik, C. R. Hagen, and T. W. B. Kibble, *Phys. Rev. Lett.*, **13**, 585 (1964).

4.11. G. 't Hooft, *Nucl. Phys.*, **35B**, 167 (1971) and **50B**, 318 (1972); G. 't Hooft and M. Veltman, *Nucl. Phys.*, **44B**, 189 (1972); B. W. Lee and J. Zinn-Justin, *Phys. Rev.*, **D5**, 3121 (1972).

4.12. J. Sakurai, *Phys. Rev.*, **D9**, 250 (1974) and references therein.

4.13. A. Pais and S. B. Treimain, *Phys. Rev.*, **D6**, 2700 (1972).

4.14. E. A. Paschos and L. Wolfenstein, *Phys. Rev.*, **D7**, 91 (1973).

4.15. D. Ljung and D. Cline, *Phys. Rev.*, **D8**, 1307 (1973).

4.16. S. L. Glashow, J. Iliopoulos, and L. Maiani, *Phys. Rev.*, **D2**, 1285 (1970).

4.17. The idea of a charm or fourth quark had a long history even before the GIM paper. The first paper seems to be by J. D. Bjorken and S. Glashow, *Phys. Lett.*, **11**, 255 (1964). Further references can be found in Ref. [4.16].

4.18. N. Cabibbo, *Phys. Rev. Lett.*, **10**, 531 (1963).

4.19. M. K. Gaillard and B. W. Lee, *Phys. Rev.*, **D10**, 897 (1974).

Early Observation of Neutrino and Antineutrino Events at High Energies*

A. Benvenuti, D. Cheng, D. Cline, W. T. Ford, R. Imlay, T. Y. Ling, A. K. Mann,
F. Messing, J. Pilcher,† D. D. Reeder, C. Rubbia, and L. Sulak
*Department of Physics, Harvard University,‡ Cambridge, Massachusetts 02138, and Department of Physics,
University of Pennsylvania,‡ Philadelphia, Pennsylvania 19104, and Department of Physics, University of
Wisconsin,‡ Madison, Wisconsin 53706*
(Received 23 April 1973)

Presented here are preliminary results of two short runs with a broad-band neutrino-antineutrino beam at National Accelerator Laboratory incident on about 120 tons of target which is part of a detector consisting of an ionization calorimeter and a muon magnetic spectrometer. These results include (i) the observed distribution in transverse muon momentum, dN/dp_μ, (ii) the average neutrino cross section at a mean neutrino energy of roughly 30 GeV, and (iii) the ratio of the antineutrino to the neutrino total cross section at a mean neutrino energy of 40 GeV.

The energy (300 GeV) and intensity ($\sim 10^{12}$ protons per pulse) of the external proton beam now available at the National Accelerator Laboratory (NAL) are sufficiently large to permit the observation of a substantial number of high-energy neutrino interactions, even without any focusing of the secondary pions and kaons which, through their decays, produce the neutrino beam. In order to provide an early description, albeit crude, of neutrino interactions in this new energy region, we report here the preliminary results of two short runs (approximately 10^{15} interacting protons on target) with such a broad-band neutrino-antineutrino beam incident on about 120 tons of target-detector.

The experimental arrangement is shown schematically in Fig. 1(a). The hadron-producing target (a collision length of iron) on which the extracted proton beam impinges is about 1400 m from the accelerator. The secondary hadrons travel unvexed through a drift region 350 m in length and 1 m in diameter. The drift region is in turn followed by a muon shield, primarily of earth, about 1000 m long, at the end of which our neutrino detection apparatus is located.

The main outlines of the target-detector are sketched in Fig. 1(b). The neutrinos are incident on an ionization calorimeter (IC) consisting of four main sections in series along the beam axis, each main section of cross-sectional area 3×3 m^2 and length along the beam of 1.8 m, and containing about 15 metric tons of mineral-oil-based liquid scintillator. Each main section is divided into four optically separated subsections viewed from two sides, as indicated in Fig. 1(b), by twelve photomultipliers (5-in. diam). There are wide-gap optical spark chambers of area 3×3 m^2 after each main section of the ionization calorimeter as shown in Fig. 1(b). Immediately downstream of the IC is a magnetic spectrometer made up of four units of toroidal iron magnets, one behind the other along the beam line, with narrow-gap optical spark chambers following each magnet unit. The toroids have an inside diameter of 0.3 m, an outside diameter of 3.6 m, and are 1.2 m long; they are driven into satura-

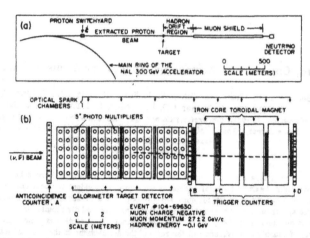

FIG. 1. (a) Experimental arrangement showing the 300-GeV accelerator, the proton beam, the hadron drift space, the muon shield, and the neutrino detector. (b) Details of the neutrino detector with quasielastic or Δ-production neutrino event superimposed.

tion and operate at an essentially constant field B of 18 kG. The narrow-gap spark chambers are 2.3×2.3 m^2 in area.

The target for neutrino interactions in the runs described here was a fiducial mass of about 60 metric tons of liquid scintillator in the ionization calorimeter plus another 60 tons of iron (part of the first unit of the magnetic spectrometer). For some of the neutrino interactions occurring in the ionization calorimeter a rough measurement of the energy of the hadron shower was made. For all events the vector momentum and sign of charge of the secondary muon was measured in the magnetic spectrometer. The detector was activated by either of the two coincidence modes CD or BC (preset minimum energy deposition in IC) [see Fig. 1(b)]. The presence of pulses in all counters and the relative times of those pulses were recorded as well as the pulse-height information from each of the sixteen subsections of the IC.

The external proton beam was obtained from the accelerator by a half-wavelength resonant extraction mode which produced a beam spill of about 300 μsec total duration. The neutrino detector was gated on by a pulse about 1 msec long coincident with the beam spill. An additional gate was opened to sample cosmic-ray events not associated with the accelerator beam. Events in coincidence with the beam were either neutrino interactions in our detector or muons emerging

from the earth in front of the detector. Almost all of the latter type were neutrino induced and are therefore of interest also but we do not discuss them here. With our relatively high useful neutrino event rate, typically a few events per hour in these runs,[1] backgrounds from (i) cosmic rays, (ii) hadrons or muons leaking through the shield, and (iii) neutrino-induced events in the shield were negligible.

The salient preliminary results that we present are obtained from the directly observed distribution in transverse momentum of the secondary muons, $dN/dp_{\mu\perp}$, both positive and negative, which is shown in Fig. 2(a). We note the striking difference between the shape of the experimental distribution in Fig. 2(a) and the exponential fall-off [as $\exp(-6p_\perp)$] of $d\sigma/dp_\perp$ observed in hadron-hadron collisions. Observe also that in Fig. 2 $\langle p_{\mu\perp} \rangle \approx 1.5$ GeV/c, which is to be compared with $\langle p_\perp \rangle \approx 0.3$ GeV/c obtained in hadron-hadron collisions. Furthermore, since $q^2 = p_{\mu\perp}{}^2 E_\nu / E_\mu$, observation of events with $p_{\mu\perp}$ as large as 6 GeV/c (the kinematic limit of the collision of a 72-GeV neutrino with a nucleon) implies values of q^2 greater than 36 (GeV/$c)^2$ and, on average,[2,3] about 72 (GeV/$c)^2$. The detection efficiency of our apparatus for such large momentum-transfer events is unrestricted.

We present in Fig. 2(b) the observed distribution in momentum, dN/dp_μ, to show that roughly $\frac{1}{4}$ of all the events have $p_\mu > 50$ GeV/c and there-

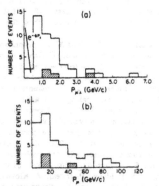

FIG. 2. (a) Distribution in $p_{\mu\perp}$. (b) Distribution in p_μ. The cross-hatched events are positive muons.

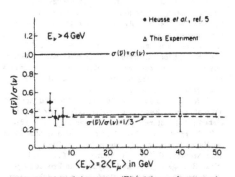

FIG. 3. Plot of the ratio $\sigma(\bar{\nu})/\sigma(\nu)$ as a function of $\langle E_\nu \rangle$. The value of $\frac{1}{3}$ is expected in the scattering of neutrinos and antineutrinos by fundamental fermions such as electrons and muons.

fore, on average,[2,3] $E_\nu > 100$ GeV. Of the twenty interactions[2] that occur in the IC, six show *visible* hadron energies[2] between 15 and 41 GeV.

It is also of interest that there are only two events in the IC that are probably quasielastic or $\Delta(1238$ MeV$)$ production, as indicated by their low value of $q^2 \lesssim 0.8$ (GeV/$c)^2$ and low value of visible hadron energy < 0.3 GeV. We take from the recent CERN–Argonne National Laboratory results[4] σ_ν(quasielastic) + $\jmath_\nu(\Delta$ production) $\approx (2.4 \pm 0.6) \times 10^{-38}$ cm^2/nucleon, essentially independent of E_ν. Then σ_ν(total) = $(24 \pm 17) \times 10^{-38}$ cm^2/nucleon at a mean neutrino energy assumed to be given roughly[2,3] by $\langle E_\nu \rangle = 2\langle E_\mu \rangle \approx 30$ GeV [see Fig. 2(b)]. This is to be compared with the value of $(21 \pm 4) \times 10^{-38}$ cm^2/nucleon expected from the relation σ_ν(total) = $(0.7 \pm 0.14)E_\nu \times 10^{-38}$ cm^2/nucleon observed at lower neutrino energies.[5,6]

Lastly, we note that there are 4 events with a positively charged muon, and 26 events with a negatively charged muon, as the outgoing lepton, all with 10 GeV/$c < p_\mu < 50$ GeV/c. Above 10 GeV/c the muon detection efficiency of our apparatus is essentially charge independent. We take the ratio of the production cross sections[7] for π^+ and π^- (of momenta greater than about 75 GeV/c) by 300-GeV protons on nucleons as 2.3 and obtain

$$\frac{\sigma(\bar{\nu}+p \to \mu^+ + \text{all}) + \sigma(\bar{\nu}+n \to \mu^+ + \text{all})}{\sigma(\nu+p \to \mu^- + \text{all}) + \sigma(\nu+n \to \mu^- + \text{all})} = 0.35 \pm 0.18$$

at $\langle E_\nu \rangle \approx 40$ GeV. This result is to be compared with the value 0.33 ± 0.1 observed[5] at $\langle E_\nu \rangle \approx 7.5$ GeV. These data are summarized in Fig. 3. It is perhaps remarkable that the ratio $\sigma(\bar{\nu})/\sigma(\nu)$ is so close to the value $\frac{1}{3}$ at $\langle E_\nu \rangle \approx 40$ GeV when one

realizes the deep inelasticity in the high-energy neutrino and antineutrino interactions that are described here. This is the numerical value of the ratio that is expected for the scattering of neutrinos and antineutrinos by fundamental fermions such as electrons and muons.

It is necessary to emphasize that our conclusions are based on few events, and therefore should not be construed as representing more than a cursory overview of a new energy region of neutrino physics. There are as yet too few events to reach any conclusion with respect to possible new phenomena, e.g., intermediate vector boson or heavy lepton production.

We wish to thank all of the dedicated people who are helping to make the National Accelerator Laboratory a scientific reality. In particular, we thank E. Bleser and Helen Edwards for their work on fast coherent extraction of the proton beam, and J. R. Orr, J. Sanford, and T. Toohig for the development of the neutrino experimental area. We are grateful to the many individuals at our respective universities whose technical skills and enthusiasm have contributed so valuably to this experimental effort. Finally, we thank Leon Lederman for the loan of the narrow-gap spark chambers.

---

*Accepted without review under policy announced in Editorial of 20 July 1964 [Phys. Rev. Lett. 13, 79 (1964)].

†Now at the University of Chicago, Chicago, Ill. 60637.

‡Work supported in part by the U. S. Atomic Energy Commission.

[1] In a subsequent short run the useful neutrino event rate has climbed to about 15 events per hour, still without focusing of the secondary hadrons.

[2] We assume that the average muon energy is $\frac{1}{2}$ of the neutrino energy. This assumption is consistent with the neutrino energy distribution determined for events observed in the calorimeter, after applying corrections for energy escape and calibration. We note that the IC has not yet been calibrated in a hadron beam. Cosmic-ray muons have been used to determine the energy loss for minimum ionizing tracks in order to calibrate crudely the visible energy measurement.

[3] E. A. Paschos and V. I. Zakharov, National Accelerator Laboratory Report No. NAL-THY-100 (to be published); J. D. Bjorken, Nuovo Cimento 68, 569 (1970).

[4] I. Budagov *et al.*, Lett. Nuovo Cimento 2, 689 (1969); A. Mann, V. Mehtani, B. Musgrave, Y. Oren, P. Schreiner, H. Yuta, R. Ammar, Y. Cho, M. Derrick, R. Engelmann, and L. Hyman, in Proceedings of the Sixteenth International Conference on High Energy Physics, National Accelerator Laboratory, Batavia, Illinois, 1972 (to be published). We explicitly assume that the quasielastic and Δ-production components remain constant with energy at high energy, as observed (approximately) in these experiments.

[5] The latest data from the CERN-Gargamelle collaboration were presented by P. Heusse, in Proceedings of the Sixteenth International Conference on High Energy Physics, National Accelerator Laboratory, Batavia, Illinois, 1972 (to be published); see also D. H. Perkins, *ibid*; I. Budagov *et al.*, Phys. Lett. 30B, 364 (1969).

[6] J. D. Bjorken and E. A. Paschos, Phys. Rev. 185, 1975 (1969).

[7] We have taken the π^+/π^- and K^+/K^- production ratio needed to determine the ratio of $\nu/\bar{\nu}$ flux from recent measurements at the CERN intersecting storage rings and proton synchrotron and interpolated the data to NAL energies. Since the π and K production approximately follow Feynman scaling in the approximate $(x, p_\perp)$ kinematic region for the NAL neutrino beam and over the s variation from proton-synchrotron to intersecting-storage-rings energies, we expect our estimate to be accurate to within $\sim 20\%$. The relevant data were taken from G. Giacomelli, to be published; H. J. Muck *et al.*, Phys. Lett. 39B, 303 (1972); J. V. Allaby, F. Benon, A. N. Diddens, P. Duteal, A. Klovning, P. Meunier, J. P. Peigneux, E. J. Sacharidis, K. Schlupmann, M. Spiegel, J. P. Stroot, A. M. Thorndike, and A. M. Wetherell, CERN Report No. 70-12, 1970 (unpublished).

Measurement of Neutrino and Antineutrino Total Cross Sections at High Energy*

B. C. Barish, J. F. Bartlett, D. Buchholz, T. Humphrey, F. S. Merritt, F. J. Sciulli,
L. Stutte, D. Shields, and H. Suter†

California Institute of Technology, Pasadena, California 91125

and

E. Fisk and G. Krafczyk

Fermi National Accelerator Laboratory, Batavia, Illinois 60510

(Received 2 September 1975)

Charged-current ν and $\bar{\nu}$ data are reported from the first application at Fermilab of a narrow-band neutrino beam for the measurement of normalized cross sections. Cross sections of about 20% accuracy were measured with a 120-GeV secondary hadron beam for ν ($\bar{\nu}$) originating from π decay ($\langle E_\nu \rangle = 38$ GeV) and Λ decay ($\langle E_\nu \rangle = 105$ GeV). The ν and $\bar{\nu}$ fluxes were determined by directly measuring the hadron flux and the $\pi/K/p$ ratios for the hadron beam.

The usual local current-current weak-interaction theory predicts that the neutrino-lepton cross section at high energy rises linearly with laboratory energy. If, in addition, the deep-inelastic structure functions scale in the dimensionless scaling variable x, the neutrino-nucleon cross section must rise linearly with laboratory energy as well.[1] The behavior of the total neutrino (antineutrino) charged-current cross section σ_ν ($\sigma_{\bar{\nu}}$) on nucleons,

$$\nu_\mu \ (\bar{\nu}_\mu) + N \to \mu^- \ (\mu^+) + \text{hadrons} ,$$

therefore provides simultaneously a directly in-

terpretable check of both weak-interaction theory and scaling.

Neutrino and antineutrino cross sections[2] previously measured at low energies ($E_\nu \lesssim 8$ GeV) are consistent with a linear rise in energy. The best-fit slopes are $\alpha_\nu = 0.74 \pm 0.03$ and $\alpha_{\bar{\nu}} = 0.28 \pm 0.01$, where all α's are in units of 10^{-38} cm^2/GeV. We describe here a measurement of σ_ν and $\sigma_{\bar{\nu}}$ at higher energy in which the neutrino flux has been measured directly in the same experiment. (Preliminary results from this experiment have been presented earlier.[3])

This is the first application of a narrow-band

neutrino beam for the measurement of normalized neutrino cross sections.[4] The Fermilab narrow-band beam, utilized in this experiment, has been described elsewhere.[5] The arrangement is shown in Fig. 1. Briefly, secondaries produced near 0 mrad by 300-GeV protons are charge and momentum analyzed and focused into a parallel beam of adjustable central momentum and approximate acceptance $\Delta\Omega\,\Delta p/p = 450$ μsr %, $\Delta p/p = 36\%$ (full width at half-maximum). This beam is directed down a 345-m evacuated decay pipe in which the decays $\pi(K)\to\mu+\nu_\mu$ yield a neutrino energy spectrum in the forward direction that contains two bands of muon neutrinos (or antineutrinos)[4-6] differing in mean energy by about a factor of 3. The pion neutrino band has a width of ±10 GeV, and the kaon band has width of ±18 GeV. (Typical measured spectra are given by Barish *et al.*[7])

The total-cross-section data for ν_μ $(\bar{\nu}_\mu)$ incident on iron nuclei were obtained with use of 300-GeV protons and 120-GeV secondaries. The neutrino data (from π^+ and K^+) and antineutrino data (from π^- and K^-) are based on 2.6×10^{16} and 4.0×10^{16} incident protons, respectively.

The scheme for directly measuring the neutrino total cross sections required measurements of the neutrino flux, detection efficiency, and event rate. The neutrino flux was determined[6] by measuring the hadron secondary-beam intensity and the $\pi/K/p$ ratio in the beam. By use of the known geometry of the detection apparatus the flux was then determined directly within the systematic errors described below. The detection efficiency, again within the stated systematic errors, was evaluated from the measured differential distributions.

The total intensity of the secondary hadron beam was continuously measured at the end of the decay pipe with a 35-cm $\times$ 50-cm ionization chamber (SIC), large enough to contain the full secondary beam (10 cm $\times$ 22 cm). This monitor was calibrated against a secondary emission monitor (SEM) in the extracted proton beam by transport-

ing the primary proton beam through the beam elements (without target) and through the entire decay region. (The SEM, previously calibrated with various standards, was checked by foil irradiation during this run and found to correspond to its calibration to better than 5%.) The hadron beam was steered into the SIC with a 2.5-cm $\times$ 5.0-cm scintillation monitor located on the axis of the SIC; this centering was continuously monitored under computer control.

The fraction of pions, kaons, and protons in this beam was determined with a Cherenkov counter[6] located downstream of the decay region where a portion of the secondary beam was allowed to exit. These measurements were taken in the secondary beam under the same conditions of targeting, focusing, and steering as for normal neutrino data taking. The incident proton intensity, however, was lowered from the normal $(1-3)\times10^{12}$ to $(2-3)\times10^{9}$ protons/pulse to allow counting of individual particles in the secondary beam. Measurements of the particle fractions were made with positive and negative secondaries at a variety of beam energies. The results of this survey have been discussed elsewhere.[6]

The flux of neutrinos into the apparatus was then calculated directly, by use of the known location of the detection apparatus relative to the decay pipe, the flux of parent particles as measured above, the known particle momenta and lifetimes, and two-body decay kinematics. Table I gives the neutrino flux into the fiducial volume with the estimated systematic error. This error comes primarily from measurements of the particle ratios, and overall particle calibration of the SIC.

The neutrino target consisted of 160 tons of Fe instrumented to detect the interaction products. The steel was segmented into 1.5-m $\times$ 1.5-m $\times$ 10-cm slabs and interspersed with scintillation counters (used as a sampling calorimeter to measure hadron energy, E_h) and spark chambers (to follow the muon trajectory). A 1.5-m-diam iron-

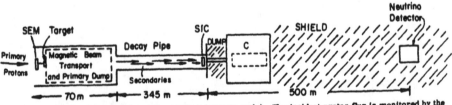

FIG. 1. Schematic of the experimental setup (*not* drawn to scale). The incident proton flux is monitored by the secondary emission monitor (SEM) immediately upstream of the target, and the secondary flux is monitored by the secondary ion chamber (SIC) at the end of the decay pipe.

core magnet, a large spark-chamber array, and trigger counters followed the target to determine the muon energy, E_μ. The apparatus was triggered by either (1) a muon traversing the magnet, or (2) significant energy deposition in the sampling calorimeter (> 99% efficient for $E_h > 6$ GeV). Roughly 90% of all neutrinos interacting inside the 127-cm × 127-cm fiducial volume satisfy at least one of these triggers.

The neutrino cross section was obtained from the restricted sample of observed events (about 50%) having a final-state muon traversing the iron-core magnet. Both muon and hadron energies were measured for this sample, yielding the total energy ($E_\nu = E_h + E_\mu$) for each event and thus determining whether the incident neutrino originated from a pion decay or from a kaon decay. The number of events, T, in each category, after small corrections for the finite (~ 25%) energy resolution, is given in Table I.

These events were then corrected by the detection efficiency, ϵ, for having the muon traverse the magnet. This efficiency was determined from the sample of all triggering events by empirically fitting the angular distribution by the following form.

$$dN^\nu/dx\,dy = CF_2^{ed}(x)[1 + a_\nu(1-y)^2], \qquad (1)$$
$$dN^{\bar{\nu}}/dx\,dy = CF_2^{ed}(x)[a_{\bar{\nu}} + (1-y)^2], \qquad (2)$$

where $x = Q^2/2ME_h$, $y = E_h/E_\nu$, and $Q^2 = 4E_\nu E_\mu \times \sin^2\theta/2$. M is the nucleon mass. θ is the muon scattering angle. and $F_2^{ed}(x)$ is the structure function measured in electron-deuteron scattering.[8] It is expected from charge symmetry that $a_\nu = a_{\bar{\nu}}$.[9] A comparison of the angular distribution of the outgoing muons for all events (including those where the muon missed the magnet) with the distribution expected from Eqs. (1) and (2) yielded $a_\nu = 0.1^{+0.4}_{-0.3}$ and $a_{\bar{\nu}} = 0.25^{+0.5}_{-0.2}$. The detection efficiency for traversing the magnet shown in Table I along with an associated systematic error was evaluated with use of the average value[10] $a_\nu = a_{\bar{\nu}} = 0.17^{+0.30}_{-0.15}$.

The total neutrino cross section per nucleon was obtained from the relation $\sigma_{tot} = T/FB\epsilon$ where T is the total number of observed interacting neutrinos with measured final muon energy, ϵ is the efficiency for the muon to traverse the magnet. F is the total number of incident neutrinos. and $B = 3.087 \times 10^{27}$ nucleons/cm^2. The resulting total cross sections with the associated statistical and total errors are shown in Table I.

In Fig. 2 the cross sections measured in this experiment are compared to the low-energy data[2] from the CERN bubble chamber Gargamelle. The data are consistent with a cross section rising linearly with energy ($\sigma_{tot} = \alpha E_\nu$). The best-fit

TABLE I. The cross sections per nucleon (σ_{tot}) for ν ($\bar{\nu}$) on an Fe target. Also shown are the values of the detection efficiency (ϵ), number of events (T), neutrino flux (F), and associated statistical and systematic uncertainties.

Parent Particle	Mean E_ν (GeV)	ϵ	$(\frac{\Delta\epsilon}{\epsilon})^{sys}$	T (events)	$(\frac{\Delta T}{T})^{sys}$	$(\frac{\Delta T}{T})^{stat}$	$F \times 10^{11}$ (neutrinos)	$(\frac{\Delta F}{F})^{sys}$	σ_{tot} (10^{-38} cm^2)	$\Delta\sigma^{stat}$	$\Delta\sigma$ total
π^+	38	.326	.066	233.6	.073	.061	7.77	.13	29.9	1.8	5.2
K^+	107	.454	.052	102.8	.078	.092	.74	.16	98.6	9.1	20.4
π^-	38	.529	.164	97.6	.049	.097	5.02	.11	11.9	1.2	2.7
K^-	102	.647	.125	10.9	.181	.29	.24	.18	22.9	6.6	9.3

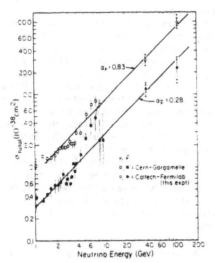

FIG. 2. Comparison of this experiment to the low-energy Gargamelle data. The inner error bars on the California Institute of Technology (Caltech)–Fermilab points correspond to statistical error only. The outer bars include the estimated systematic error added in quadrature. The slopes quoted on the figure are best fits to the Caltech–Fermilab data above.

slopes from this experiment alone are

$$\alpha_\nu = (0.83 \pm 0.11) \times 10^{-38} \text{ cm}^2/\text{GeV}. \tag{3}$$

$$\alpha_{\bar\nu} = (0.28 \pm 0.055) \times 10^{-38} \text{ cm}^2/\text{GeV}. \tag{4}$$

The Gargamelle data are in agreement with this fit. Other high-energy neutrino data,[11] obtained with the broad-band horn-focused beam at Fermilab and normalized to "quasi-elastic" events, also agree with the above slope to within a standard deviation.

The results presented here are significant in three major respects: (1) The observed linear rise in cross section supports a current-current interaction and scaling of the nucleon structure functions. (2) The ratio of neutrino to antineutrino cross sections is consistent with the value of approximately 3 expected from a predominant $V-A$ coupling to the quark component of the nucleon.[12] (3) The sum of the slopes $\alpha_\nu + \alpha_{\bar\nu}$ is related to the electromagnetic structure function measured in deep-inelastic e-d scattering [see Eqs. (1) and (2)] and in parton models measures the mean square charge of the constituent partons.[12] The slopes given above are consistent with the fractional charges expected in the sim-

plest quark model.

The first application of the narrow-band beam technique for the measurement of normalized cross sections has been extremely encouraging. The simple flux-measuring techniques applied in this initial run have provided overall accuracies of between 10 and 20% in the normalization. By improving critical areas and providing redundancy in others, we expect that the fluxes will be measured to about 5% in the near future. In the longer term, improvements in the neutrino detection apparatus and the narrow-band beam transport should ultimately give overall cross sections to match this accuracy.

This experiment, involving readout and control equipment staged over very long distances, necessarily entails rather strong interaction between experimental and accelerator personnel. We express our gratitude to the Accelerator and Neutrino Laboratory staff for help and favors, past, present, and future.

*Work supported in part by the U. S. Energy Research and Development Administration. Prepared under Contract No. AT(11-1)-68 for the San Francisco Operations Office.
†Swiss National Fund for Scientific Research Fellow.
[1]J. D. Bjorken, Phys. Rev. 179, 1547 (1969).
[2]T. Eichen *et al.*, Phys. Lett. 46B, 274, 281 (1973).
[3]F. Sciulli, in *Proceedings of the Seventeenth International Conference on High Energy Physics, London, England, 1974*, edited by J. R. Smith (Rutherford High Energy Laboratory, Didcot, Berkshire, England, 1975), p. IV-105.
[4]B. C. Barish *et al.*, FNAL Proposal No. E-21, 1970 (unpublished).
[5]P. Limon *et al.*, Nucl. Instrum. Methods 116, 317 (1974). Also see Ref. 4.
[6]T. Humphrey, Ph.D. thesis, California Institute of Technology, 1975 (unpublished).
[7]B. C. Barish *et al.*, Phys. Rev. Lett. 32, 1387 (1974); F. Sciulli, Ref. 3, and in "Neutrino '75," Proceedings of the International Conference on Particle Physics, Balatonfured, Hungary, June 1975 (to be published).
[8]A. Bodek, Ph.D. thesis, Massachusetts Institute of Technology, Laboratory for Nuclear Science Report No. LNS-COO-3069-116, 1972 (unpublished).
[9]In general a_ν and $a_{\bar\nu}$ are functions of x. The x dependence is being studied by use of a larger sample of deep-inelastic-scattering data and those results will be reported elsewhere; preliminary results agree with the values used here.
[10]Previous neutrino data taken with the narrow-band beam are in good agreement with this value and yield $a_\nu = 0.05 \,{}^{+0.11}_{-0.07}$.
[11]A. Benvenuti *et al.*, Phys. Rev. Lett. 32, 125 (1974);

R. Imlay, in *Proceedings of the Seventeenth International Conference on High Energy Physics, London, England, 1974*, edited by J. R. Smith (Rutherford High Energy Laboratory, Didcot, Berkshire, England, 1975), p. V-50.

[12]R. P. Feyman, in *Neutrinos—1974*, edited by C. Baltay, AIP Conference Proceedings No. 22 (American Institute of Physics, New York, 1974); D. C. Cundy, in *Proceedings of the Seventeenth International Conference on High Energy Physics, London, England, 1974*, edited by J. R. Smith (Rutherford High Energy Laboratory, Didcot, Berkshire, England, 1975), p. IV-131.

THE ROLE OF GAUGE THEORY, SYMMETRY-BREAKING, AND ELECTROWEAK UNIFICATION IN THE DISCOVERY OF WEAK NEUTRAL CURRENTS*

S. A. Bludman[†]

Department of Physics, University of Pennsylvania, Philadelphia, PA 19104

Abstract

The three historical and logically independent components of the Electroweak Standard Model are the exact chiral gauge theory of weak interactions, the Higgs mechanism for spontaneous symmetry-breaking, and electroweak mixing. I put into historical perspective my 1958 invention of the first gauge theory of weak interactions, predicting weak neutral currents, and show how the fundamental differences between global and gauge symmetries and between partial and exact symmetries gradually emerged.

Only renormalizability is necessary for theoretical consistency; electroweak mixing is logically independent. This extra " unification condition" distinguishes a unified theory, even in the $\sin^2 \theta_W \to 0$ limit, from a pure $SU(2)_W$ gauge theory of weak interactions. In the low-energy effective theory, the effects of unification turn out to be small ($< 11\%$). Nevertheless, *historically* the mixing $\sin^2 \theta_W \sim 0.3$ observed in the first weak neutral current experiments gave circumstantial support for the Standard Model and targetted the search for W- and Z-bosons.

1 THE ELECTROWEAK STANDARD MODEL

The electroweak sector, $SU(2)_L \times U(1)_Y$, of the Standard Model [1, 2, 3] contains three logically and historically distinct elements: (1) A chiral theory of weak interactions with an exact gauge symmetry $SU(2)_L$ symmetry [4]; (2) The Higgs mechanism [5] for spontaneous symmetry-breaking, giving some of the gauge bosons finite masses, while maintaining renormalizabilty [6]; (3) Electroweak "unification" through $W^0 - B^0$ mixing by $\sin \theta_W$ [7]. (The $SU(2)_L \times U(1)_Y$ symmetry depends on two gauge coupling constants g', g, whose ratio $g'/g \equiv \tan \theta_W$ is theoretically free. We adhere to the convention of calling this mixing "unification", although true unification would require that the symmetry group be semi-simple.)

---

*Talk given at an International Symposium on 30 Years of Neutral Currents: "From Weak Neutral Currents to the W/Z and Beyond", UCLA, February 3-5,1993.

†Supported in part by DOE Contract No. DOE-AC02-76-ERO-3071.

This report is concerned with the early history of intermediate vector meson theories and my own motivation for publishing [4] the first chiral gauge theory of weak interactions, predicting weak neutral currents of exact $V - A$ form and approximately the weak strength observed 15 years later [8]. I discuss the evolving appreciation of the fundamental distinctions between global and gauge, partial and exact symmetries, in the weak and strong interactions. Exact gauge symmetry is necessary for the Higgs mechanism for symmetry breaking, but electroweak unification is not required logically: Within the Standard Model, the electroweak mixing angle, $\sin \theta_W$, is not determined, but could have any value, including zero. The unification condition is thus non-trivial, even in the $\sin^2 \theta_W \to 0$ limit. Indeed, in a perturbative theory, unification determines the Z-boson mass within a factor ~ 2 and the W-boson mass within a factor ~ 3.

Logically, a consistent theory without electroweak mixing was conceivable. Nevertheless, *historically* the discovery of WNC with electroweak mixing angle $\sin^2 \theta_W \sim (0.2-0.3)$ provided circumstantial evidence for the unified Standard Model and drove the search for massive gauge bosons of $M_W \approx 80\ GeV$, $M_Z \approx 90\ GeV$. The ultimate discovery [29] of these gauge bosons with unequal masses then directly confirmed the Electroweak Standard Model.

2 HISTORY OF INTERMEDIATE VECTOR BOSONS

The idea of intermediate vector bosons arose naturally in the 1930's following the work of Fermi, Yukawa and many others, but the distinctions between hardronic and weak interactions, between approximate and exact symmetries, and between global and gauge symmetries were not appreciated. Different meson theories were classified, and some were found to be perturbatively renormalizeable. Nevertheless, meson-nucleon effective couplings are strong and ultimately none are fundamental. Likewise, isospin [10] and other flavour symmetries are global and approximate, deriving ultimately from the small u-, d-, s-quark masses.

We now know that the fundamental interactions are between gauge bosons and quarks and leptons, and that, for a consistent (renormalizeable) theory, massive vector meson interactions must be through exact gauge symmetries, which are spontaneously broken by the Higgs mechanism [5], giving some of the gauge bosons masses.

3 NON-ABELIAN GAUGE THEORY FOR THE WEAK INTERACTIONS

The successes of QED in the 1940's had established the importance of electromagnetic gauge invariance. Indeed, in simple enough theories, it led to minimal electromagnetic interactions that were renormalizable. For charged vector mesons, however, minimal electromagnetic interaction was ambiguous [11] and the theory was non-renormalizable. The divergences derive from the longitudinal component of the massive vector meson field and are minimal if the gyromagnetic ratio $g = 2$ and the electric quadrupole moment $Q = -e(\hbar/Mc)^2$. (In the Standard Model, the electroweak scale acts as a regulator for the longitudinal vector meson field, making vector meson electrodynamics renormalizable for just these electromagnetic moments.)

While Noether's First Theorem, that global symmetries implied well-known conservation laws, was well-known, I had always been more impressed by her Second Theorem, that *local* Lagrangian symmetries implied *new* (gauge) fields. This, together with the Yang-Mills theory [12], led to my first publication entitled "Extended Isotopic Spin Invariance and Meson-Nucleon Coupling " [13], showing that the then-current pion- nucleon interaction could not be derived directly from a gauge principle. Because I was always motivated only by exact gauge symmetries, I did not think to make the axial current partially conserved, nor the pseudoscalar pion a pseudo-Goldstone boson. We now realize that these approximate symmetries, while important, are not fundamental, but derive from the mass hierarchy of quarks in QCD.

Once the experimental situation clarified in 1957, Sudarshan and Marshak [14], Feynman and Gell-Mann [15] and Sakurai [16] each immediately presented their own derivations of the $V - A$ β-decay interaction. My own derivation [4] followed from what I called Fermi gauge invariance, generated by charge-raising and -lowering chiral Fermi charges F^+, F^-. If the algebra of generators is to close, then neutral Fermi charges $2iF^0 = [F^+, F^-]$ are required, i.e. $SU(2)_L$ is the minimal symmetry of the Fermi interactions. I went on to impose this symmetry locally and was led to an $SU(2)_L$ triplet of gauge bosons, $W^{\pm, 0}$, coupled to a triplet of chiral Fermi currents $F_\mu^{\pm, 0}$. This chiral gauge theory predicted weak neutral currents of exact $V - A$ form and the same strength as the weak charged currents. The observed strength of the Fermi interactions, $G_F/\sqrt{2} = g^2/8M_W$, then required, in tree approximation, $M_W = gv/2$, where $v \equiv (\sqrt{2}G_F)^{-1/2} = 246\ GeV$. Neither g nor M_W was predicted separately, but, if the field theory was to be perturbative, then $g < 1$, so that $M_W < 123\ GeV$ was to be expected.

This theory did not provide a mechanism for gauge bosons masses, did not unify weak with electromagnetic interactions, and did not explain the absence of flavour-changing weak neutral currents (WNC). Flavour-changing WNC were known to be absent to $\mathcal{O}(10^{-8})$ and even flavour-preserving WNC were incorrectly reported [17] to be at least thirty times weaker than charged neutral currents. (Ultimately, the absence of flavour-changing WNC at tree-level and the reduction of their radiatively-induced $\mathcal{O}(G_F\alpha)$ amplitude by a suitably small factor $(m_c^2 - m_u^2)/M_W^2$, where m_c and m_u are quark masses and M_W is the mass of the W-boson, was explained by the GIM mechanism [27].)

My 1958 paper was soon followed by proposals [18, 19, 20] to use accelerator neutrino beams to search for flavour-preserving weak neutral currents. This search was very difficult, because of backgrounds of neutrino- induced charged-current processes in which muons escape undetected, which were large and hard to estimate. Thus, neutrino experiments began only a decade later and were, for several years, preoccupied with deep inelastic scattering at SLAC and with scaling. These experimental difficulties, together with the need for a consistent theory allowing massive gauge bosons, suggest why chiral weak neutral currents needed to wait from 1958 to 1973 for experimental confirmation.

4   SPONTANEOUSLY BROKEN GLOBAL SYMMETRIES

The idea of spontaneous symmetry breaking (SSB) came to condensed matter physics to quantum field theory through the work of Heisenberg [21] and Nambu [22]. It soon led to the Goldstone Theorem [23, 24] showing how SSB could produce long-range interactions out of a short-range global interactions. Klein and I identified the Goldstone bosons expected from different levels of global symmetry- breaking and emphasized that Goldstone bosons were not present in theories with long-range interactions. Following Anderson [25], we suggested that, in an inverse Goldstone Theorem, long-range interactions might be converted into short-range. But the apparently massless neutrino could not be a Goldstone particle: the vacuum could not be macroscopically occupied by fermions. We, therefore, failed to connect the Goldstone Theorem with my earlier proposal of a *gauge* theory of weak interactions.

The 1958 work on chiral invariance was cited by Gell-Mann and by Nambu [26] and ultimately led to current algebras, soft-pion theorems, and PCAC. These successes, however, tended to gloss over the fundamental differences between global and gauge symmetries, and between partial flavour symmetries and exact gauge symmetries, in the strong and weak interactions.

Exact symmetries were useful in classifying fields and particles and were most satisfying aesthetically. For these reasons, I tended to avoid hadron physics and concentrated on weak interactions where, I was convinced, exact symmetries were to be found. At his time. I left the University of California Radiation Laboratory (now the Lawrence Berkeley Laboratory), which was then dominated by dispersion relations and S-matrix theory. I took an academic position at the University of Pennsylvania and my interests gradually shifted from laboratory to astrophysical particle physics.

5 SPONTANEOUSLY BROKEN GAUGE SYMMETRIES, WITH AND WITHOUT UNIFICATION

The Higgs mechanism [25, 5] sharply differentiates the role Goldstone bosons play in gauge theories from their role in global symmetry theories. If an exact gauge symmetry is spontaneously broken by the Higgs mechanism, so that some gauge bosons acquire masses, the symmetry is hidden, but the theory remains renormalizable. Indeed, the Standard Model has only exact gauge symmetries, so that, except for the photon, massless gauge bosons do not appear: Either the (colour) gauge symmetry is unbroken, but the massless gluons are confined, or the gauge symmetry is spontaneously broken, providing masses for the gauge bosons other than the photon.

In the minimal Standard Model, the electroweak couplings enter through the $SU(2)_L \times U(1)_Y$ covariant derivative $D_\mu = \partial_\mu - ig\mathbf{T} \cdot \mathbf{W}_\mu - ig'(Y/2)B_\mu$, so that: (1) The charged vector mesons couple to the electromagnetic field with magnetic moment $2(eh/2Mc)$ and electric quadrupole moment $-e(\hbar/Mc)^2$, where $e^{-2} \equiv g^{-2} + g'^{-2}$; (2) the WNC couple to Z^0 with coupling constant $g/\cos\theta_W$; (3) Charged currents couple to the electromagnetic field with coupling constant $e \equiv g \sin\theta_W \equiv g' \cos\theta_W$, so that both gauge couplings g, $g' \geq e$; (4) The Higgs mechanism gives the vector mesons (generally) unequal masses $M_W = M_Z \cos\theta_W$. In tree approximation,

$$M_W \sin\theta_W = M_Z \sin\theta_W \cos\theta_W = (e/2)(\sqrt{2}G_F)^{-1/2} = \sqrt{\pi\alpha}v \equiv A_0 = 37.3 \; GeV. \quad (1)$$

so that $M_W > 37 \; GeV$, $M_Z > 74 \; Gev$.

If the theory were not "unified", either $U(1)$ or weak $SU(2)_L$ could be spontaneously broken. The former theory, $U(1)$ symmetry with no weak currents, $(g = 0, g' = e)$, is Schwinger's gauge-invariant electrodynamics with massive photons [28]. The latter, purely weak interactions $(g' = 0 = e)$, is the Bludman 1958 theory [4]. These two examples illustrate that, contrary to ref. [1], consistent (renormalizable) theories without unification are logically possible. In a perturbative $SU(2)_W$ theory without unification, only the constraints $g < 1$, and therefore $M_W < 123$ Gev obtain.

In the electroweak sector of the Standard Model, besides G_F and e, there is only one free parameter,

$$\sin^2 \theta_W \equiv (M_Z^2 - M_W^2)/M_Z^2 = (1/2)[1 - \sqrt{1 - (2A_0/M_Z)^2}\,] \qquad (2)$$

in tree approximation, which measures the $SU(2)_L$ symmetry-breaking through W^0 – B^0 mixing. This mixing angle is a free parameter of the Standard Model, but cannot vanish for finite M_Z, M_W. This makes the unification condition non-trivial, even in the $\sin^2 \theta_W \to 0$ limit. Holding G_F and e constant as $\sin^2 \theta_W \to 0$, makes $g' \to e$ as g, M_W, M_Z all diverge. In a unified theory, if $\sin^2 \theta_W \to 0$, we obtain unmixed electromagnetic interactions and weak interactions that are point-like in tree-approximation, but which are unitary and renormalizable because of huge radiative corrections! But if the theory is to be both perturbative and unified, $e \le g$, $g' < 1$, $0.0836 < \sin^2 \theta_W < 0.916$, and 39 $Gev < M_W < 129$ Gev, 75 $Gev < M_Z < 135$ Gev. $\sin^2 \theta_W$ cannot vanish and $\sin^2 \theta_W, M_W, M_Z$ are all observed to lie in the middle of their theoretically allowed ranges.

In the low-energy ($\ll v = 246$ Gev) effective theory, the W- and Z-bosons have different masses and only electromagnetic gauge invariance is explicit. The mass difference between W- and Z-mesons is only 11% and, except for the top, the quark masses and mass differences are even smaller. The original $SU(2)_L$ theory reappears as a good approximation to leptonic and semi-leptonic processes, other than top quark decay. Thus, far below the weak scale v, the low-energy effective theory actually shows only small effects of electroweak mixing.

6 HISTORICAL CONCLUSIONS

The proof that, in an exact gauge theory, renormalizability would persist even as the Higgs mechanism gave masses to some of the gauge bosons, immediately attracted theorists and experimentalists to the entire Standard Model. Although a consistent theory without electroweak mixing was logically conceivable, the discovery of weak neutral currents [8] with mixing $\sin^2 \theta_W \sim 0.3$, gave circumstantial evidence for the Electroweak Standard Model and predicted $M_W \approx 80$ GeV, $M_Z \approx 90$ GeV. (Indeed, M_W and M_Z had to lie within 0.5-1.6 and 0.8-1.5 times these values for any $\sin^2 \theta_W$ observed.) The ultimate discovery [29] of these gauge bosons with unequal masses then directly confirmed the Electroweak Standard Model.

References

[1] S. L. Glashow, "Towards a Unified Theory: Threads in a Tapestry", *Rev. Mod. Phys.* **52** (1980), 539-43.

[2] A. Salam, "Gauge Unification of Fundamental Forces", *Rev. Mod. Phys.* **52** (1980), 525-38.

[3] S. Weinberg, "Conceptual Foundations of the Unified Theory of Weak and Electromagnetic Interactions", *Rev. Mod. Phys.* **52** (1980), 515-23.

[4] S. Bludman, "On the Universal Fermi Interaction", *Nuovo Cimento* **9** (1958), 433-44.

[5] P. W. Higgs, "Broken Symmetries, Massless Particles and Gauge fields, *Phys. Lett.* **12** (1964), 132-3; "Spontaneous Symmetry Breaking without Massless Bosons", *Phys. Lett.* **145** (1966), 1156-63; F. Englert and R. Brout, "Broken Symmetry and the Mass of Gauge Vector Mesons", *Phys. Rev. Lett.* **13**, (1964), 321-3; T. W. Kibble, "Symmetry Breaking in Non-Abelian Gauge Theories", *Phys. Lett.* **155**, (1967), 1554-61; G. S. Guralnik, C. R. Hagen and T. W. B. Kibble, "Global Conservation Laws and Massless Particles", *Phys. Rev. Lett.* **13** (1964), 585-7.

[6] G. 't Hooft, "Renormalizable Lagrangians for Massive Yang-Mills Fields", *Nucl. Phys.* **B 35** (1971), 167-88; G. 't Hooft and M. Veltman, "Regularization and Renormalization of Gauge Fields", *Nucl. Phys.* **B 44** (1972), 189-213; G. 't Hooft, "Combinations of Gauge Fields", *Nucl. Phys.* **B 50** (1972), 318-53; B. W. Lee and J. Zinn-Justin, "Spontaneously Broken Gauge Symmetries. I. Preliminaries", *Phys. Rev.* **D5** (1972), 3121-37.

[7] S. L. Glashow, "Partial Symmetries of Weak Interactions" *Nucl. Phys.* **22** (1961). 579-88.

[8] F. J. Hasert *et al.*, "Search for Elastic Muon-Neutrino Electron Scattering", *Phys. Lett.* **46B** (1973), 121-4; A Benvenuti *et al.*, "Observation of Muonless Neutrino-Induced Inelastic Interactions", *Phys. Rev. Lett.* **32** (1974), 800-3.

[9] W. Pauli, "Relativistic Field Theories of Elementary Particles", *Rev. Mod. Phys.* **13** (1941), 203-232.

[10] N. Kemmer, "Field Theory of Nuclear Interactions", *Phys. Rev.* **52**(1937), 906-910; E. Teller, "Scattering of Neutrons by Ortho- and Para-Hydrogen", *Phys. Rev.* **52**(1937), 286-295; G. Wentzel, "β-interaction", *Helv. Phys. Acta* **10**(1937), 107-111.

[11] J. A. Young and S. A. Bludman, "Electromagnetic Properties of a Charged Vector Meson", *Phys. Rev.* **131**(1963), 2326-2334; G. Feinberg, "Decays of μ-meson in Intermediate-meson Theory", *Phys. Rev.* **110**(1958), 1482; T. Kuo-Hsien, "Charged Vector Field of Zero Proper Mass",*C.R. Acad. Sci. (Paris)* **245**(1957), 289.

[12] C. N. Yang and R. L. Mills, "Conservation of Isotopic Spin and Isotopic Gauge Invariance", *Phys. Rev.* **96** (1954), 191-5.

[13] S. A. Bludman, "Extended Isotopic Invariance and Meson-Nucleon Coupling", *Phys. Rev.* **100**(1955), 372-375.

[14] E. C. G. Sudarshan and R. E. Marshak, "The Nature of the Four-Fermion Interaction", *Proc. Padua-Venice Conference on Mesons and Newly Discovered Particles, (Bologna, 1958)*; "Chirality Invariance and the Universal Fermi Interaction" *Phys. Rev.* **109** (1958), 1860-2.

[15] R. P. Feynman and M. Gell-Mann, "Theory of Fermi Interaction" *Phys. Rev.* **109** (1958), 193-8.

[16] J. Sakurai, *Nuovo Cimento* **7**(1958), 649-660.

[17] M. M. Block *et al*, "Neutrino Interactions in the CERN Heavy Liquid Bubble Chamber" *Phys. Lett.* **12** (1964), 281-5; Perkins group report at Sienna Conference, 1963.

[18] T. D. Lee and C. N. Yang, "Theoretical Discussions on Possible High-Energy Neutrino Experiments", *Phys. Rev. Lett.* **4** (1960), 307-11.

[19] B. Pontecorvo, "Small Probability of the $\mu \to e + \gamma$ and $\mu \to e + e + e$ Processes and Neutral Currents in Weak Interactions", *Phys. Lett.* **1**(1962), 287-8.

[20] S. S. Gershtein, N. Van Hieu and R. A. Eramzhyan, "Possibility of Observing Neutral Currents in Neutrino Experiments", *J. Exptl. Theoret. Phys. (U.S.S.R.)* **43** (1962), 1554-6.

[21] W. Heisenberg, "Theory of Elementary Particles", *Zeitschr. f. Naturf.* **14a**(1959), 441-485.

[22] Y. Nambu, "Axial Vector Current Conservations in Weak Interactions", *Phys. Rev. Lett.* **4** (1960), 380-2; Y. Nambu and G. Jona Lasinio, "A Dynamical Model of Elementary Particles Based upon an Analogy with Superconductivity. I, II", *Phys. Rev.* **122** (1961), 345-58, **124**(1961), 246-54.

[23] J. Goldstone, "Field Theories with 'Superconductor' Solutions", *Nuovo Cimento* **19** (1961), 154-64; J. Goldstone, A. Salam and S. Weinberg, "Broken Symmetries", *Phys. Rev.* **127** (1962), 965-70; J. C. Taylor, *Proc. 1962 Intl. Conf. on High-Energy Physics (CERN,Geneva, 1962)*.

[24] S. A. Bludman and A. Klein, "Broken Symmetries and Massless Particles" *Phys. Rev.* **131** (1963), 2364-2372.

[25] P. W. Anderson, "Plasmons, Gauge Invariance and Mass", *Phys. Rev.* **130** (1962), 439-42.

[26] M. Gell-Mann, *Proc. 1960 Annual Intl. Conf. on High-Energy Physics at Rochester (Interscience Publishers, Inc., New York, 1960)*; Y. Nambu and F. Lurie "Chirality Conservation and Soft Pion Production", *Phys. Rev.* **125**(1960), 1429-36.

[27] S. L. Glashow, J. Iliopoulos and L. Maiani, "Weak Interactions with Lepton-Hadron Symmetry", *Phys. Rev.* **D2** (1970), 1285-92.

[28] J. Schwinger, "Gauge Invariance and Mass", *Phys. Lett.* **125** (1962), 397-8.

[29] G. Arnison *et al.*, "Experimental Observation of Isolated Large Transverse Energy Electrons with Associated Missing Energy at $\sqrt{s} = 540 GeV$", *Phys. Lett.* **122B** (1983), 103-16; "Experimental Observation of Lepton Pairs of Invariant Mass around $95 GeV/c^2$ at the CERN SPS Collider", *Phys. Lett.* **126B** (1983), 398-410; M. Banner *et al.*,"Observation of Single Isolated Electrons of High Transverse Momentum with Missing Transverse Energy at CERN pp Collider", *Phys. Lett.* bf 122B (1983), 476-85; P. Bagnaia *et al.*, *Phys. Lett.* **129B** (1983), 130.

CHAPTER 5
THE DISCOVERY OF WEAK NEUTRAL CURRENTS

Nothing In – Nothing Out (1973 – 1974)

5.1. THE EARLY LIMITS ON THE WNC AMPLITUDES

The search for neutral current processes that did not involve a change in strangeness started in the early 1960s, soon after the initiation of high-energy neutrino beams at the BNL AGS and the CERN PS. The reactions searched for were

$$\nu + p \rightarrow \nu + p \quad , \tag{5.1}$$

and

$$\nu + p \rightarrow \nu + (\text{inelastic products}) \quad . \tag{5.2}$$

Unfortunately, these experiments probably never took the idea of the existence of WNC seriously. The spark chamber detectors were not matched to the requirements of detecting small hadronic energy deposits, and the bubble chamber groups seemed to ignore or explain away what we know today to be a 30% signal. (We attempt to trace the early period in Table 5.1.) One setback to the search was the erroneous limit published by the CERN bubble chamber group in 1964 (shown in Fig. 5.1) [5.1]. This limit was later revised but, unfortunately, discouraged some searches for WNC during this period. By 1970, the limit was revised upward, and there were searches in other channels that were less restrictive, but still null.

Therefore, the situation in 1970 was that there were extremely strong limits on FCNC and seemingly restrictive limits on WNC from neutrino experiments that convinced most particle physicists that the neutral currents, in general, did not exist in first order in Nature. It was indeed unfortunate that the neutrino experiments had somehow overlooked a 30% signal that was discovered in 1973 in the new generation of bubble chambers at CERN and the high-energy neutrino beam at Fermilab. We have attempted to provide a brief summary of the road to the discovery of WNC in Table 5.1, which covers the period of 1962 to 1973.

The most interesting aspect of the early search for WNCs was that a number of the experiments observed unusual events, which we now believe to be WNC processes but were, at that time, largely explained away as the result of neutron interactions. While there was certainly a low flux of neutrons in the neutrino beams, no one attempted to actually estimate or measure this level. This was most likely due to the belief by most theorists and experimentalists that WNCs did not exist. This belief was reinforced by the absence of FCNC to a very low level. The history of the standard model development would likely be very different if WNC had been discovered in the 1960s, and if FCNC had been searched for to a lower and lower limit. One can speculate that some clever person might have explained them both by what we now call the standard model. But theory was far ahead of experiment, and the GWS model [5.2] actually set the stage for the renewed search for WNC in the 1970s. Another way to state this is that the real importance of the absence of FCNC was only fully appreciated after WNC was discovered at the 30% level.

During this period, the Weinberg–Salam model was being developed (Glashow had developed an early version in 1961), which suggested once again that

Table 5.1. Some milestones in the search for WNC (1962 – 1973).[*]

	Milestones	Comments
1962	BNL neutrino experiment detects 2nd neutrino. Unusual events (Little Crappers) noted in PhD thesis.	Strange events are associated with neutrons from the shield.
1964	CERN bubble-chamber group reports a limit of 3% for WNC.	Reason for this incorrect limit given later in 1970 paper.
1967	E. C. Young thesis at Oxford notes that NC-like events exist in data.	
	Large, heavy-liquid bubble chamber studied by the MURA–Wisconsin–Berkeley Group to search for WNC ($v + p \rightarrow v + p$) at ANL. Later proposed for BNL then rejected.	GGM bubble chamber already approved. Incorrect CERN limits on WNC discourage further searches.
1969	HPW experiment proposed. Only mention of NCs is for leptonic processes.	1964 limit on WNC makes search for this process difficult, so leptonic processes considered.
1970	CERN group revises 1964 limit on WNC to ~ 12%.	
1972	Schwartz–Mann–SLAC black-hole experiment finds some NC-like events.	Too few events to make conclusive arguments.
	Perkins reports limits on WNC at FNAL meeting.	Limit still below currently observed values.
1973	GGM and HPWF	Struggle with large WNC signal.

[*](From [5.3].)

> *Other conclusions.* A number of other fundamental
> questions were discussed in our previous report[1],
> their present status is:
>   Violation of muon number conservation
> $(\nu_e \neq \nu_\mu) < 1\%$;
>   Violation of leptonic conservation $< 6\%$;
> $(\Delta S = 0$, neutral current coupling$/\Delta S = 0$,
>   charged current coupling$)$ $\Delta S = 0$, 3%;
>   Neutrino flip intensity/no neutrino flip intensity
> $$= \frac{K \to \mu + \nu_e}{K \to \mu + \nu_\mu} < 10\%.$$
> A boson has been postulated which produces a
> resonance of the type $\nu_\mu + n \to W_{\mu'} \to \mu^- + p$. If
> its properties are as predicted,[11] its mass is
> > 5.5 GeV.

Figure 5.1. This figure gives the text of the 1964 CERN paper that indicated that the coupling of the ($\Delta S = 0$) neutral-current coupling to charged current was less than 3%. This result was later revised upwards by a large factor.

WNC should exist but at a reduced level [5.2]. The severe limit on FCNC had to be addressed, and this was the focus of the GIM model in 1970 [5.4].

5.2. THE IMPORTANCE OF THE EARLY SEARCH FOR FCNCs

The concept of flavor was not well understood in the early and mid-1960s, and it appeared that as good a technique as any was to search for WNC by studying rare K decays. One of the earliest experiments set a limit of $\sim 10^{-6}$ on

$$K^+ \to \pi^+ e^+ e^-$$

[5.5, see article 3(B)]. The limits on FCNC were summarized in a report I wrote in 1967, which included $K \to \mu\mu$ and the search for $K^+ \to \pi^+ \nu\bar{\nu}$ from my early work [5.6, see article 3(A)].

In 1965–1970, a crucial search for $K^+ \to \pi^+ \nu\bar{\nu}$ was started [5.6]–[5.8]. This search was a pure FCNC process, whereas all other searches had charged particles in the final state and may have been affected by electromagnetic interactions, as some people claimed. The history of the search for this decay mode, which continues today, is given in Fig. 5.2.

By 1970, the incorrect CERN limit on WNC and the extremely strong limits on FCNC caused most people to believe that neutral currents *did not* exist in Nature.

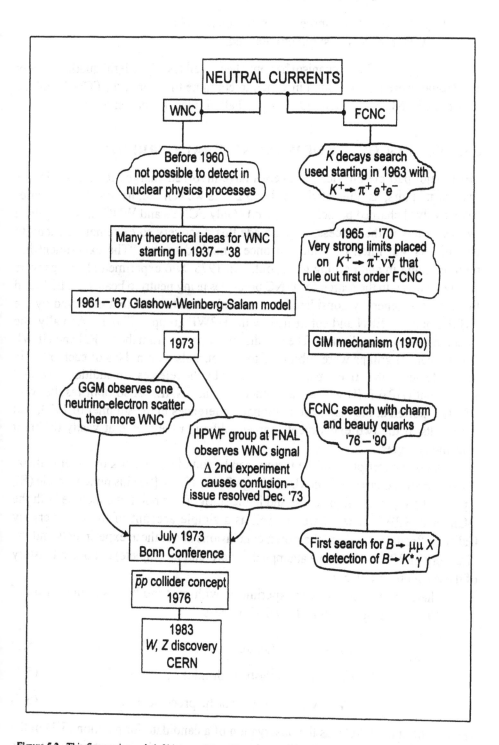

Figure 5.2. This figure gives a brief history of the efforts to discover the WNC and the search for FCNC. Article (E) by P. Galison gives a full account of the discovery of WNC, whereas Chapter 3 discusses the search for FCNC.

The GIM model was invented and showed that [5.4]

    1. FCNC could be suppressed in certain models,
    2. A hadron–lepton symmetry existed.

Today, the GIM mechanism has only been well tested for light quarks. Charm and b quarks are a rich ground in which to continue the search for FCNC, and the search for FCNC has started there, as we shall discuss in Chapter 8.

5.3. THE DISCOVERY OF WEAK NEUTRAL CURRENTS

Most major particle-physics discoveries have involved charged leptons or, in one case, $\pi^+\pi^-$ pairs, as illustrated in Table 5.2. As we can see, nearly all discoveries have involved charged particles (e, μ, *etc.*). Only FCNCs and WNCs have involved two neutrinos in the process, which made this discovery so extremely difficult to obtain. As Val Fitch of Princeton once remarked to me, "The experiment has nothing going in and nothing going out." In 1973, two experimental groups were deeply involved in the search for WNC processes using neutrino beams – CERN and FNAL. It is generally considered that neutral currents were first observed by the GGM group at CERN and confirmed by the HPWF group at FNAL. Actually, the situation is more complicated. The two discovery papers from the CERN and HPWF (Experiment I) groups were submitted to the journals within days of each other in 1973. However, the HPWF paper was delayed by the referees. In addition, a second experiment (HPWF II) was already underway at FNAL. Also both results were widely reported at the major international conferences in the summer of 1973, and the results of the two experiments agreed, although they were at vastly different neutrino energies.

A good description of the history can be found in the book by Peter Galison entitled, *How Experiments End* (U. Chicago Press, 1984). [See his review article (E), reprinted here.] For completeness and in the spirit of this book, we include both the GGM and HPWF papers here, as well as a simple account of the discovery by Galison. Readers can draw their own conclusions about these experiments and the discovery of WNCs. We have attempted a short (and incomplete) but overall history of these events in Fig. 5.2.

There were three types of experiments with neutrino beams in the period of 1973–1976 that helped define the WNC process:

$$\nu_\mu + e \rightarrow \nu_\mu + e \text{ (leptons)} , \tag{5.3}$$

$$\nu + p \rightarrow \nu + p \text{ (elastic hadronic)} , \tag{5.4}$$

$$\nu + N \rightarrow \nu + X \text{ (deep inelastic processes)} . \tag{5.5}$$

The first limit of WNC was the observation of a candidate for reaction (5.3) in the GGM bubble chamber [5.9]. While this was an extremely interesting event, one

Table 5.2. Some major discoveries in particle physics that used charged lepons.

Discovery	Observation	Group or Persons	Key Particle Detected
Weak interaction (1896)	β decay	H. Becquerel (Ch. 1)	e^-
Parity violation (1957)	β or $\pi \to \mu \to e$ decay	(see Ch. 2)	e^-, $\pi \to \mu \to e$
Free neutrino (1957)	$\bar{\nu}_e + p \to e^+ + n$	F. Reines & C. Cowen (Ch. 2)	e^+
2nd neutrino (1962)	$\nu_\mu + N \to \mu^- + p$	Lederman, Schwartz, & Steinberger (Ch. 2)	μ^-
CP violation (1964)	$K_L^0 \to \pi^+ \pi^-$	Fitch, Cronin Group	$\pi^+ \pi^-$
Quark–parton structure (1969)	$e + N \to e + x$	R. Taylor, *et al.*, SLAC Group	e^-
$J/\psi (c\bar{c})$ (1974)	$J \to e^+ e^-$ $e^+ e^- \to \psi$	S-C-C. Ting Group SPEAR Group	e^+/e^-
τ lepton (1976)	$\tau\bar{\tau} \to e\mu + ...$	Perl, SPEAR– SLAC Group	e/μ
Bare charm (1974–1976)	$\nu_\mu + N \to \mu^+ \mu^- x$ $D^0 \to$ hadrons	(see Ch. 6)	μ/μ $D^0 \to$ hadrons
$\Upsilon(b\bar{b})$ (1977)	$\Upsilon \to \mu^+ \mu^-$	FNAL Group	$\mu^+ \mu^-$
W,Z particles (1983)	$W \to e\nu_e$ $Z \to e^+ e^-$	UA1 (Ch. 7)	$e^+/e^-/\mu^+\mu^-$
Neutral Currents			
Absence of FCNC (1965)	$\begin{bmatrix} K^+ \to \pi^+ e^+ e^- \\ K^+ \to \pi^+ \nu\bar{\nu} \end{bmatrix}^*$	(see Ch. 3)	e^+/e^- $\nu/\bar{\nu}$
WNC (1973)	$\begin{bmatrix} \nu_\mu + e \to \nu_\mu + e \\ \nu + N \to \nu + x \end{bmatrix}^{**}$		$\nu/\bar{\nu}$

*See [5.6, 5.8]
**See [5.9–5.12]

event is not usually considered proof of the existence of a new class of weak interactions. While it may seem logical to use reaction (5.4) to continue the search because of the quark-like structure in the proton, this reaction has a much smaller cross section than (5.5) and was observed later (Chapter 6). The next series of events occurred in the Spring of 1973 when the GGM group observed candidates for the reaction

$$\nu + N \rightarrow \nu + X \ ,$$

where X was a hadronic final state. The GGM group then set about proving that these events were not due to neutrons entering the bubble chamber from the front or side. It was crucial for this proof that the GGM chamber be so large and the liquid freon so dense that neutron interactions could be separated structurally from the neutrino interactions [5.10].

The HPWF experiment used a different technique and had different problems. (The author was the spokesperson for this experiment.) This detector was a massive counter detector with approximately 100 tons of liquid scintillator and a massive magnetized-iron toroid to measure the muon momentum. [Paper 4(A) shows the HPWF detector.] Soon after the detector was turned on in March of 1973, many muon-less or neutral current events were detected [5.11]. The problem was to prove conclusively that these were not due to a background of charged current events with a muon ($\nu_\mu + N \rightarrow \mu + X$) that escaped the muon detector. The first data showed almost conclusively that the signal was real [5.11]. The data was submitted for publication (Experiment I). (The data and the CERN data were also discussed at the Bonn high-energy conference that year in July of 1973.) However, to obtain further proof, the group rebuilt the detector during the period of August–September 1973, took new data, and became bogged down (and confused) in the attempts to understand the new detector (Experiment II) [5.12]. In essence, to identify wide-angle muons, a nearby muon wall was constructed. However, some hadrons from the neutral current events leaked through the wall and reduced the apparent signal. Since no previous experiment had measured the hadronic distribution for deep-inelastic scattering, it was exceedingly difficult to correct for the punch-through. Nevertheless, by December of 1973, the data in Experiment II also showed conclusively that WNC had been detected (see article (E) and Ref. [5.13] by P. Galison for a comparative discussion of both experiments).

In 1974, many experiments detected WNC (Caltech, ANL bubble chamber, Columbia spark chamber detector, *etc.*) [5.14]. (These will be discussed in Chapter 6.) The next step was to detect several different types of WNCs. Several more events of type (5.3) were observed by the GGM group. The process,

$$\nu + N \rightarrow \nu + N + \pi \ , \qquad (5.6)$$

was detected. [This paper is reproduced in Chapter 6, article (A.]

A search was started for reaction (5.4) at BNL by the HPW group using a large liquid-scintillator detector (similar in some sense to the HPWF detector at FNAL), and the elastic scattering reaction in protons was clearly observed [paper (B) reproduced in Chapter 6]. This was a landmark in some ways, since the simplest neutral-current reaction, which had been incorrectly limited to 3% in 1963 [5.1], was now observed at the level of 20% in 1976. This basically completed the discovery phase of WNC. Many different channels had been detected, and there was strong evidence for a similarity in all of these channels. The next step was to determine the space-time nature of the WNC coupling.

5.4.   A BRIEF DISCUSSION OF THE PARAMETERS THAT DESCRIBE WEAK NEUTRAL CURRENTS

In the mid-1970s, several experiments had detected WNC and a uniform parameterization was required. The parameters of the GWS theory were natural to use, such as ρ and $\sin^2\theta_W$ (for the Z–γ mixing). No one knew how extremely successful this would turn out to be. The value of $\sin^2\theta_W$ was first measured by observing the ratios of

$$R^\nu = \frac{\nu_\mu + (I=0/\text{target}) \rightarrow \nu_\mu + x}{\nu_\mu + (I=0/\text{target}) \rightarrow \mu^- + x} \quad , \tag{5.7}$$

and

$$R^{\bar\nu} = \frac{\bar\nu_\mu + (I=0/\text{target}) \rightarrow \bar\nu_\mu + x}{\bar\nu_\mu + (I=0/\text{target}) \rightarrow \mu^+ + x} \quad . \tag{5.8}$$

The targets for the massive FNAL and CERN detectors were very approximately $I = 0$ systems, thus leading to the possible measurement of R^ν and $R^{\bar\nu}$. Those interested in the history of this period should read the many beautiful papers on the data analysis and techniques to obtain $\sin^2\theta_W$, some of which are reproduced in Chapter 6 [articles (C), (D), (F), (G)]. We will reproduce one set of arguments here. A careful definition of the target and muon spectrometer for the HPWF detector and neutrino beam was made, for example, to aid in the determination of ρ and $\sin^2\theta_W$.

The values of R^ν and $R^{\bar\nu}$ were measured and compared to the GGM data, showing that there was no appreciable energy dependence in the cross section. The data were then fit to a simple V–A model:

$$\frac{d\sigma_N^\nu}{dy} = a + b(1-y)^2 \quad , \tag{5.9}$$

and

$$\frac{d\sigma_N^{\bar\nu}}{dy} = b + a(1-y)^2 \quad , \tag{5.10}$$

where $y = E_{\text{hadron}}/E_{\nu}$ for a simple quark–parton model of the scaling. The data only fit a model with both V and A and, hence, was parity violating (a,b nonzero). Later the beautiful atomic physics experiments and the SLAC polarized experiments proved conclusively that the WNC was parity violating (Chapter 6). This exercise gives some insight into how the space–time parameters of the WNC were measured. The values of a and b can be related to $\sin^2\theta_W$, and by 1978, a unique value of $\sin^2\theta_W$ had been deduced. The discovery of WNCs was complete, as is discussed in Ref. [5.13] and in the papers reproduced here and in Chapter 6. A glance at Table 5.2 shows that the discovery of WNC is the only major discovery in particle physics that was made without charged leptons (or $\pi^+\pi^-$) in the final state – an indication of just how difficult this discovery was! Thus, the confusion of the period of summer through the end of 1973, when the HPWF Experiment II verified Experiment I, might somehow be understood in the context.

REFERENCES

5.1. M.M. Block et al., Phys. Lett., **12**, 281 (1964).

5.2. S.L. Glashow, Nucl. Phys., **22**, 579 (1961); S. Weinberg, Phys. Rev. Lett., **19**, 1264 (1967); A. Salam, in Proceedings of the VIII Nobel Symposium, N. Svartholm, ed. (Almquist and Wiksell, Stockholm, 1968), p. 367.

5.3. D. B. Cline, in Discovery of Weak Neutral Currents: The Weak Interaction Before and After (Proc., Santa Monica, CA 1993), A. K. Mann and D. B. Cline, eds. (AIP Conference Proceedings 300, AIP, New York, 1994), p. 175.

5.4. S. J. Glashow, J. Iliopoulos, and L. Maiani, Phys. Rev., **D2**, 1285 (1970).

5.5. U. Camerini, D. Cline, W. F. Fry, and W. M. Powell, Phys. Rev. Lett., **13**(9), 318 (1964) (Ch. 3, article (B)].

5.6. D. Cline, "Experimental Search for Weak Neutral Currents," in Methods in Subnuclear Physics, v. III (Proc., Int. School of Elementary Particle Physics, Herceg-Novi, Yugoslavia, 1967), M. Nikolić, ed. (Gordon Breach, New York, 1969), p. 355 [Ch. 3, article (A)]; also Ph.D. thesis, Univ. of Wisconsin (1965).

5.7. U. Camerini et al., Phys. Rev. Lett., **23**, 326 (1969).

5.8. J. Klems et al., Phys. Rev. Lett., **24**, 1086 (1970).

5.9. F. J. Hasert, H. Faissner, W. Krenz, et al., Phys. Lett., **46B**(1), 121 (1973) [article (A)].

5.10. F. J. Hasert, S. Kabe, W. Krenz, et al., Phys. Lett., **46B**(1), 138 (1973) [article (B)].

5.11. A. Benvenuti, D. C. Cheng, D. Cline, et al., Phys. Rev. Lett., **32**(14), 800 (1974) [article (C)].

5.12. B. Aubert et al., Phys. Rev. Lett., **32**(25), 1455 (1974) [article (D)].

5.13. P. Galison, Rev. Mod. Phys., **55**, 477 (1983); also see article (E).

5.14. B. C. Barish, J. F. Bartlett, K. W. Brown, et al., Phys. Rev. Lett., **34**, 538 (1975); also see Ch. 6, article (A).

SEARCH FOR ELASTIC MUON-NEUTRINO ELECTRON SCATTERING

F.J. HASERT, H. FAISSNER, W. KRENZ, J. Von KROGH,
D. LANSKE, J. MORFIN, K. SCHULTZE and H. WEERTS
III Physikalisches Institut der technischen Hochschule, Aachen, Germany

G.H. BERTRAND-COREMANS, J. LEMONNE, J. SACTON, W. Van DONINCK and P. VILAIN[*1]
Interuniversity Institute for High Energies, U.L.B., V.U.B. Brussels, Belgium

C. BALTAY[*2], D.C. CUNDY, D. HAIDT, M. JAFFRE, P. MUSSET, A. PULLIA[*3]
S. NATALI[*4], J.B.M. PATTISON, D.H. PERKINS[*5], A. ROUSSET, W. VENUS[*6] and H.W. WACHSMUTH
CERN, Geneva, Switzerland

V. BRISSON, B. DEGRANGE, M. HAGUENAUER, L. KLUBERG, U. NGUYEN-KHAC and P. PETIAU
Laboratoire de Physique des Hautes Energies, Ecole Polytechnique, Paris, France

E. BELLOTTI, S. BONETTI, D. CAVALLI, C. CONTA[*7], E. FIORINI and M. ROLLIER
Istituto di Fisica dell'Università, Milano and I.N.F.N. Milano, Italy

B. AUBERT, L.M. CHOUNET, P. HEUSSE, A. LAGARRIGUE, A.M. LUTZ and J.P. VIALLE
Laboratoire de l'Accélérateur Linéaire, Orsay, France

and

F.W. BULLOCK, M.J. ESTEN, T. JONES, J. McKENZIE, A.G. MICHETTE[*8]
G. MYATT[*5], J. PINFOLD and W.G. SCOTT[*5, *8]
University College, University of London, England

Received 2 July 1973

One possible event of the process $\nu_\mu^- + e^- \rightarrow \nu_\mu^- + e^-$ has been observed. The various background processes are discussed and the event interpreted in terms of the Weinberg theory. The 90% confidence limits on the Weinberg parameter are $0.1 < \sin^2\theta_W < 0.6$.

Recently many theoretical models have been postulated in an attempt to resolve the divergency of the classical current-current theory by unifying the weak and electromagnetic interactions. All these theories require neutral currents, heavy leptons or both. One of these theories, that of Salam and Ward [1] and Weinberg [2], gives specific predicitons about the amplitudes of the neutral currents which are susceptible to experimental tests.

In particular, using this model, t'Hooft [3] has calculated the differential cross sections for the purely leptonic processes

$$\nu_\mu + e^- \rightarrow \nu_\mu + e^- \qquad (1)$$

[*1] Chercheur Agréé à l 'I.I.S.N. Belgique.
[*2] Columbia University, New York, U.S.A.
[*3] On leave from the University of Milan.
[*4] Istituto di Fisica, Bari, Italy.
[*5] And at the University of Oxford, England.
[*6] Rutherford High Energy Laboratory, Chilton, England.
[*7] On leave of absence from the Sezione INFN and University of Pavia
[*8] Grant from the Science Research Council.

$$\bar{\nu}_\mu + e^- \rightarrow \bar{\nu}_\mu + e^- \qquad (2)$$

which are forbidden to first order in the conventional Feynman Gell-Mann theory. The predicted cross-sections are of the order of 10^{-41} cm^2/electron at 1 GeV, depending on the Weinberg angle θ_W, which is the only free parameter of the theory.

A search for these processes has been carried out in the large heavy liquid bubble chamber Gargamelle, useful volume 6.2 m^3, filled with freon CF_3Br, exposed to both the neutrino and antineutrino beams at the CERN PS. The large length of the chamber, 4.8 metres, compared to the radiation length of freon, 11 cm, ensured that electrons were unambiguously identified.

These interactions are characterized by a single electron (e^-) originating in the liquid, unaccompanied by nuclear fragments, hadrons or γ rays correlated to the vertex. The kinematics of the reactions are such that the electron is emitted at small angle, θ_e, with respect to the neutrino beam; the electron is expected to carry typically one third of the energy of the incident neutrino which is peaked between 1 and 2 GeV. As the neutrino interactions in the surrounding magnet and shielding produce a low energy background of photons and electrons, a lower limit on the electron energy was set at 300 MeV. This energy cut ensures that all electrons from reactions (1) and (2) will have $\theta_e < 5°$.

A total of 375 000 ν and 360 000 $\bar{\nu}$ pictures were scanned twice and one single electron event satisfying the selection criteria was found in the $\bar{\nu}$ film. This event is shown in fig. 1. The curvature of the initial part of the track shows the negative charge, and the spiralisation and bremsstrahlung prove unambiguously that the track is due to an electron. The electron energy is 385 ± 100 MeV, and the angle to the beam axis is $1.4°^{+1.6°}_{-1.4°}$. The electron vertex is 60 cm from the beginning of the visible volume of the chamber and 16 cm from the chamber axis.

The scanning efficiency for single electrons with an energy > 300 MeV was determined to be 86% using the isolated electronpositron pairs found in the chamber.

The main source of background is from the process

$$\nu_e + n \rightarrow e^- (\theta_e < 5°) + p \qquad (3)$$

where the proton is either of too low an energy to be

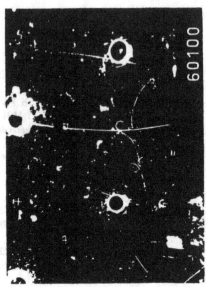

Fig. 1. Possible event of the type $\bar{\nu}_\mu + e^- \rightarrow \bar{\nu}_\mu + e^-$.

observed or is captured in the nucleus and no visible evaporation products are formed. This is due to the small (< 1%) ν_e flux present in the predominantly ν_μ or $\bar{\nu}_\mu$ beam.

This background has been determined empirically using the observed events of the type

$$\nu_\mu + n \rightarrow \mu^- (\theta < 5°) + p \qquad (4)$$

where the proton is not observed, and the ν_e flux calculated from the observed electron-neutrino events.

This is a good estimate as the two processes are kinematically similar at these energies and the ν_μ and ν_e spectra have nearly the same shape. In a partial sample of the film we have observed 450 events, occurring in a fiducial volume of 3 m^3, of the type:

$$\mu^- + m \text{ protons } (m \geqslant 0)$$

where the visible energy is > 1 GeV, and the momentum in the beam direction is > 0.6 GeV/c. These cuts eliminate the background due to incoming charged particles.

In these events, only 3 have no protons and a μ^- angle < 5°. The scanning efficiency for single μ^- has been assumed to be the same as that for the single μ^+ found in the anti-neutrino film. This was determined to be 50% using the sample of 200 single μ^+.

Hence we obtain that

$$\frac{\mu^-(\theta_\mu < 5°) + 0_p}{\mu^- + mp} = 1.3 \pm 7\%$$

This ratio is an over-estimate as the inclusion of events of energies < 1 GeV would be expected on kinematical grounds to lower it.

In the neutrino film 15 ν_e events of the type $e^- + m$ protons ($m > 0$) have been observed in the fiducial volume (3 m^3). This number is in agreement with the one expected from the estimated ν_e/ν_μ flux ratio (0.7%). Hence one deduces a background from this source 0.3 ± 0.2 events.

Another estimate using the calculated ν_e and ν_μ fluxes and expected cross-sections, gives 0.4 ± 0.2.

In the $\bar{\nu}$ film zero $e^- + m$ proton events have been observed and a background estimate is obtained as above using the calculated ν_e and ν_μ fluxes. The ν_e flux in the anti-neutrino film is an order of magnitude less than in the neutrino film. Hence the background from the above source in the $\bar{\nu}$ film is 0.03 ± 0.02 events.

The other sources of background could be due to Compton electrons or asymmetric electron pairs. Only 2 isolated electronpositron pairs having an energy greater than 300 MeV and making an angle of less than 5° with the beam direction were observed in the visible volume of the chamber in the ν film, and none in the $\bar{\nu}$ film.

Given these events and using the ratio of Compton to pair production cross-sections as well as the differential cross-section for pair production for the energy repartition among the electron and positron, this source of background is estimated to be 0.04 ± 0.02 events in ν and negligible in $\bar{\nu}$.

As the ν_e flux is less than 1% of the ν_μ flux the background from the V-A reactions

$$\begin{pmatrix} \nu_e \\ \bar{\nu}_e \end{pmatrix} + e^- \rightarrow \begin{pmatrix} \nu_e \\ \bar{\nu}_e \end{pmatrix} + e^-,$$

of which the cross-sections are of the same order as processes (1) and (2). are negligible. Similarly the lack of high energy neutrons (> 16 GeV) eliminates the background contribution from the electro-magnetic interaction $n + e^- \rightarrow n + e^-$.

To calculate the detection efficiency, i.e. the fraction of reaction (1) and (2) that would survive the selection criteria. the electron laboratory energy and angular distributions have to be known. These spectra are not uniquely predictable but depend on the model assumed to introduce the neutral currents into the weak interactions. However, the detection efficiency in the present experiment is not very sensitive to these uncertainties since the electron minimum energy accepted is small compared to the incident neutrino energy.

In the case of isotropy in the centre of mass the detection efficiency is 87%.

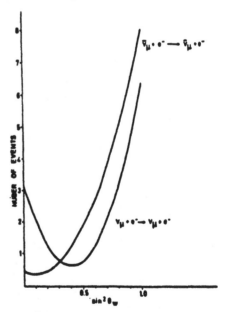

Fig. 2. Expected event rate as a function of the Weinberg parameter.

Table 1
Number of single e^- events of $E_e > 300$ MeV, $\theta_e < 5°$

Flux neutrinos/m^2	Weinberg predictions		Background	Observed
	Minimum	Maximum		
ν	1.8×10^{15}	0.6	6.0	0.3 ± 0.2	0
$\bar{\nu}$	1.2×10^{15}	0.4	8.0	0.03 ± 0.02	1

In this case the 90% confidence upper limits for the cross-sections for the processes (1) and (2) are:

$0.26 E_\nu \times 10^{-41}$ cm^2/electron

and

$0.88 E_\nu \times 10^{-41}$ cm^2/electron

respectively.

Table 1 shows the upper and lower event rates expected from the Weinberg model, taking into account the detection efficiencies, and using the measured ν_μ and $\bar{\nu}_\mu$ fluxes. The estimated backgrounds are also shown. These are to be compared with the one event found in the $\bar{\nu}$ film.

Fig. 2 shows the number of expected ν and $\bar{\nu}$ events as a function of the Weinberg parameter $\sin^2\theta_W$.

In order to combine the neutrino and anti-neutrino results a maximum likelihood method has been used, taking into account the fluxes and backgrounds. The 90% confidence limit gives:

$0.1 < \sin^2\theta_W < 0.6$.

It may be remarked that, in the context of the Weinberg theory, the proportion of electrons with $E_e > 1$ GeV is much lower in neutral current events than in the ν_e background, and hence our quoted background is over-estimated. We conclude that the probability that the single event observed in the $\bar{\nu}$ film is due to non-neutral current background is less than 3%.

It is a pleasure to expresss our thanks to the members of the CERN TC-L group who have carried the technical responsibility for the experiment. We also thank the CERN PS operational staff, and the scanning the programming personnel in the various laboratories.

References

[1] A. Salam and J.G. Ward, Phys. Lett. 13 (1964) 168.
[2] S. Weinberg, Phys. Rev. Lett 19 (1967) 1264.
[3] G. t'Hooft, Phys. Lett. 37B (1971) 195.

OBSERVATION OF NEUTRINO-LIKE INTERACTIONS WITHOUT MUON OR ELECTRON IN THE GARGAMELLE NEUTRINO EXPERIMENT

F.J. HASERT, S. KABE, W. KRENZ, J. Von KROGH, D. LANSKE, J. MORFIN,
K. SCHULTZE and H. WEERTS

III. Physikalisches Institut der Technischen Hochschule, Aachen, Germany

G.H. BERTRAND-COREMANS, J. SACTON, W. Van DONINCK and P. VILAIN[*1]

Interuniversity Institute for High Energies, U.L.B., V.U.B. Brussels, Belgium

U. CAMERINI[*2], D.C. CUNDY, R. BALDI, I. DANILCHENKO[*3], W.F. FRY[*2], D. HAIDT,
S. NATALI[*4], P. MUSSET, B. OSCULATI, R. PALMER[*4], J.B.M. PATTISON,
D.H. PERKINS[*6], A. PULLIA, A. ROUSSET, W. VENUS[*7] and H. WACHSMUTH

CERN, Geneva, Switzerland

V. BRISSON, B. DEGRANGE, M. HAGUENAUER, L. KLUBERG,
U. NGUYEN-KHAC and P. PETIAU

Laboratoire de Physique Nucléaire des Hautes Energies, Ecole Polytechnique, Paris, France

E. BELOTTI, S. BONETTI, D. CAVALLI, C. CONTA[*8], E. FIORINI and M. ROLLIER

Istituto di Fisica dell'Università, Milano and I.N.F.N. Milano, Italy

B. AUBERT, D. BLUM, L.M. CHOUNET, P. HEUSSE, A. LAGARRIGUE,
A.M. LUTZ, A. ORKIN-LECOURTOIS and J.P. VIALLE

Laboratoire de l'Accélérateur Linéaire, Orsay, France

F.W. BULLOCK, M.J. ESTEN, T.W. JONES, J. McKENZIE, A.G. MICHETTE[*9]
G. MYATT[*] and W.G. SCOTT[*6],[*9]

University College, London, England

Received 25 July 1973

Events induced by neutral particles and producing hadrons, but no muon or electron, have been observed in the CERN neutrino experiment. These events behave as expected if they arise from neutral current induced processes. The rates relative to the corresponding charged current processes are evaluated.

We have searched for the neutral current (NC) and charged current (CC) reactions:

$$NC \quad \nu_\mu / \bar{\nu}_\mu + N \to \nu_\mu / \bar{\nu}_\mu + \text{hadrons}, \qquad (1)$$

$$CC \quad \nu_\mu / \bar{\nu}_\mu + N \to \mu^-/\mu^+ + \text{hadrons} \qquad (2)$$

which are distinguished respectively by the absence of any possible muon, or the presence of one, and only one, possible muon. A small contamination of $\nu_e / \bar{\nu}_e$ exists in the $\nu_\mu / \bar{\nu}_\mu$ beams giving some CC events which are easily recognised by the e^-/e^+ signature. The analysis is based on 83 000 ν pictures and 207 000 $\bar{\nu}$ pictures taken at CERN in the Gargamelle bubble chamber filled with freon of density 1.5×10^3 kg/m^3 [*]. The dimensions of this chamber are such that most

[*1] Chercheur agréé de L'Institut Interuniversitaire des Sciences Nucléaires, Belgique.
[*2] Also at Physics Department, University of Wisconsin.
[*3] Now at Serpukhov.
[*4] Now at University of Bari.
[*5] Now at Brookhaven National Laboratory.
[*6] Also at University of Oxford.
[*7] Now at Rutherford High Energy Laboratory.
[*8] On leave of absence from University and INFN-Pavia.
[*9] Supported by Science Research Council grant.

[*] A more detailed account of the analysis of this experiment appears in a paper to be submitted to Nuclear Physics.

hadrons are unambiguously identified by interaction
or by range-momentum and ionisation. Any track
which could possibly be due to a muon has consigned
the event to reaction (2).

Analysis of the signal. To estimate the background
of neutral hadrons coming from neutrino interactions
in the shielding and simulating reaction (1). events
where a visible charged current interaction produces
an identified neutron star in the chamber (associated,
AS, events) were also studied. To obtain a good estimate of the true neutral hadron direction from the
direction of the observed total momentum a cut in
visible total energy of `>` 1 GeV was applied to the
NC and AS events, as well as to the hadronic part of
the CC events.

We have observed, in a fiducial volume of 3 m^3,
102 NC, 428 CC and 15 AS in the ν run and 64 NC,
148 CC and 12 AS in the $\bar{\nu}$ run. Using these numbers
without background substraction the ratios NC/CC
are then 0.24 for ν and 0.42 for $\bar{\nu}$, whilst the NC/AS
ratios are 6.8 and 5.3 respectively.

The spatial distributions of the NC events have
been compared to those of the CC events and found
to be similar. In particular, the distribution along the
beam direction of NC (fig. 1) has the same shape as
the CC distribution. In contrast the observed distribution of low energy neutral stars shows a typical exponential attenuation as expected for neutron background. The distributions of radial position, hadron
total energy, and angle between measured hadron total momentum and beam direction are also indistinguishable for NC and CC.

Using the direction of measured total momentum
of the hadrons in NC and CC events, a Bartlett method has been used to evaluate the apparent interaction
mean free paths, λ_a, for NC and CC which are found
to be compatible with infinity. For the NC events we
find λ_a `>` 2.6 m at 90% CL; this corresponds to 3.5
times the neutron interaction length for high energy
(`>` 1 GeV) inelastic collisions in freon.

Evaluation of the background. Since the outgoing
neutrinos cannot be detected in reaction (1), the NC
events may be simulated by neutral hadrons coming
from the ν beam or elsewhere.

As a check for cosmic ray origin, the up-down
asymmetries of NC events in vertical position and momenta have been measured and found to be $(3 \pm 8)\%$
and $(-8 \pm 8)\%$ respectively. In addition, a cosmic ray

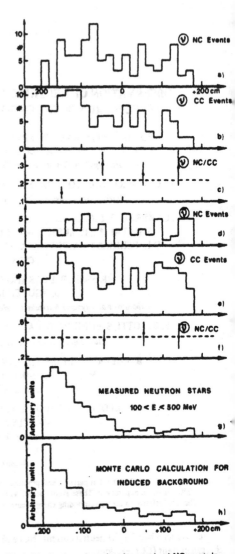

Fig. 1. Distributions along the ν-beam axis. a) NC events in ν.
b) CC events in ν (this distribution is based on a reference
sample of ~ 1/4 of the total ν film). c) Ratio NC/CC in ν
(normalized). d) NC in $\bar{\nu}$. e) CC events in $\bar{\nu}$. f) Ratio NC/CC
in $\bar{\nu}$. g) Measured neutron stars with $100 < E < 500$ MeV
having protons only. h) Computed distribution of the background events from the Monte-Carlo.

exposure of 15 000 pictures shows no NC type event satisfying the selection criteria. We conclude that the cosmic background is negligible.

The low energy muons (< 100 MeV/c) captured at rest in the ν run could be mistaken as protons. A study of the observed muon spectrum in CC events, as well as a theoretical estimate of the low end of this spectrum shows that the correction to be applied is 0 ± 5 events.

Interactions of neutral hadrons produced by the primary protons up to and including the target should produce events at an equal rate in ν and $\bar{\nu}$ runs. On the contrary, we observe an absolute rate 4 times larger in the ν run than in the $\bar{\nu}$ run. If the neutral hadrons are due to defocussed secondary pions and kaons, the disagreement is larger since we expect $1-2$ times more events in $\bar{\nu}$ than in ν. Since the whole installation is shielded from below by earth we should again expect up-down asymmetries in the NC events. This is not observed.

The most important source of background is the interaction of neutral hadrons produced by the undetected neutrino interactions in the shielding. The high elasticity (0.7) of the neutrons causes a cascade effect in propagation through the shielding. The neutron energy spectrum at production can, in principle, be obtained from the AS events together with available nucleon-nucleus data. Due to the limited statistics in the AS events we make the extreme assumption that all the NC events are neutron produced an use their observed energy spectrum to calculate the neutron spectrum from neutrino interactions. This gives an energy dependence described by E^{-2}. The effective interaction length λ_e of neutrons in the shielding is then found to be 2.5 times the inelastic interaction length, λ_i. A smaller effective interaction length is found for K_L^0 although the background from this source must be negligible since we find no examples of Λ^0 hyperon production among the NC events.

From the absolute value of the number of AS events, we can calculate the number of background events. This has been done by Monte-Carlo generation of events in the shielding surrounding the fiducial volume according to the radial intensity distribution of the beam. The ratio of background events (B) to AS events is found to be B/AS = 0.7 for $\lambda_e = 2.5 \lambda_i$.

If the NC sample has to be explained as being entirely due to neutral hadrons, the Monte-Carlo requires $\lambda_e / \lambda_i > 10$, instead of the best estimate of 2.5. Both ratios would predict distributions along the beam direction in the chamber in strong disagreement with those observed.

Another evaluation of this type of background has been made using the simple assumption that an equilibrium of neutral hadrons with neutrinos exists throughout the entire chamber/shielding assembly. For a radially uniform ν flux it gives B/AS < 1.0 which confirms the Monte-Carlo prediction.

Conclusion. We have observed events without secondary muon or electron, induced by neutral penetrating particles. We are not able to explain the bulk of the signal by any known source of background, unless the effective interaction length of neutrons and K_L^0 is at least 10 times the inelastic interaction length. These events behave similarly to the hadronic part of the charged current events. They could be attributed to neutral current induced reactions, other penetrating particles than ν_μ and ν_e, heavy leptons decaying mainly into hadrons, or by penetrating particles produced by neutrinos and in equilibrium with the ν beam.

On subtraction of the best estimate of the neutral hadron background, and taking into account the $\nu(\bar{\nu})$ contamination in the $\bar{\nu}$ (ν) beam, our best estimates of the NC/CC ratios are

$$(NC/CC)_\nu = 0.21 \pm 0.03$$

$$(NC/CC)_{\bar{\nu}} = 0.45 \pm 0.09$$

where the stated errors are statistical only. If the events are due to neutral currents, these two results are compatible with the same value of Weinberg parameter, $\sin^2 \theta_W$ [1–3] in the range 0.3 to 0.4.

References

[1] S. Weinberg, Phys. Rev. D5 (1972) 1412.
[2] A. Pais and S.B. Treiman, Phys. Rev. D6 (1972) 2700.
[3] E.A. Paschos and L. Wolfenstein, Phys. Rev. D7 (1973) 91.

Observation of Muonless Neutrino-Induced Inelastic Interactions

A. Benvenuti, D. C. Cheng,* D. Cline, W. T. Ford, R. Imlay, T. Y. Ling, A. K. Mann, F. Messing,
R. L. Piccioni, J. Pilcher,† D. D. Reeder, C. Rubbia, R. Stefanski, and L. Sulak

*Department of Physics, Harvard University,‡ Cambridge, Massachusetts 02138, and Department of Physics,
University of Pennsylvania,‡ Philadelphia, Pennsylvania 19174, and Department of Physics, University of
Wisconsin,‡ Madison, Wisconsin 53706, and National Accelerator Laboratory, Batavia, Illinois 60510*
(Received 3 August 1973)

We report the observation of inelastic interactions induced by high-energy neutrinos
and antineutrinos in which no muon is observed in the final state. A possible, but by no
means unique, interpretation of this effect is the existence of a neutral weak current.

We report here additional results of our study[1] of high-energy neutrino interactions produced by the broad-band unfocused neutrino beam of the National Accelerator Laboratory. In this note we concentrate on reactions which are distinguished from the "ordinary" processes, $\nu_\mu(\bar{\nu}_\mu)+$ nucleon $\to \mu^-(\mu^+)+$ hadrons, by the absence of a muon in the final state. Data obtained with proton energies $E_p = 300$ and 400 GeV are presented here.

The experimental setup is shown in Fig. 1(a). The light from each of the sixteen segments of the target-detector was collected to generate a pulse proportional to the energy deposited in each segment.[2] The sixteen signals were combined to generate an event trigger whenever the total energy[3] exceeded a specific threshold (6 GeV at $E_p = 300$ GeV, 12 GeV at $E_p = 400$ GeV). The detector was gated on during two equal periods, one coincident with the machine burst, the other delayed to detect cosmic-ray events exclusively. At $E_p = 300$ GeV (400 GeV) the effective beam-spill duration was 100 μsec (15 μsec). For each

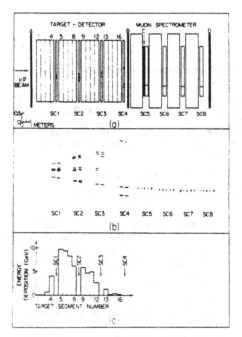

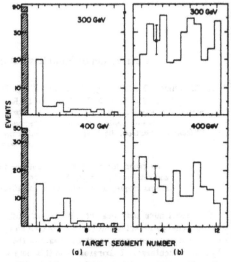

FIG. 2. (a) Distributions of event vertices along the neutrino beam path for events in which counter A fired. The crosshatched bins contain all events with a vertex upstream of the detector. (b) Events that did not fire A.

FIG. 1. (a) Plan view of experimental apparatus. The target-detector consists of liquid-scintillator segments (1–16) with wide-gap spark chambers (SC1–SC4) interspersed, each with two gaps. The muon spectrometer consists of four magnetized iron toroids whose axes coincide with the beam line. After each toroid are narrow-gap spark chambers (SC5–SC8) each with six gaps. Auxiliary scintillation counters are labeled A, B, C, and D. A typical inelastic neutrino event with an associated muon is sketched into the spark chambers. Its enlarged photograph appears in (b) and the energy deposition in each segment is shown in (c).

event we triggered all spark chambers and recorded on magnetic tape the energy deposited in each target segment, and the pattern and relative timing of the auxiliary scintillation counters A, B, C, and D [Fig. 1(a)].

The subdivision of the target-detector and the good multispark efficiency of the wide-gap chambers, coupled with the high multiplicity of the interactions, led to unambiguous identification of the hadron cascade. The position of the vertex of the primary interaction was determined by the distribution of the energy depositions in the target-detector and by extrapolation of the spark-

chamber tracks with a precision of ± 5, ± 15, and ± 25 cm in the vertical, horizontal, and longitudinal coordinates, respectively.

Results are based on 1116 triggers at $E_\nu = 300$ GeV and 368 triggers at 400 GeV. None of the 92 triggers within the cosmic-ray gate had a clear vertex within or just outside the target volume. The longitudinal distribution of vertices for events within the beam gate and with a good vertex is shown in Fig. 2, divided into two groups depending on whether or not counter A fired. As Fig. 2(a) shows, the majority of events with a pulse in A have vertices either upstream of or in the first segment of the target-detector. The vertices of those events occurring inside the target are attenuated in 1.5 target segments, about 1 strong-interaction absorption length. The residual flat contribution is consistent with accidental counts in A. Vertices for events in which A did not fire [Fig. 2(b)] have a distribution uniform over 7 absorption lengths, indicating that interactions of neutrons entering the front of the apparatus are insignificant.

The vertex distribution in the plane perpendicular to the beam is shown in Fig. 3 for events with no count in A. Hadrons entering from the

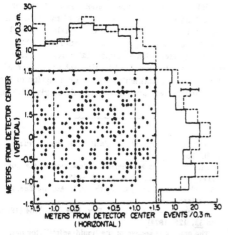

FIG. 3. Distribution of event vertices in the plane perpendicular to the beam (data for E_p = 300 GeV only). Filled circles depict events with a muon and open circles events without a muon. Projections onto the horizontal and vertical axes represent muon events by solid lines and muonless events by dashed lines. Vertices within the dashed square satisfy the final fiducial-volume requirement.

sides would have a transverse attenuation length equal to 1 absorption length times their mean angle with respect to the beam direction. The angles of the observed hadron showers have been measured and have a distribution centered on 0 mrad projected angle with a rms deviation of 25 mrad. The absence of any substantial enhancement near the edges of the target indicates that there is no appreciable background of hadrons penetrating from the sides.

To ensure good containment of the hadronic cascade, moderately high muon detection efficien-

cy, and rejection of possible hadrons incident from the sides, fiducial-volume cuts are imposed: The vertex must occur within the first twelve target segments and be at least 0.5 m away from the edges. After these cuts the total number of events is 236 at 300 GeV and 94 at 400 GeV.

The presence of an associated muon in the spectrometer was identified independently by a count in C or by observation of a track in $SC5$ [Fig. 1(a)]. The only correction is for wide-angle muons that miss the muon identifier. The geometrical acceptance of the apparatus has been calculated using the uniform vertex distributions in longitudinal and transverse position (Figs. 2 and 3), and assuming azimuthal symmetry of the muon angular distribution; it remains substantial (≥ 10%) out to 380 mrad for events occurring in segments 5-12 of the calorimeter (and out to 600 mrad for interactions in the first section of the iron magnet which we also observe). Observe that the raw ratio of events without and with muons is higher outside the horizontal-vertical fiducial boundary (Fig. 3) than it is inside, which is consistent with the calculated dependence of the geometrical acceptance on vertex position. Furthermore, the muon detection efficiency has been obtained from two independent Monte Carlo calculations, which assume scale invariance and form factors consistent with low-energy neutrino data[4] and electroproduction data.[5] These calculations are in agreement with the measured muon angular distribution and other properties of events with muons, as reported previously.[1]

The numbers of events with and without observed muons, before and after correction for muon detection efficiency, are given in Table I. We divide the events into four sets according to whether the vertex occurs in the first or last half of the target-detector at each of the two proton energies. The four independent sets of data

TABLE I. Summary of data analysis

Target segments	Proton energy (GeV)	Visible muon events	No visible muon (1)	Undetected muon events (calc)	Excess muonless events (2)	Purity of sample (2)/(1)	Ratio of cross sections, R
1-6	300	52	59	41	18	25%	0.20 ± 0.12
	400	27	23	20	3		0.06 ± 0.14
7-12	300	72	53	28	25	50%	0.25 ± 0.10
	400	21	23	10	13		0.42 ± 0.23
7-12	300 + 400 combined	93	76	38	38	50%	0.29 ± 0.09

give consistent values for R, the corrected ratio of cross sections without and with muons.[6] Since the geometric acceptances of the two halves of the target-detector are substantially different, the internal consistency of the results gives confidence in the calculated correction for unobserved muons. For the combined segments 7-12, we obtain $R = 0.29 \pm 0.09$, where the error is statistical only. We estimate that the uncertainty in R due to a possible systematic error in the calculated detection efficiency is smaller than the statistical error.

The simplest explanation of this result is the existence of a neutral weak current. If so, our measurement is not in disagreement with the Weinberg model,[7] which predicts a value of R between 0.22 and 0.55 for our mixture of ν and $\bar{\nu}$. Muonless events have also been reported in an experiment done at CERN[8] at much lower neutrino energies. However, other origins of the effect we observe cannot as yet be excluded. Among these are (i) a ν_e ($\bar{\nu}_e$) contamination of the ν_μ ($\bar{\nu}_\mu$) beam more than an order of magnitude larger than the 2% estimated; (ii) an anomalous excess of events in the vicinity of $y = E_h/E_\nu \simeq 1$, due, for example, to the production of a new particle, which would substantially affect the correction for undetected muons; (iii) some novel process resulting from a new type of lepton. Any of these phenomena would indicate a new feature of weak interactions.

We thank the staff of the National Accelerator Laboratory and gratefully acknowledge the help of many individuals at our universities, especially William Haught, Kirk Levedahl, Alan McFarland, Wesley Smith, James Strait, Steve Summers, and Robert Wagner.

Note added in proof.—Further study of the calculated correction for undetected muon events indicates that measurement error in the transverse position of event vertices leads to an uncertainty in the boundary of the fiducial volume, which results in a small reduction of the muon detection efficiency. Reanalysis of the data with this revised detection efficiency yields $R = 0.23 \pm 0.09$. In addition, the entire experiment has been repeated with an appreciably larger muon detection efficiency and a different admixture of neutrinos and antineutrinos. A description of the new experiment and its result will be given soon.

---

*Now at the University of California, Santa Cruz, Calif. 95060.

†Alfred P. Sloan Foundation Fellow, now at the University of Chicago, Chicago, Ill. 60637.

‡Work supported in part by the U. S. Atomic Energy Commission.

[1]A. Benvenuti *et al.*, Phys. Rev. Lett. 30, 1084 (1973), and 32, 125 (1974).

[2]The energy response of the liquid-scintillator detector is uniform to ±15% over the active area of a segment, and the sensitivity is 50 keV per collected photoelectron. A minimum-ionizing particle is well resolved in each segment. A detailed report on the detector is in preparation.

[3]The visible energy E_{vis} has been measured relative to the energy deposition of a minimum-ionizing particle. Because of scintillator saturation and nuclear-binding losses we expect $E_{vis} \sim 0.7E_h$, where E_h is the energy of the final-state hadrons.

[4]T. Eichten *et al.*, Phys. Rev. Lett. 46B, 274, 281 (1973).

[5]We use an empirical fit to the data of G. Miller *et al.*, Phys. Rev. D 5, 528 (1972); V. Barger, private communication.

[6]The ratio R is averaged for neutrino and antineutrino contributions in which the final muons are in the ratio $\mu^+/(\mu^+ + \mu^-) = 0.17 \pm 0.02$. The ratio of antineutrinos to neutrinos is then 0.59 (see Ref. 1).

[7]S. Weinberg, Phys. Rev. D 5, 1412 (1972), and Phys. Rev. Lett. 19, 1264 (1967); A. Salam and J. C. Ward, Phys. Lett. 13, 168 (1961); A. Pais and S. B. Treiman, Phys. Rev. D 6, 2700 (1972); E. A. Paschos and L. Wolfenstein, Phys. Rev. D 7, 91 (1973); C. H. Albright, NAL Report No. NAL-PUB-73/23-THY, 1973 (unpublished); L. H. Sehgal, Nucl. Phys. B65, 141 (1973).

[8]F. Hasert *et al.*, Phys. Lett. 46B, 138 (1973).

Further Observation of Muonless Neutrino-Induced Inelastic Interactions*

B. Aubert,† A. Benvenuti, D. Cline, W. T. Ford, R. Imlay, T. Y. Ling, A. K. Mann, F. Messing,
R. L. Piccioni,‡ J. Pilcher,§ D. D. Reeder, C. Rubbia, R. Stefanski, and L. Sulak

*Department of Physics, Harvard University, Cambridge, Massachusetts 02138, and
Department of Physics, University of Pennsylvania, Philadelphia, Pennsylvania 19104, and
Department of Physics, University of Wisconsin, Madison, Wisconsin 53706, and
National Accelerator Laboratory, Batavia, Illinois 60510*
(Received 19 March 1974)

We report here additional positive results of a search for muonless neutrino- and antineutrino-induced events using an enriched antineutrino beam and a muon identifier of
relatively high geometric detection efficiency. The ratio of muonless to muon event
rates is observed to be $R = 0.20 \pm 0.05$. We observe no background derived from ordinary
neutrino or antineutrino interactions that is capable of explaining the muonless signal.

The investigation reported here is a search for inelastic neutrino and antineutrino interactions at a mean energy of 40 GeV that differ from the usual processes by the absence of a muon in the final state.[1] A previous search for such events yielded a positive signal. Muonless events have also been reported in a CERN-Gargamelle experiment at a mean energy in the range of 2 to 4 GeV.[2]

The experiment was carried out at the National Accelerator Laboratory, where collisions of 300-GeV protons with an aluminum target produced secondary hadrons that were focused by a single magnetic horn to provide a beam enriched in antineutrinos.[3] The experimental apparatus [Fig. 1(a)] is a modified version of the arrangement described previously.[1,4] The modifications are (1) the addition of 35 cm of iron immediately downstream of the ionization calorimeter to form a muon identifier (μ_1) consisting of counter B and spark chamber 4, and (2) doubling of the area of counter C and replacement of the 5.3-m^2 narrow-gap spark chambers in the magnetic spectrometer with 8.4-m^2 wide-gap chambers to increase the solid angle of the second (original) muon

identifier (μ_2).

The experiment was triggered by the deposition of energy (E) in the ionization calorimeter greater than a preset minimum value,[5] with counter A in anticoincidence. The hadron cascade of an actual event is illustrated in Fig. 1(b) and the track pattern observed in spark chambers SC1-SC8 is illustrated in Fig. 1(c) which shows one of the three stereoscopic views of the event. Events were verticized by extrapolation of tracks in the two ± 7.5° stereo views and in the 90° stereo view, and consistency was required between the z position obtained from the visual reconstruction of the vertex and the calorimeter pulse-height information.

The z dependence of all triggers was observed to be uniform except for a small excess of events in the first segment (module) of the target detector, which is consistent with the small (5%) geometrical inefficiency of counter A. There is no evidence of neutrons or photons, unaccompanied by charged particles, entering the front of the target. Since each segment of the calorimeter corresponds to approximately 0.6 nuclear collision lengths, to provide additional protection

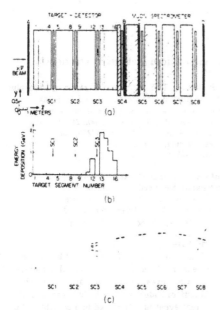

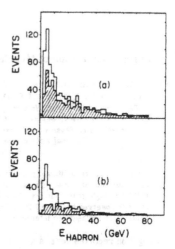

FIG. 2. (a) The E_H distribution of all $\overline{A}EB$ triggers (951) and of the corresponding verticized events (627, cross-hatched). (b) The E_H distribution of all $\overline{A}E\overline{B}$ triggers (459) and the corresponding verticized events (173, cross-hatched).

FIG. 1. (a) Plan view of the modified experimental apparatus. The target detector consists of liquid scintillator segments (1–16) with wide-gap spark chambers (SC1–SC4) interspersed. The muon spectrometer includes SC5–SC8. Auxiliary scintillation counters are A, B, C, and D. An inelastic neutrino event with an associated muon is sketched into the spark chambers (c) and the energy deposition in each segment is shown in (b).

from penetrating hadrons, events occurring upstream of module 5 were excluded. A further restriction, $z \lesssim 12$, was imposed to reserve 4 modules for detection of the hadronic cascade and as additional absorber for μ_1.

Events can be tentatively classified as having a muon or not according to the response of scintillation counter B, and are labeled $\overline{A}EB$ and $\overline{A}E\overline{B}$, respectively. In Fig. 2 are shown the distributions in hadron energy, E_H, for the 1410 triggers in the z fiducial region. The shaded histograms show those events for which a hadron shower with a reconstructed vertex was present in the spark chambers. Except for very low-energy $\overline{A}E\overline{B}$ triggers, about two-thirds of the events have a reconstructed vertex.[6] The remainder can be accounted for in the main by neutrino interactions in the calorimeter beyond the spark

chamber boundaries. The larger excess of low-energy $\overline{A}E\overline{B}$ triggers appears to be diffuse showers that originate outside the target, of which only traces are seen in the spark chambers.

The distribution of shower vertices in the transverse plane is nearly uniform, with a small depletion toward the target boundary consistent with the neutrino beam rms radius of ~1.5 m. A fiducial boundary limit of ±120 cm in x and y was imposed to avoid regions of low detection efficiency and to insure complete rejection of hadrons entering the sides.[7] A total of 535 events passed the threshold on E_H and satisfied the criteria on vertex location, event quality, and time of occurrence within the beam gate.

To use counter B or SC4 as a muon identifier it is necessary to measure the probability ϵ, that hadrons penetrate the μ_1 absorber. A sample of events with muons identified by the counter configuration $\overline{A}EBC$ were used to measure the penetration (punch-through) probability of the accompanying hadrons as a function of the z position of the event vertex [Fig. 3(a)], and as a function of the energy of the hadronic shower [Fig. 3(b)]. The shapes of these dependences are consistent with other measurements[8] of hadron penetration as indicated in Figs. 3(a) and 3(b).

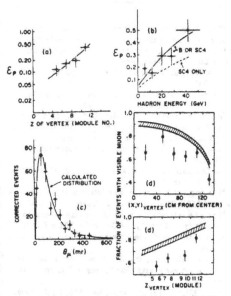

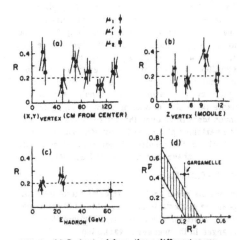

FIG. 4. (a) R obtained from three different muon identifiers as a function of the transverse distance from the center of the calorimeter. (b) The z variation of R obtained using three different muon identifiers. (c) The E_H variation of R from three muon identifiers. (d) The allowed region of R^ν and $R^{\bar\nu}$ from this experiment compared with R^ν and $R^{\bar\nu}$ obtained in the CERN measurement (Ref. 2).

FIG. 3. (a) The measured punch-through probability of hadrons accompanying $\overline{A}EBC$ events (for all hadron energies) as a function of z, and the expected shape of the distribution. (b) The measured punch-through probability (for z between 5 and 12) as a function of E_H, compared with the expected variation. (c) The corrected muon angular distribution measured in SC4 compared with the predicted distribution. (d) Comparison of the observed fraction of events with a muon for the μ_1' identifier (SC4 alone) and ϵ_μ as functions of transverse position and z position. The cross-hatching indicates the uncertainty in ϵ_μ arising from the statistics of the data in (c).

The angular distribution of muons[9] identified by a spark in SC4 (for about $\frac{2}{3}$ of the muon sample) is shown as the histogram in Fig. 3(c), after correction for the geometric acceptance of SC4. The geometric acceptance of SC4 is calculated using only the observed distributions of event vertex positions and assuming azimuthal symmetry of the primary neutrino interaction. The raw data divided by the calculated acceptance yield directly the intrinsic muon angular distribution in Fig. 3(c). Calculations which use the measured neutrino-antineutrino spectrum and a model of the interaction dynamics[4] yield the intrinsic muon angular distribution shown by the solid curve in Fig. 3(c). We conclude from the good agreement in Fig. 3(c) that these latter calcula-

tions reproduce the observed angular distribution out to the maximum detected angle of 500 mrad. Only 4% of the muons are predicted to lie at angles greater than 500 mrad. The muon detection efficiency ϵ_μ is then obtained from the measured angular distribution, corrected for the loss of events with $\theta_\mu > 500$ mrad.[10]

The ratio of muonless events to events with muons is obtained from the formula

$$R = \frac{[\epsilon_\mu + \epsilon_p - \epsilon_\mu \epsilon_p](1 + R_m) - 1}{1 - \epsilon_p(1 + R_m)}, \qquad (1)$$

where R_m is the measured ratio of events without and with a count in a given muon identifier. In Fig. 3(d) we plot the observed fraction of events with a muon as a function of z and of (x, y), after correction for hadron punch-through, which should be equal to ϵ_μ if $R = 0$ [Eq. (1)]. The data were obtained using SC4 alone as a muon identifier (μ_1'), and are to be compared with ϵ_μ, shown as the cross-hatched regions in Fig. 3(d). A clear discrepancy is indicated, providing evidence that R is different from zero.

Figures 4(a), 4(b), and 4(c) present R as a function of (a) the transverse position of the event vertex, (b) the z position of the event vertex, and (c) the hadron energy E_H. We have included in Fig. 4 results for SC4 alone (μ_1'), as well as for μ_1 and μ_2. The best solid angle is achieved with μ_1, but for μ_1' the measured values of ϵ_s are smaller than for μ_1 [Fig. 3(b)], and for μ_2, $\epsilon_s = 0$. Furthermore, the dependence on z, (x, y), and E_H of the ϵ_μ for μ_1, μ_1', and μ_2 is significantly different, which serves to test the internal consistency of the data.[11] Apart from statistical fluctuations, Fig. 4 indicates that the same value of R is obtained from each of the muon identifiers over the entire range of each of the variables plotted and provides evidence for stability of the results. These results integrated over z, transverse position, and E_H yield

$$R = 0.20 \pm 0.05,$$

where the error includes an estimate of 0.03 for possible systematic effects, which follows from an exhaustive study of the sensitivity of R to changes in the measured variables.

The value of R measured for the combined neutrino-antineutrino beam used in this experiment is related to the values of R^ν and $R^{\bar\nu}$, for pure neutrinos and antineutrinos, respectively, by

$$R = aR^\nu + (1 - a)R^{\bar\nu},$$

where $a = 0.63 \pm 0.11$ is the corrected observed ratio of the negative muon event rate to the total muon event rate. The allowed values of R^ν and $R^{\bar\nu}$ are presented in Fig. 4(d) which also shows the recent results for R^ν and $R^{\bar\nu}$ from Gargamelle.[2]

It is a pleasure to acknowledge the aid and encouragement of the National Accelerator Labora-

tory staff and the efforts of Robert Beck and Hans Weeden.

---

*Work supported in part by the U. S. Atomic Energy Commission under Contract No. AT(11-1)-881-401.

†On leave of absence from Laboratoire de L'Accelerateur Lineaire, Orsay, France.

‡Now at Stanford Linear Accelerator Center, Stanford, Calif. 94305.

§Alfred P. Sloan Foundation Fellow, now at the University of Chicago, Chicago, Ill. 60637

[1]A. Benvenuti *et al.*, Phys. Rev. Lett. **32**, 1025 (1974).

[2]F. J. Hasert *et al.*, Phys. Lett. **46B**, 138 (1973), and CERN Report No. TC-L/Int. 74-1 (to be published).

[3]The horn was constructed under the supervision of F. Nezrick.

[4]A. Benvenuti *et al.*, Phys. Rev. Lett. **30**, 1084 (1973), and **32**, 125 (1974).

[5]The triggering circuitry was determined from the data to be fully efficient for hadron energy $E_H > 4$ GeV.

[6]For $\bar{A}E\bar{B}$ triggers the ratio of verticized events to all events is 0.22 for $E_H < 12$ GeV and 0.57 for $E_H > 12$ GeV, for $\bar{A}EB$ events the corresponding ratios are 0.57 and 0.74.

[7]As Fig. 3(d) shows, the data do not indicate any excess of muonless events near the edges (142 cm) of the detector.

[8]R. W. Ellsworth *et al.*, Phys. Rev. **165**, 1449 (1968).

[9]The angle of the muon is determined from the angle of the 10-cm-long track segment in SC4 and from the line joining the track centroid to the shower vertex.

[10]The detection efficiency ϵ_μ includes a small correction ($\sim 0.3\%$) for muons within the angular acceptance that range out.

[11]For muon identifiers μ_1, μ_1', and μ_2 the average ϵ_μ are 0.89, 0.81, and 0.79.

THE DISCOVERY OF NEUTRAL CURRENTS

Peter Galison
Program in History of Science; Department of Physics and
Department of Philosophy, Stanford University, Stanford CA 94305

For the better part of the 1970's and 1980's, gauge theories
gripped the imagination of the physics community the way Newton's
gravity fascinated natural philosophers in the early eighteenth
century. In both periods the new theory opened a great number of
new phenomena to calculation; gauge unification of the
electromagnetic and weak interactions promised to constitute a
template for further unification of the remaining forces just as
central force laws in the eighteenth centuries held, for thinkers
like Laplace, the hope that all other forces could be unified.

By the mid-1980's the "new" gauge physics had already exer-
ted a sufficient influence to justify historical analysis. Physicists had written scores of textbooks, produced thousands of
articles and organized tens of conferences. Accelerator
designers had grounded plans for billion-dollar machines on its
basis, and the Nobel Prize Committee had several times recognized
contributions to the experimental and theoretical development of
gauge theories. In an important historical sense it does not
matter what the eventual fate of the theories are. It may be
that, like Newtonian gravity, gauge theories will be replaced by
a theory as different from the physics of the 1970's and 1980's
as the inverse square law is from curved space. Or it may come
to pass that the gauge theories will prove to be but low rungs on
a long hierarchical ladder of other effective gauge theories.

Given the importance of the theoretical shift that has
accompanied gauge physics, a natural historical inquiry would
trace its conceptual history. Such an investigation would (aside
from its recentness) fit easily into a long tradition in the
history of physics: examine the history of theory and then
survey its experimental "confirmations." In the case of the
unified gauge theory of the weak and electromagnetic
interactions, the first experimental check would be of neutral
currents. After all, first the experimental prediction seems
straightforward: can neutrinos scatter without turning into
charged particles -- that is, do neutral currents exist? The old
theory of the weak interactions said no, the new theory said yes.

In this talk I will sketch the history not from the theorists' perspective but from the experimentalists', a history from
below, not from above. I will focus on two experimental groups,
the Gargamelle collaboration which found neutrinos at CERN and
the E1A collaboration which measured theirs at Fermilab. Though
a fuller discussion is presented elsewhere [1], I hope here to
communicate a sense of the experimentalists' reasoning. It is
too simple to view the experiments purely as an outgrowth of the

electroweak theory. This was <u>not</u> the case. Theorists and
experimentalists did not suddenly turn to the gauge physics as a
result of the Glashow-Weinberg-Salam model. E1A and Gargamelle
began in the "old" physics and only gradually traversed the
conceptual and practical boundaries that marked the beginning of
the "new" physics. Changes in interpretive techniques,
experimental goals and even in the physical apparatus, reflected
the difficulties of effecting this transition. Like rocks caught
in the fault line between two tectonic plates, these experiments
exhibited the strains and stresses of having been pulled in two
directions. Fault geology teaches the geologist much about both
plates; following the analogy, these experiments have a great
deal to tell us about physics both before and after the "gauge
revolution."

By following the progress of the two experiments we can see
how evidence came to persuade the experimentalists themselves
that they were peering at a real effect, not an artifact of the
machine or of the environment. We can also see how evidence was
mustered for presentation to the community at large. Finally, the
case study of this discovery offers a glimpse at how a modern
particle physics experiment proceeds when the theoretical stakes
are very high.

Before continuing, it is worthwhile to review briefly some
of the governing assumptions of the "old" weak interaction
theory of the 1960's. By recalling the basic tenets of the
earlier physics, we can understand why E1A and Gargamelle were
built, and why they undertook their original physics program.
More importantly, the full drama of the neutral current work will
only be apparent when set against the routine assumptions of the
older weak interaction theory.

Of all weak interaction processes, the best known (indeed
the prototype for all other weak interactions) was beta decay: a
neutron disintegrates into a proton, an electron and an
antineutrino. In 1932 Fermi had put forward a quantum field
theory description of this process that had the great virtue of
accounting for many experimental results, while avoiding any
explanation of the internal mechanism of the decay.[2] Arguing
along a line analogous to Fermi's it was possible to describe
related weak interactions such as inverse beta decay, in which a
neutrino and neutron interacted to produce a proton and an
electron. The two processes, with their "black box" interactions,
are depicted in Figs. 1a, and 1b.

Throughout the 1940's and 1950's, physicists refined Fermi's
black box treatment into a successful phenomenological
description of the weak interactions of many particles. More
specifically, theorists came to view all weak interactions as the
product of two currents, by analogy with the interaction of two

electrons in which each constituted a microscopic electric
current through space. In this language, Fig. 1b depicts the
interaction of a current of strongly interacting, heavy
particles (the hadronic current) with a current composed of
light particles immune to the strong force (leptonic current).
One could also have weak interactions involving two hadronic
currents, or two leptonic currents.

When currents keep the same charge, as they do in the
interaction of two electrons, they are called neutral-currents,
of which the electromagnetic interaction of figure 2b is an
example. When the currents change charge as in 1a and 1b the
currents are dubbed charged-currents. Before the work of E1A and
Gargamelle, all experimentally observed weak interactions wi1e
thought to have been composed of weak charged-currents. For our
purposes, the significant point is that both E1A and Gargamelle
initially had as a principal goal the exploration of charged-
current interactions in more detail than previously. It was not
the only purpose, however.

Since the time of Yukawa a great number of theorists had
speculated that charged-current interactions could be better
understood as the result of the exchange of an intermediate
particle, the singly charged W. Yukawa and his many followers
were impressed by the possibility of imitating quantum
electrodynamics. In that theory the electromagnetic force is
explained as resulting from the exchange of a photon between
currents of charged particles. Analogously, in the weak
interactions a charged particle would be emitted by one current
and absorbed by another. (See Figs. 2a, 2b.) Possibly, the
theorists reasoned, this could make the high energy behavior of
the theory sensible; the old Fermi theory only yielded finite
predictions at relatively low energies. In its most obvious
application, the W particle would break up beta decay into two
subprocesses. First the neutron would turn into a proton and
emit a negatively charged W, then the negatively charged W would
decay into an electron and an antineutrino.

Beginning in earnest in the 1960's a host of experiments and
theories sought to define more precisely the properties of the
elusive W. For the experimentalists, the most promising route
appeared to be the following: protons would be rammed into a
target producing assorted particles including pions and muons.
Most particles would be filtered out, leaving a beam of pions and
muons. The muons would be stopped by large quantities of earth,
and the pions would decay into a beam of neutrinos. The hope was
that this beam of neutrinos would produce a charged W and either
an electron or a muon. E1A and Gargamelle were but two in a long
series of neutrino experiments that were designed to use this
technique to make W's. If the accelerators were not powerful
enough to produce the W, then the experimentalists hoped at least

to set lower bounds on the W's mass. To the extent that most
physicists considered it even a possibility that weak neutral-
currents existed, they expected that such effects would be tiny
compared to charged particle interactions.

Gargamelle's proximate origin lies in a letter written by
André Lagarrigue, then an accomplished bubble chamber physicist
at Ecole Polytechnique in Paris, to Victor Weisskopf, CERN's
Director General.[3] With Polytechnique's successful bubble
chamber program behind him, Lagarrigue could offer CERN a proven
collaboration of engineers, technicians and physicists. Moreover,
the Polytechnicien could set out a financial offer that was
 difficult to refuse: the French government would pay a large
fraction of the costs and there was at least an indication that
Italian and English laboratories were interested in the project.
CERN would still have to bear the fiscal responsibility of
installing and running the chamber, and most importantly the CERN
administration would have to balance Gargamelle's appetite for
beam time against the needs of competing detector groups.
Weisskopf responded by reminding Lagarrigue that the proposal
would have to pass through the usual channels.[4]

Gargamelle was to be huge, ten times larger than the biggest
existing heavy liquid chamber. Construction problems occurred on
a commensurate scale. Even the administration of a detector of
this size was formidable. In mid - 1966 an Orsay engineer, Jean
Lutz was put in charge of the project, with Paul Musset as deputy
for physics and experimentation. Another deputy held
responsibility for coordination and planning, while Lagarrigue
remained scientific advisor with André Rousset as his assistant.
Below this administrative level stood fourteen principal groups,
each of which had responsibility for a special project. Any one
of these tasks was, in its own right, a substantial endeavor
involving coordination with other projects, contractors and
subcontractors. Separate groups aimed at the assembly of the
magnet, chamber body, expansion and piping, optics, lighting,
cameras, electronics, thermal regulation, command and control,
safety, study bureau, supplies, installation at Saclay (for
tests) and installation at CERN.[5]

When planning began for the physics to be done with
Gargamelle, neutral-currents, when they were mentioned at all,
were left as a distinctly low priority. Neither theorists nor
experimentalists had much use for the neutral-currents which, if
they existed, should be characterized by particle interactions
differing from the typical beta-decay -like processes. In beta-
decay the charges on the leptons (electron and neutrino) add up
to one. In a weak neutral-current decay, one should be able to
have the pairs of leptons adding to zero charge, for example two
neutrinos or an electron and a positron. If there were a neutral
weak intermediate particle such a process would clearly be

possible. See Figs. 3a, 3b. Such phenomena were not observed. "It
is a remarkable fact," one author wrote in 1964, "without known
exception that the two leptons in a weak current always consist
of a charged and a neutral particle... this could only imply
that neutral [weak] currents must be absent."[6]

The absence of processes such as the one depicted in Figure
3 was especially striking to experimentalists. In this process
the hadronic current consists of a positive kaon which turns into
a positive pion. The charge stays the same, both are positively
charged, so the current is neutral. However, the current does
change in another respect: the kaon is a <u>strange</u> particle while
the pion is not. Consequently, the process is called
"strangeness-changing." Almost all of the evidence against
neutral-currents came from such "strangeness-changing" processes,
and for good reason. In both strong and electromagnetic
  interactions strangeness does not change, so strangness-changing
events seemed to present a perfect laboratory for the study of
neutral-currents. Now comes the essential point: in the 1960's
there seemed to be no reason to expect neutral currents to behave
fundamentally differently in the strangeness-changing and
strangeness-conserving cases. It seemed perfectly reasonable to
assume that neutral-currents -- of any type -- were ruled out.

The situation resembles that described by Locke in his
<u>Essay Concerning Human Understanding</u>. It seems that after a very
pleasant visit with the Dutch ambassador, the King of Siam
inquired after the habits of people from the Low Countries. As
King of such a great seafaring nation, the Siamese Royalty
undoubtedly felt rather familiar with water in its various forms.
One can imagine his suspicion upon being told that cold water in
the Dutchman's part of the world would occasionally become "so
hard, that men walked upon it, and that it would bear an
elephant, if he were there. To which the king replied, "<u>Hitherto
I have believed the strange things you have told me, because I
look upon you as a sober fair man</u>, but now I am sure you
<u>lie</u>."[emphasis Locke's][7] Physicists, having repeatedly
measured strangeness-changing neutral-currents' million-fold
suppression, were no more likely to expect an absence of
suppression in the strangeness-conserving case than the King of
Siam expected elephants to walk on cold water. There was simply
no good reason to believe that the distinction would make such a
difference.

Further evidence against neutral-currents came from
experiments designed to measure the scattering of neutrinos from
protons. Results in the early 1960's seemed to indicate that
neutrino scattering in which no charged particle was exchanged
comprised only a tiny fraction of similar neutrino scattering
experiments in which the neutrino changed into a charged particle
such as a muon. G. Bernardini commented on similar results in his

Fig. 1

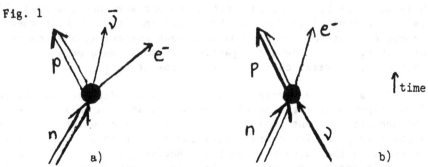

↑ time

a)

b)

Fig. 1a. Beta Decay in Fermi Theory. The black dot indicates that no internal mechanism for the interaction was presupposed.
Fig. 1b. Inverse Beta Decay in Fermi Theory.

Fig. 2

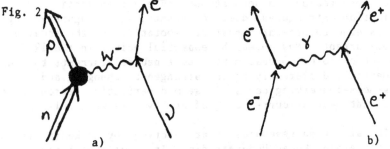

a)

b)

Fig. 2a. Beta Decay with Hypothetical W-exchange.
Fig. 2b. Electromagnetic Neutral Current. Scattering of an electron from a positron by a photon exchange.

Fig. 3

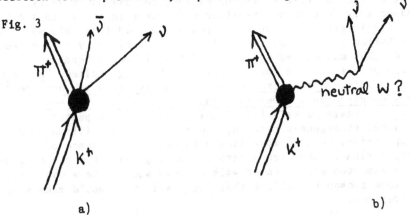

a)

b)

Fig. 3a. Strangeness-Changing Weak Neutral Current. This process in which a "strange" particle decayed into nonstrange particles was known to be extremely rare in comparison to charged-current processes such as 1a that did not change strangeness.
Fig. 3b. Same as 3a but with hypothetical neutral W exchange.

introductory lecture to the 1964 Enrico Fermi Summer School
arguing that, "neutral leptonic currents if they do exist are
coupled with hadronic currents more weakly by several orders of
magnitude than the charged ones." In a widely used textbook,
R.W.E. Marshak, Riazuddin, and C.P. Ryan included a section
entitled, "Absence of Neutral Lepton Currents," in which they
concluded that results resembling those just mentioned "support
the absence of neutral lepton (or at least neutrino) current(s)
..." As late as 1973, E. Commins referred to the selection rule
of "no neutral-currents."[8]

        To the extent that one can speak of consensus within the
physics community in the 1960's regarding the weak interactions
it would be this: the theory of weak interactions could be
written as the interaction of charged-currents. In order for
these interactions to give interpretable predictions at high
energies, an intermediate particle, the charged W, had to be
found. It was therefore natural, in February 1964, for
Lagarrigue, Musset and Rousset to use their preliminary proposal
for Gargamelle to advocate a W search. Of course there was no
assurance that the CERN accelerator would provide sufficient
 energy to find the particle. Still, as indicated by comments at
the Sienna conference of 1963, it was considered reasonable at
the time to expect the W to weigh in at a few GeV. A W that light
was within the grasp of Gargamelle operating on the CERN Proton
Synchrotron.

        Musset and Rousset discussed other projects, but even when
the builders of Gargamelle expanded their preliminary discussion
into a major project proposal in 1964, they disposed of neutral-
currents in three sentences.[9] In 1970, with Gargamelle nearing
completion, D.H. Perkins rewrote the proposed physics program
for Gargamelle.[10] At the top of his list of goals came the
hunt for the intermediate vector boson. Then came the study of
various processes predicted by the older theory of weak
interactions which did not depend on the W's existence.

        Still, not everything was the same. Since the original
project proposals had been submitted, a group at SLAC had
conducted an experiment in which they scattered electrons off
nuclei with a large transfer of momentum. Just as Rutherford had
shown that atoms contained hard nuclei by scattering alpha
particles from gold foil, so the SLAC team exhibited the inner
structure of nucleons to be granular by the scattering of
electrons from a target of nucleons. The SLAC result -- that
electron scattering was simply proportional to ordinary
Rutherford scattering -- was known as "scaling."[11] Perkins
added that tests of scaling ought to be carried out with
neutrinos in Gargamelle. SLAC had shown that the seat of
electromagnetic forces within the nucleons was in the granules
that Feynman dubbed "partons." Was the seat of the weak force

also pointlike? In addition to these exciting issues, Perkins pointed out, "there are, of course many other topics of interest, for example, neutral currents,.... However, these problems can also be investigated with [other] chambers. On the other hand Gargamelle is, we claim, a unique instrument for investigating problems like [the W and parton hypotheses]."[12]

Suddenly, in the spring of 1971, the theoretical community took notice of the almost forgotten gauge unification scheme of Glashow, Salam and Weinberg. In a _tour de force_ of mathematical physics, Gerard 't Hooft proved that the gauge symmetry of the theory guaranteed that it would be renormalizable.[13] As Sidney Coleman put it, 't Hooft's kiss tranformed Weinberg's frog into an enchanted prince.[14] SU(2) X U(1) was now as much of a theory as quantum electrodynamics. The question remained, would the predictions of the model be seen? Beginning in November 1971 the Gargamelle collaborators had contact with theorists advocating the new gauge theories and who pushed the neutral current search. Among these were Bruno Zumino, J. Prentki, and Mary K. Gaillard. Zumino explained to the experimenters why the theorists were so excited. Musset recalls being somewhat discouraged by the test the theorists advocated: scattering a muon neutrino from an electron. (See Fig. 4a.) Though extremely "clean" of background effects (because no strong interactions were involved) the cross section (or likelihood) of such an event was extremely small. By contrast the cross section for neutrino-hadron scattering ought to have been much higher. (See Fig. 4b.) For the theorists, however, the hadrons with their newly discovered partons remained an enigma: were they quarks? It seemed safer to stay away from relying on calculations based on such speculative entities.

Many of the bubble chamber experimentalists concurred with the theorists' hesitancy to concentrate on neutrino-hadron weak neutral currents, though on different grounds. Perkins and Faissner, to give but two examples, knew at first hand how treacherous it was to try to disentangle this type of neutral-current event from the background. They had tried to do so in earlier work. The problem was: theory predicted that a genuine neutral-current event would consist of a neutrino emitting a neutral exchange particle, the Z-zero, which would be received by the hadron, whereupon the hadron would fragment. Since the neutrino is uncharged, and reacts so rarely, it would leave no tracks. The hadronic fragments would deposit fan-like tracks in the liquid, all of which would stop before penetrating too far. (Hadrons are absorbed quickly since they interact by the strong force.) In addition to the fan of hadron tracks, in a charged-current event the neutrino would become a muon. Since muons do not interact strongly, their tracks would extend the length of the chamber and then exit.

Severe problems of background had plagued earlier bubble chamber attempts to examine hadronic neutral-currents. The difficulty was that in ordinary charged-current events, some of the fragments would be high energy neutrons. If the charged-current event took place outside the visible volume -- for example in the concrete shielding -- the neutrons could enter the chamber and fragment a nucleon in the visible volume. (See Fig. 5.) Since such an event would have no exiting muon, one could easily mistake its occurrence for a bona fide neutral-current event. One would have to find a way to separate the wheat from the chaff if the hadronic neutral-currents were to be the proving ground for the new unification schemes.

Among the first members of the Gargamelle collaboration to take part in the neutral-current search were Musset, (then at CERN), Antonio Pullia (from Milan), Bianca Osculati (a student from Milan), Ugo Camerini (from Wisconsin), William Fry (from Wisconsin) and Robert Palmer (from Brookhaven). Their earliest attempts to attack the neutral-current background problem lacked the force of their later demonstrations. Nonetheless, these first analyses helped shape the convictions and methods that guided subsequent stages of the experiment.

Palmer began one approach to the problem early in 1972 after discussions with Fry and Camerini, even before many measurements of the events had taken place.[15] In this work Palmer tried to establish the rough event positions within Gargamelle by using the images of the eight cameras to divide the chamber into eight zones. He could then plot the number of events as a function of distance from the front of the chamber. As an initial test of the hypothesis that the events were due to neutrinos and not neutrons, he compared his plot with one constructed from a run in which a cascade of neutrons was dumped into the bubble chamber. Both plots showed approximately the same distribution of neutral-current candidates clustered near the front end of the chamber. In short, there was no evidence for neutral-currents.

Next Palmer sought to plot only events from higher energy interactions. Real neutral-current interactions would still be plentiful at high energies whereas the background events ought be scarcer. After plotting only the high energy events, he found a more surprising result: "we note that the distribution ...shows a very significant number of events [without muons] in the downstream end. It is very different from that seen in the beam dump run [where most particles were neutrons] and very different from that expected for events from external neutrons."[16] It remained possible, however, that the events were caused not by directly entering neutrons but by neutrons produced in the walls and shielding of the chamber by neutrinos. To treat this possibility, Palmer made some very rough approximations. He supposed, for example, that the neutrinos interacted in the walls

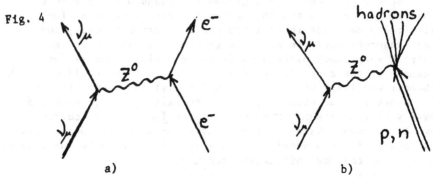

Fig. 4

a)   b)

Fig. 4a. Purely Leptonic Weak Neutral Current. Such an
interaction was "clean of background effects but was rare.
Fig. 4b. Hadronic Neutral-Current. Interactions in which the
neutrino scattered from hadrons was expected to occur often but
could not be faked by background from the process depicted
in Fig. 5.

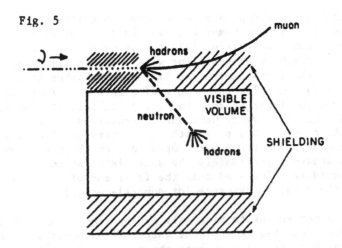

Fig. 5

Fig. 5. Neutron Background. Ordinary charged-current event in
shielding produces neutron which causes "fake" neutral-current
event in chamber.

just as they did in the Freon. Under such simplifying assumptions
he concluded that there remained a significant number of real
events, at least 56 +/- 15 of them.[17]

While Palmer was sketching his crude estimates, Musset,
Camerini and Pullia were working to develop a methodical
procedure for the scanning, measuring and data processing of the
film. At a Paris meeting on 2-3 March 1972 a few members of the
CERN collaboration gathered to discuss event rates, the program
for examining hyperons, and the establishment of subgroups.[18]
Pullia announced some preliminary results from the neutral-
current search. Cosmic rays, he concluded, could not account for
the neutral-current candidates because there were simply too many
candidates, an average of five per roll. With sixteen neutrino
rolls measured, he found that the spatial distribution of
candidates was constant along the length of the chamber. This,
Pullia surmised, "seems to exclude secondary interactions from
[charged-current] neutrino interactions in the magnet. ..."[19]
Gargamelle's keepers were still a long way from claiming that
they had seen neutral-currents.

In addition to the study of the spatial distribution of
events, Camerini, Musset, and Pullia attempted as much as
possible to treat the charged- and neutral-current candidates on
an equal footing. That is, criteria for selection of the hadronic
shower (location, energy, etc.) were chosen to be precisely the
same for charged and muonless events. Furthermore, since the
primary question at the time was whether neutral-currents existed
(not yet in what proportion), only completely unambiguous events
were used so as not to confuse charged- with neutral-currents.
Finally, to reduce the effects of any remaining biases, and to
make the measurement less sensitive to calculations of the
neutrino flux, the group chose to express their results in terms
of the ratio of neutral-currents to charged-currents. This also
 rendered their results independent of energy since both neutral-
and charged-current cross sections increased linearly with
energy.

Such innovations in the choice of variables in which to
express the results were crucial. In earlier neutrino experiments
the total visible energy deposited in charged events was compared
to the total visible energy in muonless events. Reinterpreting
these experiments after the fact one sees that some fraction of
the muonless events were neutrino events with unseen neutrinos
emerging from the vertex. Thus the earlier experimentalists had
thought that they were comparing the number of equally energetic
neutral-current and charged-current events. In fact, they were
comparing the number of high energy neutral events to low energy
charged events. Since the number of either type of event
decreased quickly with energy, the team persuaded themselves that
there was a very small proportion of neutral-current events.

Sufficient interest had developed in the neutral-current problem by September 1972 for those interested to gather apart from the rest of the collaboration in a second Paris meeting. Members of this <u>ad hoc</u> working group had to devise a standardized format for the computer data cards as well as a common nomenclature for the many scanning groups. Procedures established at the Paris meeting would guide the participating laboratories as they sorted through hundreds of thousands of images. Finally, from the memoranda preliminary to the meeting, it is evident that different members of the collaboration were concerned about different possible backgrounds. Fry was especially interested in the possibility that K-zeroes would create fake neutral-current events. Others were worried that muons, stopping in the chamber, might be mistakenly counted as hadrons. Still others were particularly vexed by the neutron background.

At a Fermilab meeting that same month, Perkins voiced his own opinion that the only good test of the Glashow-Weinberg-Salam theory would come from the scattering of neutrinos from electrons:

"As far as the Weinberg theory is concerned, the most definitive and unambiguous evidence, for or against, must come from the purely leptonic reactions...since the hadronic processes involve details of the strong interactions which might contain unknown suppression effects..."[20]

By contrast, Perkins noted, certain purely leptonic interactions were clearly forbidden in the old phenomenological theory. One would not have to make assumptions about the properties of the still recently discovered partons. Perkins added that "In the CERN Gargamelle experiment to date, the expected number of events was between 1 and 9, and none was observed...If none were observed [in the remainder of the experiment], this would be fairly conclusive evidence against the Weinberg theory."[21]

During the fall of 1972 each of the subgroups conducted their respective data analyses, a long, often frustrating task in which hundreds of events had to be analysed, categories of events modified, and criteria for selection adjusted. By January 1973, Musset and the others had gathered sufficient data to present their findings to the American Physical Society meeting in New York. Musset devoted his talk almost entirely to the neutron background problem. His data included a plot of the number of events (charged, neutral, associated) plotted against longitudinal and radial position in the bubble chamber. From the data, Musset argued that the events occurred evenly throughout the volume of the chamber, as would "real" neutrino events. Neutron induced events ought to have been clustered near the walls.[22]

All of the arguments hinged on the neutrons entering the chamber. Very little was known about them. No one was sure what their energies or angular distributions were. To get a direct grip on the problem, the group began to study <u>associated events</u>. These were events in which a regular charged-current event occurred inside the chamber, and ejected a neutron which also interacted inside the chamber (see Fig. 6). By studying these associated events the Gargamelle team could measure the interaction length, energy and angular distributions of neutrons ejected from the initial charged-current events. With this information in hand, computers could simulate the effect of neutrons ejected in charged-current events in the walls, magnets, and shielding around the bubble chamber. In all previous chambers this kind of measurement had been impossible. Neutrons travelled too far to interact inside the small visible volumes of the old detectors. Gargamelle's size thus proved essential.

At the time of Musset's talk there were not enough associated events to do a serious statistical analysis. There were even fewer when the team restricted their interest to neutral-current candidates above a cutoff energy of 1 GeV in order to exclude the typically low energy neutrons. After hearing Musset's presentation, Paschos called Musset to report some new theoretical results and to report their relevance to the Gargamelle experiment. Only a few weeks earlier, Paschos and Wolfenstein had completed an analysis of precisely the interaction in which the hadronic group at CERN was interested: neutrino + nucleon goes to neutrino + anything.[23] Their result was striking. Neutral-current interactions of this type should occur at a rate of over 18% of the charged-currents. If Paschos and Wolfenstein were right Gargamelle would imminently reveal a huge effect.

Musset's excitement over the new theoretical results was heightened by another piece of good news. In early January, just a few days before he had left for the United States, a candidate for a neutrino - electron scattering event had been found at Aachen. During a routine re-scan of some photographs at Aachen, a picture had been taken that could be interpreted as a single, high energy electron apparently originating in the center of the chamber. The Aachen electron event satisfied all the criteria that the electron group had imposed on their search. It was isolated and well within the measurable visible volume of the chamber. This ruled out the possibility that the electron had been hit by a photon, since high-energy photons could only travel a few centimeters in the bubble chamber liquid. The electron had been knocked in the beam direction making it consistent with the hypothesis that it had been hit by one of the beam neutrinos. Finally, the electron had such a high energy that it could not have been hit by one of the relatively low-energy neutral hadrons ejected in charged-current events.

The recognition of the Aachen event and its positioning as a real "discovery" took place in four stages, as the "golden event" rose through the hierarchy. At the first level, one of the women scanning the bubble chamber negatives (H. von Hoengen) noticed an unusual event (see Fig. 7). She classified it as an extremely rare muon plus gamma ray. While checking the scanners' work, one of the research students, Franz Hasert, grew curious about such an odd occurance. He went back to the film and recognized the spiralling particles as electrons. (Again see Fig. 7.) The next day Hasert carried the picture another rung up the ladder to Deputy Group Leader Jürgen von Krogh. Krogh concurred that the picture was of considerable interest. He brought it to his Group Leader, Helmut Faissner, who later wrote:

"I got to realize that this event was a "Bilderbuch example" of what we had been expecting...[for] months to show up: a candidate for neutrino electron scattering. But the crucial point to assess was [the] background."[24]

The dominant background to a single electron was just inverse beta decay. Ordinary beta decay is,

$$n \longrightarrow p + \text{e-minus} + \text{anti-neutrino}; \qquad (1)$$

inverse beta decay is either:

$$\text{neutrino} + n \longrightarrow \text{e-minus} + p, \qquad (2)$$

or,

$$\text{anti-neutrino} + p \longrightarrow \text{e-plus} + n. \qquad (3)$$

Only neutrinos produce e-minuses. Since the photograph definitely involved an e-minus and the film had been shot on an anti-neutrino run, the only background was from the tiny admixture of neutrinos that invariably found their way into the anti-neutrino beam. Whence Faissner's excitement.

Faissner packed the "picturebook" example and set off for England to show it to Perkins. Judging from his Batavia talk a few months earlier, Perkins seems to have suspected that the neutral currents did not exist at a high level since none of the coveted lepton events had thus far been found. With Faissner's Aachen photo in hand, Perkins' attitude changed: "I had only one question. Was it in neutrino or antineutrino film...On learning it was an antineutrino event, I led the charge to the bar, to celebrate. At that time, the expected background was ˜.01 events, so this one case was enough to convince me (although clearly not the world at large)."[25] Faissner wrote to Lagarrigue on 11

Fig. 6

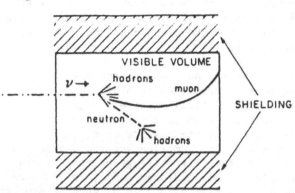

Fig. 6. Associated Event. These are events in which both a charged-current event and a neutral-current candidate occur in the same frame. They were interpreted as follows: in the charged-current event a neutron is released causing the "fake" neutral-current event. By studying the angular distribution and penetration length of the neutrons, the experimentalists hoped to estimate the number of neutron-induced "fake" neutral-current events from neutrons produced in the shielding (see Fig. 5).

Fig. 7

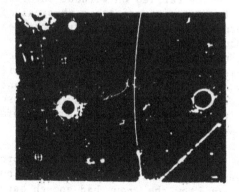

Fig. 7. Candidate for Neutrino-electron Scattering. This photo, analyzed at Aachen, was interpreted as a neutral-current candidate corresponding to the process depicted in Fig. 4a. The Gargamelle experimenters concluded that the most likely explanation for this event was that an unseen neutrino had entered along the beam direction (indicated by arrow) then scattered the electron to the right. Haloed spots are flashes for bubble chamber photography.

January 1973: "The event has excited us a great deal; it is in
effect a lovely candidate for an example of the neutral current."
Lagarrigue wrote back asking for more background studies, but
most of the electron group was converted. Faissner expressed his
optimism in a letter to W. Jentschke, the director general of
CERN, declaring that the discovery "would be a great one not
just for Aachen."[26]

An event like the Aachen electron was very compelling to
experienced bubble chamber experimentalists. Their specialty was
famous for several critical discoveries grounded on a few well-
defined instances. The omega-minus had been accepted on the basis
of one picture, as had the cascade zero. Emulsion and cloud
chamber groups had also compiled arguments based on such "golden
events" including the first strange particles, and a host of K-
decays. In a letter of 11 January 1973, Faissner reminded
Lagarrigue of their discussions about such shining examples: "I
still vividly remember your declaration of twelve years ago, that
a single distinct electron would be sufficient to demonstrate the
identity of the muon-neutrino and the electron-neutrino."[27]
Nonetheless too much was at stake to publish immediately.
Besides, the American competition seemed far behind.

The effect on the collaboration's research priorities was
immediate. Until January 1973, work on neutral-currents had been
far from dominant. One advantage of the single electron search
was that it could be be carried on without interfering with
routine scanning for a variety of other physical effects. By the
time Musset returned from the United States the hesitancy to
commit major resources to the neutral-current work had begun to
fade. Partly the new outlook was shaped by the increasing number
of hadronic muonless events. More directly the shift was effected
by the discovery of the Aachen electron. With mounting evidence
from both hadronic and leptonic quarters, A. Rousset issued a
memorandum on 19 February 1973 reflecting the newfound confidence
of the neutral-current investigators.[28]

By this time Lagarrigue and Rousset were thoroughly involved
in the search. Neutral currents now stood at the center of
everyone's attention. Now the group had enough data to demand
that the neutral-current candidates be even more energetic than
before. The CERN and Orsay groups reanalyzed the old data tapes.
The Orsay group set up a separate group to perform an independent
check. The two concurred that the ratio of neutral- to charged-
current events was .24 for neutrinos and .43 for
antineutrinos.[29] This, they commented, was perfectly
compatible with the Weinberg model, but the nagging question
remained: Could the background account for the numbers?

Musset put out a plea for help with the extraordinary amount
of work required to study each neutral-current candidate in

detail. Scanning groups across Europe were working to sort,
classify and measure the event types. Technicians prepared huge
enlargements of the appropriate photographs so that the group as
a whole could judge their validity. Physicists gathered around
each photo to argue over its proper analysis. Records from these
meetings contain long lists of such judgments: "OUT possible
[muon],"--that picture was no longer a candidate because one of
the "hadron" tracks might actually be a stopping muon; "OK one
track badly measurable,"-- if the event could not be properly
measured, then conclusions drawn from it would be unreliable;
"OUT cosmic [ray]," "OUT entering track,"-- if a particle other
than a muon came into the chamber it could indicate charged-
current events were occuring in the walls, releasing neutrons or
other dangerous particles; "OUT outside fiducial volume"-- if the
track was not in the best part of the chamber measurements were
not reliable; "OUT possible mu-kink" -- a "muon" track with a
kink might in fact be a pion.[30]

These debates took place at the University of Milan, at
CERN, at Orsay and at other laboratories across Europe. By April
of 1973 the question of finding an upper limit for neutral-
current processes was abandoned. The question became: what was
the ratio of neutral-currents to charged-currents? In the single
electron search a consensus had also emerged. Donald Cundy's
minutes from a March 1973 meeting began by declaring, "There was
general agreement that a paper should be published as soon as
possible concerning the electron search and the one event
found."[31]

Not everyone was persuaded, and those that were based their
assent on a variety of arguments. On 17 May 1973, the neutrino
collaboration held a meeting at CERN, most of which was devoted
to the neutron problem.[32] Pullia presented a method for
estimating the neutron background entirely from the
characteristics of events in the visible volume of the chamber.
This avoided any assumptions about the distribution of matter
around the machine or the flux of neutrinos into the experimental
area. At the same meeting, R. Baldi and Musset offered another
kind of self-contained analysis involving an imaginary division
of the chamber volume into inner and outer shells. Under the
supposition that all of these candidates were due to neutrino-
induced neutrons, they could estimate the probability that the
"parent" charged-current event would be in an outer shell that
was still inside the visible volume. Their conclusion: "...the
events cannot be explained by [neutrons]."[33]

Other methods were applied as well. Rousset advocated a
technique involving a thermodynamic analysis that put the
neutrinos and neutrons in equilibrium.[34] Such methods had been
used in earlier neutrino analysis but had been stymied because
the small size of the older chambers prevented the acquisition of

reliable data about neutron interactions.[35] The computer could
be used to simulate physical processes including the neutron
induced "fake" neutral current events. Several subgroups applied
"Monte Carlo" methods to the background problem, in order to
estimate the number of background events to be expected.
Unfortunately in many cases the result depended on one's choice
of physical paramaters, such as how far one assumed a neutron
could penetrate before interacting. Different computer subgroups
therefore ran Monte Carlo routines to be sure that the
background would remain small even if one varied the parameters
significantly.

None of these approaches satisfied William Fry and Dieter
Haidt. Above all they were worried that neutrons could cause a
cascade of other neutrons in the chamber and the shielding and,
by so doing, increase the number of neutrons that could cause
fake neutral-current events. In particular, no one had
systematically explored how neutrons were scattered from nuclei
at different energies. When would a neutron knock another neutron
free? What energies would the secondary neutrons have? Could some
variation of these variables give rise to a much higher neutron
background than anyone had suspected? For some members of the
collaboration, including F. Bullock, it was Fry and Haidt's
careful computer-aided analysis of the neutron cascade background
that persuaded them that the neutral-current effect was real.[36]

During the last week of June and the first two weeks of July
1973 members of the collaboration suggested argument after
argument for and against the neutral-currents. Vialle, for
example, remembers Lagarrigue storming into his office almost
daily with a new source of possible background to be
investigated.[37] Only days before Musset's seminar announcing
the discovery of the neutral-currents, Ettore Fiorini became
concerned that neutral kaons might oscillate back and forth
between their two forms, extending their dangerous range far
enough to cause fake neutral-current events. Shortly afterwards
he wrote that he had satisfied himself that it was not a
problem.[38]

In the end, the group mustered a great number of arguments
to persuade themselves that their neutral-current effect would
not disppear into the background. The final catalyst, though,
came not from the computer teletype, nor the back of an envelope,
nor the blackboard. Carlo Rubbia, who also held a position at
CERN, let it be known that the Fermilab group was close on
Gargamelle's heels. According to many of the participants, this
tipped the already tilting balance, and the decision to publish
was made. Not all were entirely happy with the arguments
presented in the final draft, but they believed that they had the
backgrounds largely under control.[39] The single electron paper
was received by <u>Physics Letters</u> on 2 July 1973.[40] On 19 July

1973, Musset gave a seminar at CERN announcing the discovery; four days later, on 23 July the paper was sent to <u>Physics Letters</u>.[41] (See Fig. 8.)

American efforts to exploit neutrino physics on a scale that could compete with that of Gargamelle began in the late 1960's. To follow this side of the neutral-current search we need to go back to the Summer Study Program held by Fermilab in Aspen, Colorado during the summer of 1969. Neutrino physics had been one of the original justifications for building the National Accelerator Laboratory. By now the reader will not be surprised to find that during the Aspen program many physicists advanced proposals to unveil the stubbornly undiscovered W. One of the program participants, A.K. Mann, presented a report to the school on the possibility of producing the W by means of a high-energy neutrino source incident on a dense target.[42] He would then detect the particle's decay products using spark chambers between segments of earth. From his preliminary calculations, Mann argued that such a search could be effected up to a W mass of about 5 GeV.

To strengthen his case against the competition, Mann chose to work with David Cline, a younger physicist whom Mann already knew. Ever since Cline received his Ph.D. he had been interested in the problem of neutral weak currents in kaon decay. As mentioned earlier, such strangeness-changing neutral-currents had quite severe upper limits placed upon them by several experiments. Almost his entire career in physics, Cline had been involved in these determinations using bubble chambers. For example, with Camerini, Fry, and Powell in 1964 he had shown that the number of neutral-current decays:

$$K+ ---> pi-plus + e+ + e-$$

was less than 10^{-6} of all K+ decays.[43] Characteristic of this and related projects was Cline's scrupulous analysis of small numbers of significant photographs. In the letter reporting this work an important section is devoted to the discussion of three critical events. As we will see, Cline's style of physics played an important role in the discovery of neutral-currents.

A few years later, Cline and other researchers compiled an impressive array of data indicating that neutral-currents did not exist, at least not at the same level as charged-currents. Summing up a 1967 summer school talk, Cline argued that, "the successful explanation of the absence of neutral lepton couplings (and possibly of primitive neutral hadron couplings) will undoubtedly be a very significant factor in the ultimate theory of the weak interactions."[44]

Cline was an important addition to the project. Together he
and Mann rewrote the old proposal, reiterating their primary
physics goals. These remained the production of the W and the
detailed study of charged-current neutrino physics. By this time
(December 1969) it was clear that their device would have to be
more complicated than Mann's original subterranean spark chamber.
They would need to magnetize the iron blocks so that they could
determine the muon momenta. Moreover, they would have to add a
liquid scintillator calorimeter that would measure the energy of
the nucleon that was blasted apart by the neutrino.[45]

With Cline on board the proposal was stronger, but in Mann's
opinion, not strong enough to sway the planning committee. Mann
turned to Carlo Rubbia (then at Harvard) whom he knew from a
leave of absence Mann had spent at CERN. Rubbia brought with him
his experience designing and building the kind of large and
sophisticated detector that the team would need. Once again the
principal investigators had to rewrite their proposal. This time
it included testing the parton hypothesis, as well as searching
for the W and studying charged-current interactions at high
energies.[46]

In this superior incarnation the device was specified more
fully. When the neutrinos entered the detector they would
encounter a series of spark chambers alternately placed between
thick containers of liquid scintillator. Because of its large
mass, this stage served as a target in which the neutrinos could
interact. The liquid scintillator formed a calorimeter which
would measure the amount of energy deposited (by collecting light
from the scintillating oil) and the spark chambers created a
visual display of charged particle tracks as they crossed the
spark chambers. (See Fig. 9.) Most of the particles produced in
neutrino interactions would be hadrons. Because they interact
strongly, hadrons typically stopped after a short flight in the
target-detector stage of the device. Some of the particles, the
muons, would penetrate through into a second stage that
alternated the spark chambers with thick iron slabs. Any hadron
energetic enough to penetrate past the target-detector would be
stopped by the iron slab that stood immediately downstream from
counter B (see Fig. 9). Muons would sail through the iron,
although they would be bent by a strong magnetic field. The
curvature of their tracks would allow a determination of their
momenta.

As the device was originally designed, neutral-currents
could not have been found. Along with much else, the electronics
of the detector had been borrowed from the Schwartz-Steinberger-
Lederman et al work at Brookhaven.   The Brookhaven
collaboration had grown out of the test of the two-neutrino
hypothesis. By only photographing events in which the neutrino
turned into a charged particle, the group saved themselves the

Fig. 8

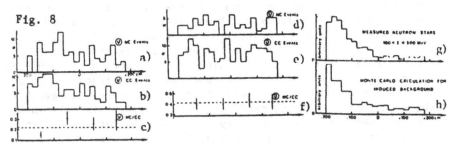

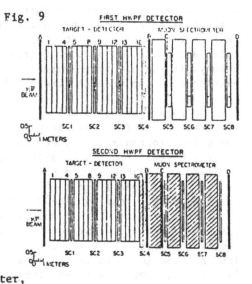

Fig. 8.   Summary of Gargamelle Collaboration's Arguments for
Hadronic Neutral-Currents.   8a and 8b indicate that events found
during neutrino runs gave neutral- and charged current candidates
distributed relatively evenly over the radius of the chamber.   Such
distributions were expected for neutrino events but not for
neutron-induced events which would tend to cluster near the walls.
This clustering could be observed in the computer-simulation
reproduced in 8h.   8d and 8e show the spatial distribution
of neutral- and charged-current events for antineutrinos.   The
Glashow-Weinberg-Salam theory predicted that the ratio of neutral-
current to charged-current events (NC/CC) would be different in
neutrino and antineutrino beams and that this ratio would be
larger for antineutrinos.   This was confirmed in figures 8c and 8f.
From Hasert et al., Phys. Rev. 46 (1973): 138-140.

Fig. 9.   Two Versions of
Experiment 1A.
Top: First Apparatus.   Beam
enters from left and interacts in the first stage where
the charged reaction products
are detected.   In the second
stage, the muon spectrometer,
thick blocks of iron filter
out hadrons allowing only
muons to penetrate.
Bottom: In order to avoid
the wide-angle muon problem,
the E1A group replaced
counter C and spark chambers
SC5-SC8 with larger chambers
and inserted a 13" steel
plate upstream of SC4.   Thus
SC4 and counters B and C were
supposed to serve as the first
elements of the muon spectrometer,
catching wide-angle muons.

trouble of looking over hundreds of events in which "nothing"
happened. Folklore has it that the Brookhaven collaboration
dubbed the "uninteresting," muonless events "crappers." In any
case the Harvard - Wisconsin - Pennsylvania - Fermilab (HWPF)
group at first followed suit, indicating what little importance
they had initially attached to the neutral-current search.
Further corroboration of this point can be found in the published
literature. In 1970 Cline, Mann and Rubbia published an article
on a method by which they could use their apparatus to search for
the W in their detector, providing that the particle fell into
the 5-10 GeV range.[47] Once again we see how far the original
experiment was from being a test of the electroweak theory. In
the Glashow-Weinberg-Salam theory --which was never mentioned--
the W weighs near 80 GeV.

For a period in early 1972, E1A and several groups began
skirmishing over who would run the first neutrino
experiments.[48] Meanwhile 't Hooft published his
renormalization proof of of the Glashow-Weinberg-Salam theory.
As in Europe, the proof piqued the theorists' interest. After a
hiatus of almost five years, Weinberg himself became intrigued
by his old work and began   calculating consequences of the
theory. Later, he recalled,

"Now we had a comprehensive quantum field theory of the weak
and electromagnetic interactions that was physically and
mathematically satisfactory in the same sense as quantum
electrodynamics -- a theory that treated photons and
intermediate vector bosons on the same footing, that was based
on an exact symmetry principle, and that allowed one to carry
calculations to any desired degree of accuracy. To test this
theory, it had now become urgent to settle the question of the
existence of the neutral currents."[49]

Weinberg published calculations of the cross sections to be
expected for neutral-current production. He then phoned Rubbia at
the old cyclotron at 44 Oxford Street to persuade him to look for
neutral-currents in his Fermilab experiment.[50] After a brief
hesitation Rubbia agreed that the neutral-current search ought
to figure prominently in the experiment 1A's program. The physics
was fundamental, and not incidentally, adding the neutral-
current search to their proposal would provide yet another
reason for the steering committee at Fermilab to choose E1A to
run as the first neutrino experiment. The senior collaborators
lost no time in pointing this out to the Director of the
laboratory, Robert Wilson: "There has recently been increasing
awareness of the need of more sensitive searches for neutral
weak currents and neutral weak intermediate bosons. The
existence of a neutral weak current or a neutral-weak propagator
would cast additional light on the connection between weak and
electromagnetic interactions.... We might now stand in a

position analogous to that of Oersted, Ampère and Faraday 150
years ago as they attempted to elicit the connection between
electricity and magnetism."[51]

The first consequence for the experiment's hardware was the
installation of a trigger that would allow events to be recorded
if sufficient energy was deposited in the calorimeter (indicating
a neutrino event candidate) even if no charged muon emerged. This
task fell to Larry Sulak. Soon after he installed
the energy trigger, photographs started to show up indicating
events without muons. But, as Mann has indicated, this did not
make instant believers of the team. "You can say, well, we came
to the conclusion immediately that we had seen weak neutral-
currents. But you'd be surprised, that was the last conclusion we
came to. Our first conclusion was that we were making some
mistake and that these muons were somehow escaping the apparatus
or being missed by us in some way and that no effect of that
magnitude could exist."[52] The reader must bear in mind that
both Cline and Mann had earlier spent years in careful efforts
that showed that kaon neutral-currents did not exist in some
processes above one part in a million. Now something that looked
like evidence for neutral-currents appeared in the experimental
data at a rate of one part in three. It had to be a quirk,
perhaps from some poorly understood feature of their new and as
yet unproven detector.

As a consequence of their skepticism, Mann and Cline were
principally concerned during the spring of 1973 with
understanding charged-current physics and other projects
originally set out as goals for their experiment. They reasoned
that a better grasp of the more traditional physics would inform
them about their machine's idiosyncracies. Simultaneously they
would certainly be able to explore the old physics at new
energies. These first efforts culminated in the collaboration's
paper, on studies of neutrino and antineutrino events at high
energies. [53] Meanwhile Sulak and a coterie of undergraduates
began sifting through the spark chamber photographs in a fourth-
floor projection room at Harvard's Lyman Laboratory. Their
principal goal was to determine whether or not muons could be
escaping from the chamber before they reached the muon
spectrometer. (see Fig.10) If the muons were escaping, ordinary
charged-current events would appear to be neutral-current events:
no muons would reach the muon spectrometer.

Two Monte Carlo computer programs were established, one at
Harvard, and one at Wisconsin. Data from both laboratories
indicated that the muonless signal was too big to be ascribed to
fleeing muons. Sulak wrote an article for the collaboration and
brought the manuscript to Mann who was confined to his bed with
back problems. Mann, Ford, Cline and Rubbia agreed that the
paper should be submitted for publication.

All of this work in the late spring of 1973 was done knowing
that the Europeans were fast accumulating evidence on neutral-
currents. The Americans´ sources of information were
several. Rubbia was commuting regularly between CERN and the
U.S., and several members of the Gargamelle collaboration had
scheduled visits to Fermilab. In mid-July, Rubbia, independently,
wrote a letter to Lagarrigue telling him of the recent HWPF work:

"I have heard from several people at CERN that your neutrino
experiment in Gargamelle in additon to the beautiful electron
event has now a growing evidence for neutral – currents. We have
observed at NAL approximately one hundred unambiguous events of
this type and we are in the phase of final write-up of the
results. In view of the significance of the result I am
addressing to you this note in order to know if announcing our
result we should mention the existence of your work on the
hadronic process (and if so in which form). In this case I hope
you will take a similar attitude toward our work."[54]

Lagarrigue declined Rubbia´s offer the next day, suggesting that
the announcements be made independently without mentioning the
other´s results. He added that the CERN announcement would be
made in twenty-four hours, on 19 July 1973.[55] It was.

On the basis of their Monte Carlo demonstration that
neutral-currents existed above the background, Rubbia and Sulak
began to prepare for the summer conferences at Aix-en-Provence
and Bonn, where they would announce their findings. In late
August 1973, Sulak brought the data over to Europe, where Rubbia
had remained since leaving the U.S. Along with Jim Pilcher and
Don Reeder they then headed to Bonn for the International
Conference on Electron and Photon Interactions at High Energies.
By this time deadlines had long passed, so the Americans could
not schedule their paper in a parallel session. Instead George
Myatt, who was presenting a summary talk, offered to read a brief
handwritten report composed shortly before his presentation.[56]
After he spoke, one of the questioners from the floor queried how
the new results could be reconciled with earlier limits on
strangeness-changing limits. "That," Myatt commented, "is a major
obstacle to the Weinberg-type theories."[57] It cannot be over-
stressed how strongly people believed that the earlier results
had killed neutral-currents.

Reconvening in the south of France, representatives of
Gargamelle and E1A once again displayed their results. Again
Musset insisted that the evidence from the compatibility of
neutrino and antineutrino events, the consistency of the ratio of
neutral- to charged-current events over a range of energy, and
the general similarity of the hadronic showers in neutral and
charged events all conspired to suggest that neutral-currents

were present.[58] Weinberg himself cautiously endorsed the
neutrino experimenters´ conclusions. "It is perhaps premature
to conclude from all this that neutral currents have really at
last been observed. There may be some mysterious source of
background contaminating all these experiments. It is certainly
too early to conclude that the old model of leptons is really
correct. However, there is now at last the shadow of a suspicion
that something like an SU(2) X U(1) model...may not be so far
from the truth."[59] Thus encouraged, by late summer it seemed
to the Harvard group that the experiment had accomplished its
new goal.

At Fermilab the experiment was just beginning. Four
considerations contributed to a certain distrust Cline and Mann
felt about the paper their collaboration had already submitted to
Physical Review Letters. First, the data analyzed at Madison had
begun to indicate a very low ratio of neutral- to charged-
currents. Second, from previous work, Cline and Mann came to the
experiment inclined to think that neutral-currents were
suppressed in much the same way as strangeness-changing neutral-
currents. Third, given the uncertainty in the use of the new
apparatus, compounded by the wide-angle muon problem, it was
natural for them to seek a further check on the new results.
Finally, Mann felt that the whole experiment could be redone
rapidly in a much improved way. As a result, the full attention
of Cline, Mann, and others at Fermilab turned to the
rearrangement of the detector. For the moment, believing the
conference reports to be a sufficient description of their work,
they postponed responding to the referees´ reports on their first
paper.[60]

To counter the wide-angle muon problem the collaboration
decided to modify their apparatus in two ways. First, they placed
a new 13-inch thick steel shield between calorimeter segment 16
and spark chamber 4 (SC4). See Fig. 9b. In principle, the
combination of the steel shield and the downstream sections of
the calorimeter were supposed to stop any hadrons from reaching
SC4. Thus, with the shield in place, only muons could get to SC4.
SC4 was thereby transformed from the last spark chamber of the
target-calorimeter stage into the first spark chamber of the muon
spectrometer. Second, counter B could now be used as a register
of muons since no hadrons were supposed to penetrate the new
steel plate. Third, the small wide-gap spark chambers SC5 -SC8
were replaced with larger, narrow gap chambers. Fourth, counter C
was enlarged. All of these changes were designed to capture muons
leaving the target area at wider angles that previously would
have been lost. For these innovations, the price the physicists
had to pay seemed relatively low. True, the hadron shield had
previously been much thicker -- four feet of iron instead of 13
inches of steel. But a four foot thick object simply could not be
wedged between calorimeter segment 16 and SC4 without pushing

SC4 and counter B so far back as to lose their usefulness as wide angle-muon detectors. Cline commented on the change in a memorandum shortly after the new device was tested: "The new iron [recte steel] placed behind the calorimeter is very effective in reducing the hadron penetration to...[spark chamber 4]. Some small number of events do show penetration, but the fraction is very likely less than 20%...More study of the data is needed to make this a reliable conclusion."[61]

Unfortunately, though this would not be understood for several months in a quantitative way, the shield was not thick enough to be a very effective hadron filter. Moreover the team did not know how ineffective it was. This was a crucial problem. For if the hadrons penetrated through the iron, even if no muon emerged from the vertex, the event would be recorded as a charged-current event. (See Fig. 11). Ironically, in an effort to prevent wide-angle muon escape which made charged-current events look like neutral-current events, the collaboration had inadvertantly introduced punchthrough, which made neutral-currents look like charged-currents.

Because the experimenters had not compensated adequately for the punchthrough, the neutral-current signal seemed to vanish. The reason precise predictions could not be calculated for the hadron punchthrough is related to the Gargamelle group's difficulty in calculating the neutron interaction length. Both the neutrons and the punching hadrons interact strongly. Strong interactions render hadrons' passage through matter much more complicated to calculate than a muon's flight. In the case of the muon only electromagnetic interactions are significant. Other factors complicated their work. E1A was probing phenomena at an energy higher than that explored in any previous neutrino experiment. They therefore could not rely on data compiled in earlier work about the reaction products of the neutrinos. Finally, since punchthrough had not been a prominent problem in earlier work, it was not at first realized that the thinner shield now made it so.

In part Cline's hesitancy to rely on demonstrations based on detailed Monte Carlo simulations of punchthrough or wide-angle muons was tied to a wider issue of experimental reasoning. In his earlier work Cline had successfully used the search for "gold plated events" in his bubble chamber work. In this respect his approach was similar to the electron group at Gargamelle: both Cline and the electron group much preferred a few clean shining examples to hundreds of events embedded in a shakier statistical argument. From Cline's 1 October 1973 memorandum (see Fig. 12) we see an example of this approach. He wrote, "It is amusing to investigate how improbable the central (x,y) event is..(the other two events are too close to the edge of the fiducial region to be gold plated). ...we expect to find...1 events. [sic] Thus,

Fig. 10

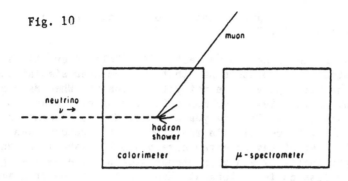

Fig. 10. Wide-Angle Muon. While using the first HWPF apparatus (see Fig. 9a) the collaborators' greatest fear was that muons produced at wide angles with respect to the beam could escape from the apparatus before registering their presence with the muon spectrometer. In such cases charged-current events would look like neutral-current events.

Fig. 11

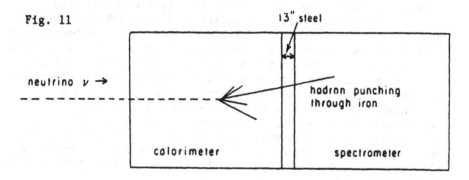

Fig. 11. Punchthrough. Hadrons unexpectedly penetrated the 13" steel plate entering into the muon spectrometer. For several weeks after modifying their apparatus the E1A group interpreted these hadrons as muons and so concluded that all neutrino events were charged-current events.

unfortunately this event is not improbable and we have not found a gold-plated event."[62]

Reinforcing Cline's predilection for "clean" events was the feeling other collaborators had that the computer simulations used in the first paper were not yet persuasive. When Mann came back from having viewed the Monte Carlo programs in action, his reaction was strong. Several years later, Mann's student Frank Messing wrote, "I recall Al's reaction, on his return, was similar to that of having eaten some bad shellfish. It doesn't taste right initially and your stomach feels worse as time goes on."[63] In less culinary language Mann felt that the programs were simply too sensitive to small changes in experimental parameters to be reliable.

On 10 October 1973 Cline distributed a memorandum speculating on the production of intermediate vector bosons -- a supposition incompatible at the available energy with the Glashow-Weinberg-Salam theory. The following day, in a memorandum of 11 October, Cline gave the first indication that their experiment was no longer giving results compatible with the Gargamelle publication.[64] Calculations by Reeder and Ling indicated that muons would be detected with an 83% efficiency, and punchthrough would not exceed 13%. This latter number was but half of the number the group later established. Since more pions were penetrating through the steel than they thought, many real muonless events were counted as charged-current events. Neutral-currents appeared to vanish. "Taken at face value," Cline concluded, "these results are inconsistent with the CERN measurement of R' = 0.28+/-0.03 for a mixed beam [of neutrinos and antineutrinos]. Clearly, it could still be that we did that one in 100 experiments or something else is wrong."[65] Something else was wrong, but it would take the group two more months to be sure what it was.

The pressure was building. Cline recalls getting less and less sleep as the team accelerated the project to provide a definite answer to the neutral-current question. On 16 October 1973, Cline prepared a new memorandum: "Because of the importance of the neutral-current question, the fact that we have extended our necks previously on the subject, and that other groups around the world are moving fast to check our results and the CERN result, I propose that a rapid, unified analysis of the [muonless] events be carried out early in November [1973] at NAL....(ii) The schedule of our run has changed, with the laboratory now inserting running time for E21 at the end of November. I suspect that this time will be used for a...[muonless] search, since they are likely submitting a proposal for this experiment in the next week or two. Again this proves the need for us to move fast in our analysis and to settle the question before others get to it."[66]

By mid-November Mann and Cline were convinced that the newer
results failed to give evidence for neutral-currents. Rubbia
concurred. Mann then drafted a letter to this effect for the
Physical Review Letters, pending the outcome of the revised
experiment. Though the "No Neutral-Currents" paper was never
submitted, it captures the spirit of the time. Its abstract read
in part as follows: "The ratio of muonless events to events with
muons is observed to be 0.05 ± 0.05 for the specific case of an
enriched antineutrino beam. This appears to be in disagreement
with recent observations made at CERN and with the predictions of
the Weinberg model."[67]   There was some division in the HWPF
group over the question of how and when the new results should
be released. Mann wanted to wait. Rubbia and Cline at different
times discussed the situation with people outside the group. When
Rubbia went back to CERN in December 1973, he spoke with a
variety of people, including Musset, Lagarrigue, Rousset,
Jentschke, and others. By this time the Gargamelle group had, of
course, already published their result defending neutral-
currents. Naturally they were somewhat distressed.

Jentschke, then Dirctor General of CERN, convoked a meeting
of the Gargamelle group to cross-examine them on the experiment.
He was afraid that CERN would be publicly embarrassed by the
forthcoming American bombshell that neutral-currents did not
exist. The Gargamelle group held their ground. Still, they were
shaken. Musset circulated a memorandum advising the subgroups to
deemphasize the Weinberg theory and to redouble effort on the
study of associated events. The memorandum began, "Dear Friends,
After our last neutral-current meeting, all of you have probably
heard rumours about new results ...at Batavia...The efficiency
for [muon] detection is better than previously and the result is
an apparent lack of neutral-current-type events. ...In the near
future, we can expect to be heavily questioned about the
reliability of our experiment. ...Independently from these new
rumours, it is much more important to know if neutral-current-
type events can be simulated by a trivial background such as
[neutrons].... than to measure accurately a [weak mixing
angle.]"[68]

Rumors were soon followed by documents. When the HWPF
collaboration had completed their "No Neutral-Currents" draft,
the principals prepared a letter to Lagarrigue bearing the bad
tidings. See Fig. 13. [69]   Mann, Cline and Rubbia signed it,
but did not mail it. Rubbia, independently, brought an unsigned
copy to Lagarrigue, where it remains among his scientific
papers. Apparently it was then copied and widely distributed,
since many members of the Gargamelle collaboration had it in
their files in 1980. "Dear Professor Lagarrigue," it began, "We
write to inform you of the preliminary result of our recent
experiment to search for neutrino interactions without final

Fig. 12

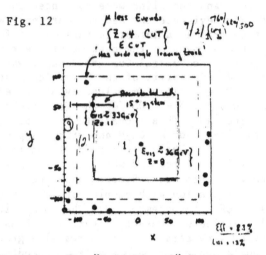

Fig. 12. Candidate for "Gold-Plated" Neutral-Current Event.
From Cline, Wisconsin TM 1 October 1973.

Fig. 13   November 13, 1973

Professor A. Lagarrigue, Director
Linear Accelerator Laboratory
University of Paris - SUD
Centre D'Orsay
Batiment 200
91405 Orsay
France

Dear Professor Lagarrigue:

We write to inform you of the preliminary result of our recent experiment to search for neutrino interactions without final state muons.
As you know, our apparatus was modified to provide a much larger detection
efficiency for muons relative to the apparatus that was used in our earlier
search for muonless events. We also improved our ability to locate accurately vertices of observed neutrino interactions, and lowered the
threshold on the total energy of the hadrons in the final state.

From about one half of the data obtained in our recent run, we find
the raw ratio R_{raw} = 0.18 ± 0.03. We estimate the muon detection

efficiency of the apparatus for the enriched antineutrino beam that was
used in this experiment to be approximately 0.85. Taking into account
small backgrounds produced by incident neutrons and by ν_e in the incident

beam, the corrected ratio is R_{corr} = 0.02 $^{+0.05}_{-0.03}$, where the error includes

an estimate of the uncertainty in the calculated detection efficiency. We
are continuing to process the remainder of the data and to improve our
understanding of the experiment.

We have written a paper intended for Physical Review Letters which
will soon be submitted. A copy will, of course, be sent to you but for
obvious reasons we wanted to convey our result informally to you before
its publication.

With kindest regards

Yours sincerely,

D. Cline

A. K. Mann

D. D. Reeder

AKM/rs   C. Rubbia

Fig. 13. Letter: Cline, Mann Reeder and Rubbia to Lagarrigue,
13 November, 1973, Mann files and files of many participants in
Gargamelle collaboration. This letter was never sent but unsigned
copies were widely distributed.

state muons." The letter went on to add that the corrected ratio
of neutral- to charged- currents was statistically
indistinguishable from zero. It closed by announcing the
imminent submission of the paper to Physical Review Letters.
Back at Fermilab background calculations dominated all other
work. Richard Imlay did another punchthrough study. Using his
analysis, Cline estimated the corrected neutral- to charged-
current ratio to be much lower than Gargamelle or the standard
model would permit. On 6 December 1973 Cline gave a talk at
FNAL. His concluding transparency delivered a clear verdict. The
corrected ratio was "very likely too small to be consistent with
Weinberg model and lower bounds deduced by Paschos and
Wolfenstein for this model -- also CERN data if due to Weinberg
model." The slide continued, R´[ratio of neutral- to charged
current events after correction]"suggested by first E1A
experiment is not confirmed in the present experiment..."[70]
Perhaps, Cline added, the discrepancy was due to an inadequate
position measuring system in the earlier experiment.

One has to imagine these tense months to understand how
things must have appeared to the participants. The HWPF
collaboration was under a great deal of pressure to announce
their findings. Demands on the group to discuss their results
came not just from other experimentalists, but also from the
theorists who were getting informal progress reports from various
participants. As Mann later recounted: "As the results began to
emerge, we were being pressed harder and harder for some kind of
decisive answer from people. It is very hard to communicate to
you how [things are], when you are in the center of the stage at
a time like that, particularly in high-energy physics where you
do not quite have control over your own destiny. You have to work
with collaborators, with the lab, with the director, with the
program committee, and with all the people who do the chores that
allow the experiment to be done. You´re being leaned on over and
over again to produce, whether you´re ready to produce or not."[71]

Throughout the months of September, October, November and
December of 1973 each of the participants was struggling to
integrate the various calculations and measurements. Each had to
persuade himself of the reality or artificiality of the effect.
Every measurement and calculation had its strengths, and its
weaknesses. But no one person could be fully conversant with
every different argument. One might have a thorough grasp of a
Monte Carlo calculation, another would be completely familiar
with the measurements associated with the punchthrough. Consider
an example that exhibits the difficulties associated with
interpreting the preliminary data. In November of 1973 Sulak had
plotted the ratio of neutral-current to charged-current
candidates as a function of longitudinal position in the
chamber.   He found that the ratio fell to zero as one approached
the last spark chamber in the target-calorimeter stage of the

detector.[72] For him this strongly suggested that punchthrough
was making neutral-currents near the end of the chamber look
like charged- current events. For other members of the
collaboration, Sulak's results confirmed their worst fears. They
thought that the plot indicated that when one took events in the
last chamber -- where wide-angle muons could not escape -- then
one got reliable data. If no neutral-currents appeared near the
end of the target- calorimeter, it meant that there were no
neutral-currents.

As a result of both outside pressure and new evidence,
opinions were beginning to change. On 13 December 1973, Cline
released a memorandum with a new tone: "Three pieces of evidence
now in hand point to the distinct possibility that a[muonless]
signal of order 10% is showing up in the data. At present I don't
see how to make these effects go away."[73]   These arguments
were: first, the Monte Carlo model now yielded a neutral-current
to charged-current ratio of 0.1. Second, the spatial
distribution of events looked like that of neutrinos. But the
third reason seems to have been most persuasive. It was the only
type of evidence that was qualitatively different from all
earlier arguments. Five events had shown up that were clear,
central in the chamber, and "had no hint of wide-angle tracks."
Reasoning from these golden events, Cline found that they were
"certainly consistent with a true[muonless]signal of [.08]."
This was the kind of argument Cline liked: a small selection of
events, clean of possible edge effects, that could be treated
without recourse to Monte Carlos.

About the same time Mann too was converted. On 26 January
1974 Mann issued a memo indicating that he had found eight events
that were "good to look at."[74] Also in December, Aubert, Ling,
and Imlay completed a rigorous study of the punchthrough, which
could then be used to subtract away the background. When this was
done the neutral- to charged-current ratio rose to 12-15%. The
second version of the 1A experiment thus neared completion, and
after several meetings during January and February the
collaboration decided to publish both their new results (see
Fig. 14) and the original Harvard-based paper with the comment
that additional work had confirmed their earlier findings.

By the end of February 1974, the 1A group had essentially
completed two separate experiments. The accelerator physicists
had changed the beam characteristics; the E1A collaborators had
shifted their detector's geometry; they had replaced the narrow
gap spark chambers; they were concerned with a new kind of
background. Even the participants in the two "experiments" did
not fully overlap. The style of experimentation in the two
subgroups was different, their expectations were at variance.
Above all, the type of evidence that finally convinced the

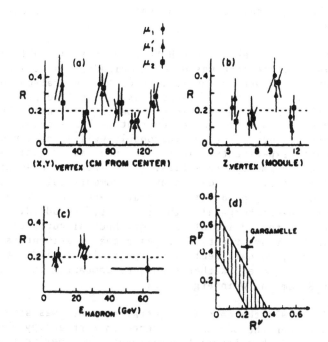

Fig. 14.   Summary of E1A's Arguments for Neutral-Currents.
(a) R, the ratio of neutral- to charged currents as a function of
the distance of the event vertex from the center of the calorimeter.
The three different markers correspond to different muon identifiers.   (b) R as a function of longitudinal position in chamber
(z is the calorimeter module number see Fig. 9).   (c) Variation
of R as a function of the total hadron energy of the event.
(d) The shaded area designates the allowed region of R (for
neutrinos) and R (for antineutrinos) from E1A's conclusions.
Gargamelle's result falls slightly outside this area. These
and five other plots of data are from Aubert et al., "Further
Observation of Muonless Neutrino-Induced Interactions," Phys. Rev.
Lett. 32 (1974): 1454-1457, on p. 1456.

experimentalists that neutral-currents were real was different in the two efforts.

Within a few weeks of the HWPF group's first publications [75] in April 1974, a conference was held in Philadelphia (26-28 April) on the topic of neutrino physics. Naturally neutral-currents figured prominently in the discussions. One participant reflected the general trend of thought in his opening remarks: "The existence of a hadronic neutral-current in high-energy neutrino experiments is, by this time, reasonably well established by four "independent" experiments. Clearly our next major goal is to establish the symmetry properties."[76] This was also Richard Feynman's opinion, as evidenced by his summary talk. He contended that the successful prediction of neutral-currents was a credit to the Glashow-Weinberg-Salam theory. "But," he added, "I should like to follow the advice of Mr. Mann. Neutral currents should be studied in their own right. That means the experimenters should say, all right, we have neutral-currents, let's find out what their properties are. (Rather than just comparing them to the theory of Salam and Weinberg.)"[77]

In June 1974 the case for neutral-currents was strengthened yet further when the London conference on high energy physics convened. Cal Tech's Fermilab experiment, an experiment from the Argonne-Concordia-Purdue 12-foot bubble chamber and the Columbia-Illinois-Rockefeller-Brookhaven collaboration all announced that they had corroborated the existence of neutral-currents.[78] In the early days of the HWPF collaboration's punchthrough crisis, the group was bemusedly said to have discovered not neutral- but "alternating" neutral-currents. By the spring of 1974 the physics community's consensus was that neutral-currents would alternate no longer.

*   *   *

Experiment 1A and Gargamelle have much to teach us, not just about the discovery of neutral currents, but about the nature of large-scale experimentation in the late twentieth century. These particular projects illustrate a frequent phenomenon in modern physics: experiments often straddle several theoretical issues. In the present case both E1A and Gargamelle began as part of a general inquiry into the nature of weak interactions. Two issues dominated their early agendas, tests of the V - A theory at high energies and searches for the intermediate W particle. At the time neutral-currents were a minor issue for both theorists and experimentalists. When they were considered at all, they were expected to occur through higher order processes at rates no greater than one percent of charged currents. After the startling discoveries of the 1969 SLAC experiment, both E1A and Gargamelle shifted their sights to the parton model. Would the model hold up at higher energies? Would the pointlike electromagnetic scatterers also act as pointlike weak scatterers? These were open and compelling issues. Even though the Glashow-Weinberg-Salam theory was available for application (at least to leptons) in 1967, its influence was not felt until much later. With 't Hooft's 1971 proof that the long neglected electroweak theory was in fact renormalizable, its fortune changed. Theorists pushed the experimentalists to include neutral currents in their list of high-priority projects. Even then it took some time for both groups to readjust their equipment, scanning and analysis proceedures and interests. Between 1971 and 1973 the experiments became tests of neutral currents.

As this essay suggests, we would be seriously mistaken to represent E1A and Gargamelle as neutral-current investigations from the start. Not only would such a depiction contradict all evidence, it would make it impossible to understand the course of the neutral-current investigation itself. We would be unable to understand why, for example, no energy trigger had been included in the earliest experiments on E1A. It would be impossible to understand why both groups felt so reluctant to accept neutral currents at the 20% level. Finally, we would be rewriting the history of experimental physics to fit the eventual triumph of gauge theories. Experiments have their own history.

In the final analysis my hope for this study is that it will help shift our historical attention away from the somewhat "theorocentric" approach that has governed its writing. When we focus on experiments we can see a variety of issues in the history of physics that would otherwise be obscured. For example, we can see how instruments affect the construction of arguments. In E1A we saw, for example, how experience with "golden events" in bubble chamber physics affected Cline's approach to data. Within the Gargamelle collaboration a similar stress on finding "clean" electron events was particularly important for the physicists who previously had had such difficulty dealing with the statistical problems of the neutron background. Following the careers of other members of the HWPF collaboration, one can understand how long experience with electronic detectors would lead more naturally to a trust in Monte Carlos and statistical argumentation. One can explore these issues for other periods, indeed for other disciplines. What role do image-producing and statistics-producing devices play in astrophysics, geophysics or molecular biology?

Comparisons between the types of evidence favored by different collaborators lead us to a crucial point. In both groups the process by which the first photographs were transformed into persuasive evidence took place over an extended period of time. Background studies played a fundamental role in effecting that change. One cannot point to the first muonless photographs and equate their production with the discovery of

neutral-currents. Knowing precisely what the photographs were
not was an inextricable part of knowing what they were. One is
reminded of the old joke about Michaelangelo's sculptorial
masterpiece: to create it, all the artist had to do was to
remove everything that wasn't David.

The task of assessing what the photographs were not fell
into the hands of various subgroups. One Gargamelle subgroup
prosecuted the neutron cascade problem, another followed a
different type of Monte Carlo. Some collaborators were
principally concerned with developing a thermodynamic analysis
îthat gave rough numbers analytically. Others tried to find
methods that made no reference to events outside the visible
volume. In the HWPF collaboration a similar division of labor
took place, though they had different backgrounds to investigate.
Bit by bit the evidence was marshalled. Theorists added studies
of the parton scattering involved in neutral current processes;
the various subgroups began to compare notes. One sees in the
dynamics of these experiments a phenomenon new in the large-scale
experimentation of high energy physics. Tasks that previously
would have taken place as separate experiments are now
incorporated into the single large-scale experiment.

Biologists have long spoken of "ontogeny recapitulating
phylogengy." Metaphorically, our individual experiment is
recreating the development of the community. Internal
publication, competition, peer evaluation, proposals and
counterproposals are all taking place inside experiments. Having
followed these two experiments in detail we can begin to
understand why. No two detectors the size of E1A or Gargamelle
are identical. Yet their particular features are precisely at
issue in the arguments surrounding the neutral-currents. In E1A
the two most important background processes issued from the
geometry of the counters and spark chambers and the thickness of
the plate between calorimeter segment 16 and spark chamber 4. The
Gargamelle collaboration faced analogous backgrounds that
depended directly on the configuration of concrete shielding,
chamber walls and magnets. The proper audience for studies of
these features of the detectors and their impact on background
physical processes had to be, in the first instance, members of
the collaborations. The sociological division of work within the
large collaboration is part and parcel of the construction of an
argument for or against the reality of a physical effect.

The neutral-current particle physics experiments that I have
discussed in this talk raise a question easily asked, perhaps
less simply answered: "What is a physics experiment?" Is the
Gargamelle collaboration conducting one continuous experiment
from the planning of Gargamelle to the day it was shut down in
the 1970's? Are there three experiments, one a search for the W,
one a study of partons and the last on neutral currents? Do the

neutral-current experiments themselves divide into leptonic and
hadronic experiments? Are each of the unpublished background
studies to be counted as separate experiments? Should individual
computer simulations themselves become experiments? In short:
how do we parse the events surrounding Gargamelle?

In order to grasp the historical continuity of experimental
work one needs to study more than the varied experimental
responses to a particular theoretical question. We need a
history of experimental skills. Skills are essential not simply
because they are means to accomplish a theory-determined end, but
because the development of experimental skills is fundamentally
interwoven with the construction of persuasive experimental
arguments. Do the experimenters trust Monte-Carlos? Do they
rely on neutron background calculations? In the case of neutral
currents we saw how crucial these questions were and how their
answers depended on the experimenters' past experience.

The histories of skills and machines are important for other
reasons as well: very frequently experiments begin with one set
of goals and end with another. We saw in this talk how many of
the attitudes and techniques from the "older" neutrino physics
carried over to the early work on "new" neutral-current
phenomena. Here, as in many other experiments, we can learn a
great deal by studying how goals are shifted and by locating the
continuities that link the different projects. The construction
of machines and the analysis of data cut across theoretical
boundaries, forcing us to consider together techniques and tools
created for one purpose, and used repeatedly across very
different questions. In the course of developing an account of
how persuasive arguments issue from the skills of
experimentation, we will come to understand how data can be
transformed from a collection of curiosities -- such as the first
muonless photographs -- into one of the empirical cornerstones
of our view of unified field theory.

1 This talk builds on and compresses a detailed study: P. Galison, "How the First Neutral Current Experiments Ended," Rev. Mod. Phys. 55 (1983): 477-509. A fascinating, opposing view is: A. Pickering, "Against Putting the Phenomena First," Stud. Hist. Phil. Sci. 15 (1984): 85-117. Pickering argues that "particle physicists accepted the existence of the neutral current bcause they could see how to ply their trade more profitably in a world in which the neutral current was real." For a comparison of our viewpoints see P. Galison, How Experiments End (Chicago Univ. Press, forthcoming).

2 E. Fermi, "Versuch einer Theorie der [beta]-Strahlen," Z. Phys. 88 (1934):161.

3 Lagarrigue, letter to Weisskopf 14 February 1964; Weisskopf to Lagarrigue 6 March 1964. CERN archives (hereinafter CERN-arch) DG 20568.

4 See for example CERN documents SPC/195; SPC/576; FC/770; CERN archives Diradm. F434. Further documentation in Galison, Experiments, op. cit., ch. 2.

5 Gargamelle notebook "Organization," document: Vernard, "Repartition des Tâches." Musset papers.

6 G. Feinberg, in Brandeis Summer Institute in Theoretical Physics. Lectures on Astrophysics and Weak Interactions, 1963, vol. 2, pp. 278-375 on 282.

7 John Locke, An Essay Concerning Human Understanding (New York: Dover, 1959), vol. II, p.367.

8 G. Bernardini, in Proceedings of the International School of Physics "Enrico Fermi" Course XXXII, pp. 1-51, on p. 1; Marshak, Riazuddin, and Ryan, Theory of Weak Interactions in Particle Physics. (New York: Wiley Interscience, 1969), p. 319; Commins, Weak Interactions (New York: McGraw-Hill, 1973), p. 239.

9 J.F. Allard, "Proposition d'une grande chambre à bulles à liquides lourds destinée à fonctionner auprès du synchrotron à protons du CERN," Printed Report, Paris and Saclay, Summer 1964.

10 D.H. Perkins, "Draft Gargamelle Proposal," typescript, 4 February 1970.

11 M. Briedenbach; J.I. Friedman; and H.W. Kendall, "Observed Behavior of Highly Inelastic Electron-Proton Scattering," Phys. Rev. Lett. 23 (1969): 935-939.

12 Perkins, "Draft Gargamelle," p.1.

13 't Hooft, "Renormalization of Massless Yang-Mills Fields."
Nucl. Phys. B 33 (1971): 173-199.

14 Coleman, "The 1979 Nobel Prize in Physics," Science 206
(1979): 1290-1292.

15 Palmer, "Very Preliminary Results of Neutral Current Search in
the Neutrino-Freon Exp." Typescript, May 1972, Palmer papers.
Letter to author 8 June 1983.

16 Palmer, ibid.

17 ibid.

18 D.C. Cundy and M. Haguenauer, "Resumé of the Gargamelle
Collaboration Meeting Held in Paris on March 2 and 3 1972," CERN-
TCL.

19 ibid.

20 Perkins, in Proceedings of the XVI International Conference on
High Energy Physics, 6 September - 13 September 1972, vol. 4,
Plenary Sessions: Mostly Currents and Weak Interactions, pp. 189-
247 (Batavia: National Accelerator Laboratory, 1972), on p.208.

21 ibid.

22 Musset, Transparencies from presentation to American Physical
Society Meeting, January 1973. Musset papers.

23 E.A. Paschos and L. Wolfenstein, "Tests for Neutral Currents
in Neutrino Interactions," Phys. Rev. D 7 (1973): 91-95.

24 Faissner, letter to author 7 December 1981.

25 Perkins, letter to author 9 September 1983.

26 Faissner, letter to Lagarrigue 11 January; Lagarrigue to
Faissner 16 January 1973; Faissner to Jentschke 9 February 1973.
Faissner papers.

27 Faissner, letter to Lagarrigue 11 January 1973.

28 Rousset, "Schedule of Neutrino Experiments in Gargamelle in
1973," CERN-TCL, Memorandum to Professor Cresti, 19 February
1973.

29 Musset, "Study of Hadronic Neutral Currents." CERN-TCL, 19
March 1973.

30 Musset, "List of the NC Events > 1 GeV Controlled at CERN on
the 12 & 13 April 73." CERN-TCL, 17 April 1973.

31 Cundy, "Minutes of the Neutrino Collaboration Meeting Held on 21st March 1973 at CERN." CERN-TCL, 26 March 1973.

32 Cundy, "Minutes of the Neutrino Collaboration Meeting Held at CERN on 17th May 1973 at CERN [sic] CERN-TCL, 21 May 1973.

33 Baldi and Musset, "Analysis of the NC candidates in [neutrino] and [antineutrino] Run by a Self-Contained Method." CERN-TCL, 16 May 1973.

34 Rousset, e.g. "Calcul du Bruit de Fond de Neutrons," CERN-TCL, 22 May 1973.

35 E.C.M. Young, "High-Energy Neutrino Interactions," CERN Yellow Report 67-12.

36 See Fry and Haidt, "Evaluation of the Neutron Flux in Equilibrium With the Neutrino Beam," CERN-TCL, Technical Memorandum, 22 May 1973; idem., "Calculation of the Neutron-Induced Background In the Gargamelle Neutral Current Search," CERN Yellow Report 75-1 (1975).

37 Vialle, interview with author 28 November 1980.

38 Fiorini, letter to Musset 11 July 1973. Musset papers.

39 Cundy, among others, did not feel the paper was in a suffiently persuasive form at this time. Interviews with Cundy, Musset, and Vialle, 1980. Pullia too was nervous about the conclusions in July 1973 and felt fully confident only when the results of E21 at Fermilab were published. Interviews with Pullia, Fiorini, 1984 .

40 Hasert et al., "Search for Elastic Muon Neutrino-Electron Scattering," Physics Letters 46 (1973): 121-124.

41 Hasert et al., "Observation of Neutrino-Like Interactions Without Muon or Electron in the Gargamelle Neutrino Experiment," Phys. Lett. B 46 (1973):138-140. An expanded version of this paper was subsequently published as Hasert et al., "Observation of Neutrino-Like Interactions Without Muon or Electron in the Gargamelle Neutrino Experiment," Nucl. Phys. B 73 (1974):1-22.

42 Mann, "W Searches with High-Energy Neutrinos and High-Z Detectors," in National Accelerator Laboratory, 1969 Summer Study. vol. 4 Experiments, pp. 201-207. (Batavia: National Accelerator Laboratory, no date)

43 Camerini, Cline, Fry and Powell, "Search for Neutral Leptonic Currents in K-plus decay," Phys. Rev. Lett. 13 (1964): 318-321.

44 Cline, "Experimental Search for Weak Neutral Currents,"
reprint from the school held in Heceg Novi, Yugoslavia, 1967.

45 E.W. Beier, D. Cline, A.K. Mann, J. Pilcher, D.D. Reeder,
and C. Rubbia, "Harvard-Pennsylvania-Wisconsin Collaboration NAL
Neutrino Proposal," Draft HWPF Proposal, 1970.

46 Ibid.

47 Cline, Mann, and Rubbia, "Detection of the Weak Intermediate
Boson through its Hadronic Decay Modes," Phys. Rev. Lett. 25
(1970): 1309-1312.

48 Representatives from E1A, 21 and 38 were all invited by Wilson
to make presentations in which they would defend their "unique"
advantages and disadvantages. R.R. Wilson to B. Barish, 9
September 1970. Program Planning Office, Fermilab.

49 Weinberg, "Conceptual Foundations of the Unified Theory of
Weak and Electromagnetic Interactions," Rev. Mod. Phys. 52
(1980):515- 523, on 518.

50 Rubbia interview, 3 October 1980.

51 Benvenuti, Cline, Imlay, Ford, Mann, Pilcher, Reeder, Rubbia,
and Sulak, letter to R.R. Wilson 14 March 1973. Mann papers.

52 Mann interview, 29 September 1980.

53 Benvenuti et al., "Early Observation of Neutrino and
Antineutrino Events at High Energies," Phys. Rev. Lett. 30
(1973):1084-1087.

54 Rubbia, letter to Lagarrigue 17 July 1973. Lagarrigue papers.

55 Lagarrigue, letter to Rubbia 18 July 1973. Lagarrigue papers.

56 Myatt, letter to author 3 November 1981.

57 Myatt, "Neutral Currents," in Proceedings of the 6th
International Symposium on Electron and Photon Interactions at
High Energies. Physikalisches Institut, University of Bonn,
Federal Republic of Germany, August 27-31 1973, edited by H.
Rollnik and W. Pfeil, pp. 389-406 (Amsterdam-London: North-
Holland Publishing Company, 1974).

58 Musset, "Neutrino Interactions," in 2nd International
Conference on Elementary Particles, Aix-en-Provence, 6-12
September 1973, published in Journal de Physique 34 (1973):C1-23-
42, supplement to no. 10, Colloques.

59 Weinberg, "Recent Progress in Gauge Theories of Weak, Electromagnetic, and Strong Interactions," in ibid.:C1-45-67, on 47.

60 Anonymous, Referee Reports, Physics Review Letters, 16 October 1973. Imlay papers, Sulak papers.

61 Cline, "Performance of Revised E1A Detector for [muonless] Event Search," Univ. of Wisconsin, Madison, Technical Memorandum, 1 October 1973.

62 ibid.

63 Messing, letter to author 22 May 1984.

64 Cline, "Statistical Analysis of the [muonless] Events from the Test Run," Univ. of Wisconsin, Madison, Technical Memorandum, 11 October 1973.

65 ibid.

66 Cline, "Unified Analysis of the [muonless] Events in the November-December runs," Univ. of Wisconsin, Madison, Technical Memorandum, 16 October 1973.

67 Typescript draft entitled, "Search for Neutrino Induced Events without a Muon in the Final State." The manuscript is undated, but it is referred to in a letter dated 13 November (discussed below), and was therefore probably written during the second week of November 1973.

68 Musset and Vialle, "Gargamelle collaboration, 20 November 1973," CERN-TCL, 20 November 1973.

69 Letter from D. Cline, A.K. Mann, D.D. Reeder, and C. Rubbia to A. Lagarrigue, 13 November 1973, signed by Cline, Mann and Rubbia. Signed version in Mann's files, unsigned copy in Lagarrigue's scientific papers at Orsay. Copies in files of other members of the Gargamelle collaboration. I would like to thank Mme. Lagarrigue and Professor Morellet for permission to see these papers.

70 Cline, "Data reported at NAL Talk, December 6, 1973," Univ. of Wisconsin, Madison, Technical Memorandum, 13 December 1973 with photocopied transparencies, 6 December 1973.

71 Mann interview, 29 September 1980.

72 Sulak, "Early Study of [antineutrino] Run Nov. '73," Harvard University Technical Memorandum, 17 November 1973.

73 Cline, "Are We Seeing a [muonless] Signal at the Level of 10%?" University of Wisconsin, Madison, TM, 13 December 1973.

74 Mann, "Summary of 300 GeV [antineutrino] Run for N/C Experiment," University of Pennsylvania, Technical Memorandum with spark chamber photographs, 26 January 1974.

75 Benvenuti et al., "Observation of Muonless Neutrino-Induced Inelastic Interactions," Phys. Rev. Lett. 32 (1974): 800-803; Aubert et al., "Further Observation of Muonless Neutrino-Induced Inelastic Interactions," Phys. Rev. Lett. 32 (1974): 1454-1457.

76 Sakurai, "Remarks on Neutral Current Interactions in Neutrino-1974. American Institute of Physics Conference Proceedings, April 1974, edited by C. Baltay, no. 22 Particles and Fields, subseries no. 9, pp. 57-63, on p. 57, (New York: American Institute of Physics, 1974).

77 Feynman, "Conference Summary," in Neutrino-1974, ibid., pp. 299-327 on p. 315.

78 B.C. Barish, "Results from the Cal Tech FNAL Experiment," presented at the Parallel Session on Neutral Currents and Heavy Leptons. In XVII International Conference on High Energy Physics, London, 1974, edited by J.R. Smith, pp. IV-111-113. (Chilton, Didcot, Oxon: Rutherford Lab, 1974); P. Schreiner, "Results from the Argonne 12-foot Bubble Chamber Experiment," ibid., pp. IV-123 126; W. Lee, "Observation of Muonless Neutrino Reactions," Columbia, Illinois, Rockefeller, Brookhaven Collaboration in ibid., pp. 127-128.

ACKNOWLEDGEMENTS

Above all I would like to thank those members of the E1A and Gargamelle collaborations for helpful discussions and the provision of documents: B. Aubert, C. Baltay, E. Belotti, F.W. Bullock, U. Camerini, D. Cline, D.C. Cundy, H. Faissner, E. Fiorini, W. Fry, D. Haidt, P. Heusse, R. Imlay, T.W. Jones, A.M. Lutz, A.K. Mann, F. Messing, D. Morellet, P. Musset, G. Myatt, R. Palmer, D.H. Perkins, C. Peyrou, D. Reeder, M. Rollier, A. Rousset, C. Rubbia, L.R. Sulak, J.P. Vialle and J.K. Walker. In addition I had many helpful conversations with: F. Everitt, H. Georgi, I. Hacking, E. Hiebert, G. Holton, A.I. Miller, C. Prescott, K.H. Reich, C. Quigg, S. Schweber, W. Sletten, S. Weinberg, V. Weisskopf and T. Yamanouchi. I would also like to thank the CERN and Fermilab archives for access and the Pew Memorial Foundation and the National Science Foundation for support and Rev. Mod. Phys. for permission to excerpt from ref. 1.

CHAPTER 6
OTHER WEAK-NEUTRAL-CURRENT PROCESSES, PARITY VIOLATION IN WNCs, AND $\sin^2\theta_W$

A Force in the Mirror (1975 - 1978)

6.1.   THE STUDY OF OTHER REACTIONS OF WNCs

While the deep inelastic interactions studied in 1973–1974 were the key to the proof of the existence of a WNC interaction, the observation of other more specific interactions was of crucial importance in establishing the underlying structure of the WNC [6.1]-[6.3]. Some of the key processes that were observed were

$$\nu + N \rightarrow \pi + N + \nu \quad , \tag{6.1}$$

$$\nu + p \rightarrow \nu + p \quad , \tag{6.2}$$

$$\nu_\mu + e^- \rightarrow \nu_\mu + e^- \quad . \tag{6.3}$$

These processes are low rate and rather difficult to separate from various backgrounds. We reproduce here three of the papers (A)-(C) that were either the first, or perhaps the most convincing, to proclaim the observation of reactions (6.1)-(6.3). The reactions are listed in the order of increasing difficulty of observation (and the increasing year), as can be seen by studying the papers reprinted here [articles (A)-(C)] and Refs. [6.4]-[6.6]. Once these reactions were all observed, it was possible to state that the WNC was a universal interaction.

These reactions were also important to help differentiate between the various modes for WNC that had been proposed. Early on, the data favored the SU(2) × U(1) model of GWS, but it was the issue of the parity violation in the WNC process that was to be the crucial test of the GWS model, as we will now see. (For lack of space, we will not discuss the competing theories that were ruled out by these studies.)

6.2.   THE OBSERVATION OF PARITY VIOLATION IN THE WEAK-NEUTRAL-CURRENT PROCESS

In 1975, the existence of WNC had been fully confirmed, and it was then time to attempt to learn more about the space–time structure of the interaction. In the case of the charged current interaction, this took about 25 years (1933 to 1958). In the case of the WNC, it only took three years (1975 to 1978), because of the possibility of comparing (1) the WNC to charged currents, and the possibility of observing effects of the WNC in (2) the atomic system and in (3) polarized electron scattering. By 1978, all three measurements yielded results consistent with the GWS model and, in the process, proved that the WNC leads to the violation of parity symmetry.

One way to test for parity nonconservation in the WNC interaction was to express the WNC cross section as

$$\frac{d\sigma^\nu}{dy} = a + b(1 - y)^2 \quad , \tag{6.4}$$

$$\frac{d\sigma^{\bar{v}}}{dy} = b + a(1 - y)^2 \quad , \tag{6.5}$$

where $y = E_{hadron}/E_v$. The HPWF experiment reported in 1976 that $a \neq b$, thus indicating parity nonconservation [6.7].

The actual comparison of WNC and the charged current interaction took the form of measuring the ratios [6.7] (and can be expressed in terms of $\sin^2\theta_W$ as)

$$R^v = \frac{\sigma(v_\mu + N \rightarrow v_\mu + x)}{\sigma(v_\mu + N \rightarrow \mu^- + x)} = \frac{1}{2} - \sin^2\theta_W + \frac{20}{27}\sin^4\theta_W \quad , \tag{6.6}$$

$$R^{\bar{v}} = \frac{\sigma(\bar{v}_\mu + N \rightarrow \bar{v}_\mu + x)}{\sigma(\bar{v}_\mu + N \rightarrow \mu^+ + x)} = \frac{1}{2} - \sin^2\theta_W + \frac{20}{9}\sin^4\theta_W \quad , \tag{6.7}$$

where the $\sigma(vN)$ refers to the cross section of a mostly $I = 0$ target. In the case of the charged current interactions, the difference between the $v_\mu N$ and $\bar{v}_\mu N$ is due to the presence of both V and A interactions, which results in maximal parity violation. [See the CMR article, Chapter 2(E), for a simple clarification of this concept.]

The first reported measurement that was made of R^v, $R^{\bar{v}}$ (and which clarified the difference) was made by the HPWF group using a restricted region of the detector, as shown in Fig. 6.1 [6.7]. The results reported in 1976 [paper (D)] showed that the WNC could not be pure A or V and must have a mixture of A and V, implying parity violation. While this was strong evidence for parity nonconservation, the issue was of such extreme importance to the understanding of the WNC that it was essential to determine this issue in a non-neutrino interaction.

The observation of parity violation in the atomic system of Bismuth was reported [paper (E)] by L. Barkov and M. Zolotorev in March 1978 [6.8]. In a classic piece of scientific work, the SLAC team of C. Prescott et al. were the first to observe parity violation in deep-inelastic electron scattering [[6.9], paper (F)]. The schematic layout of the experiment is shown in Fig. 6.2.

By 1978, it was clear that the WNC had components of V and A interactions and was consistent with the GWS model. Thus, the major breakthrough of determining the space-time structure of the WNC took less than five years!

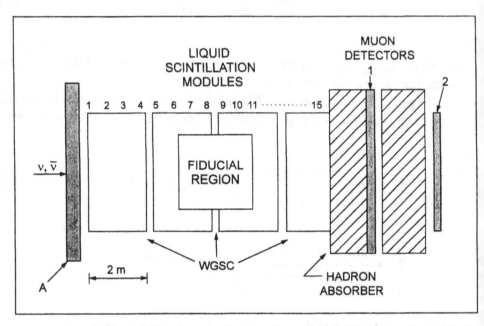

Figure 6.1. The HPWF detector at FNAL in 1975 that measured the modern value of $\sin^2\theta_w$ and indicated parity violation in the WNC (from [6.1].

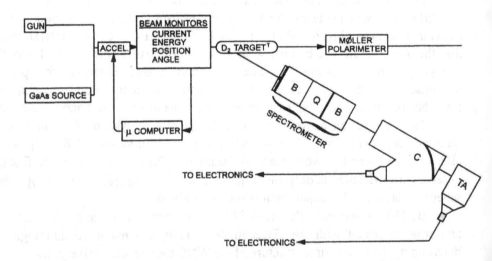

Figure 6.2. Schematic layout of the experiment. Electrons from the GaAs source or the regular gun are accelerated by the linac. After momentum analysis in the beam transport system, the beam passes through a liquid deuterium target. Particles scattered at 4° are analyzed in the spectrometer (bend–quad–bend) and detected in two separate counters (a gas Čerenkov counter and a lead–glass shower counter). A beam monitoring system and a polarization analyzer are only indicated, but they provide important information in the experiment (from [6.9]).

6.3. EARLY DETERMINATION OF $\sin^2\theta_W$ IN NEUTRINO INTER-ACTION AND COMPARISON WITH THE GWS MODEL

The key to proving that the GWS model was correct was measuring the mixing angle, $\sin^2\theta_W$, in as many different reactions as possible. If the same value (within errors) were obtained, this would be strong evidence for the validity of the GWS model, and this is exactly what happened during this period of time. In addition, as we will see later, the value of $\sin^2\theta_W$ set the mass scale for the W and Z particles and was crucial in the choice of the $\bar{p}p$ collider as a means to discover the W and Z particles later on!

One of the most convincing early determinations of $\sin^2\theta_W$ was carried out by the CDHS group at CERN in 1977 [see paper (G)]. This group carried out a precision measurement of R^ν and $R^{\bar\nu}$ and then, using formulas (6.6) and (6.7), determined $\sin^2\theta_W$ [6.10]. In Fig. 6.3, we show the history of the determination of $\sin^2\theta_W$. (We have taken the liberty of also using the 1983 W/Z data in Fig. 6.3 to show how the various determinations of $\sin^2\theta_W$ agreed at the time.)

6.4. OBSERVATION OF DIMUON EVENTS AND CHARM PRODUCTION BY NEUTRINO BEAMS

In this book, we are focusing on the observation of WNC and the electroweak force. However, the neutrino detectors that yielded the first observation of the WNC also observed a curious set of events with two muons, or a muon and an electron in the final state, which caused significant confusion in the period of 1974 to 1976. Early on, Gaillard, Lee, and Rosner [6.11] suggested that these events were due to the production of bare charm, and we now know this to be true. This finding led to a unified explanation of all of the phenomena observed in the neutrino interactions at that time.

The first group to observe the dimuons was the HPWF group [6.12],[6.13]. Figure 6.4 shows one of the first events to be reported by the HPWF group in early 1974 [6.13]. The reaction is now known to be

$$\nu_\mu + N \rightarrow \mu^- + \text{charm} + X \rightarrow (\mu^+ + \nu + \text{strange particle}) \ . \qquad (6.8)$$

There were other observations of charm in neutrino interactions at BNL and at CERN, for which we provide references [6.14]–[6.17]. Of course, the famous J/ψ particle is now known to be a charm–anticharm bound state.

The direct observation of charm particles, where the mass was directly determined, was carried out just at SPEAR at SLAC [6.18]. We should point out that Dr. Niu's group at Nagoya University observed events in nuclear emulsions exposed to cosmic rays, which also suggested a new particle. Subsequently, the

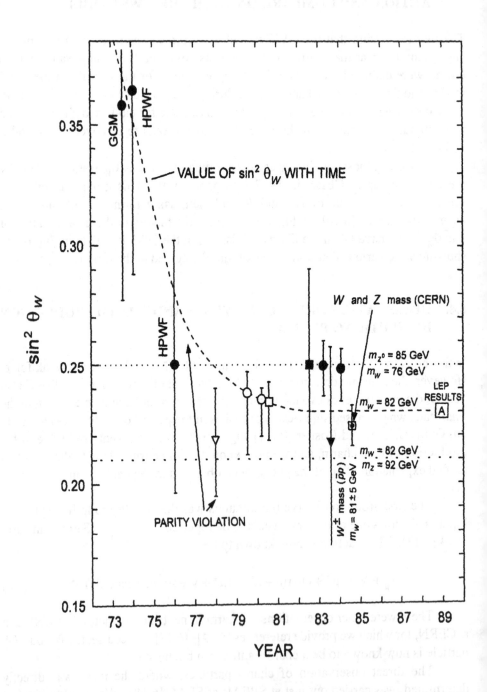

Figure 6.3. History of the value of $\sin^2\theta_w$ and m_w/m_z. Note that the first values of $\sin^2\theta_w$ below 0.3 were determined in 1976 by the HPWF group. This led to the expectation that the mass of the W and Z particles would be above 70 GeV or so.

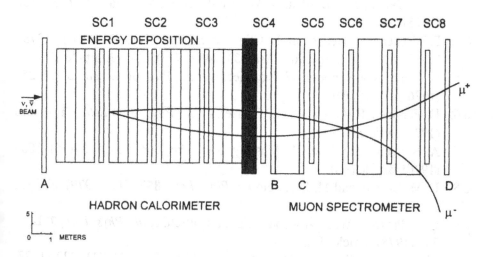

Figure 6.4. Sketch of a muon-pair event in the HPWF detector at FNAL in 1974, which starts in module 5 of the ionization calorimeter and deposits 21.8-GeV ionization energy. The muon momenta are $\rho_{\mu^+}\rho = 14.7$ GeV and $\rho_{\mu^-} = 8.4$ GeV (from [6.13].

dimuons, SPEAR charm, and emulsion events have been interpreted as the production and decay of charmed particles, thus leading to the identification of the fourth (or charm) quark.

In summary, during the period of 1974 to 1978, the following major advances had been made:

1. The WNC was shown to be universal in that in the reactions where it was expected, it was observed.
2. Parity nonconservation had been observed for the WNC interaction.
3. The first "precision" value of $\sin^2\theta_W$ was determined, and it was shown to be less than 0.25, which is consistent with the expectations of item 2.
4. The value of $\sin^2\theta_W$ determined from different processes was the same within errors – in excellent agreement with the GWS model.
5. The fourth (charm) quark had been clearly identified in e^+e^- and neutrino interactions, as was expected for the GIM mechanism.

This was indeed a remarkable period in science and particle physics.

REFERENCES

6.1. A. Benvenuti, D. C. Cheng, D. Cline, *et al.*, *Phys. Rev. Lett.*, **32**, 800 (1974).

6.2. B. C. Barish, J. F. Bartlett, K. W. Brown, *et al.*, *Phys. Rev. Lett.*, **34**, 538 (1975).

6.3. P. Wanderer, A. Benvenuti, D. Cline, *et. al.*, *Phys. Rev.*, **D17**, 1679 (1978).

6.4. S. J. Barish *et al.*, *Phys. Rev. Lett.*, **33**, 448 (1974); also see article (A).

6.5. D. Cline, A. Entenberg, W. Kozanecki, *et al.*, *Phys. Rev. Lett.*, **37**(5), 252 (1976) [article (B)].

6.6. H. Faissner, H. G. Fasold, E. Frenzel, *et al.*, *Phys. Rev. Lett.*, **41**(4), 213 (1978) [article (C)].

6.7. A. Benvenuti, D. Cline, F. Messing, *et al.*, *Phys. Rev. Lett.*, **37**(16), 1035 (1976) [article (D)].

6.8. L. M. Barkov and M. S. Zolotorev, *Phys. Lett.*, **85B**, 308 (1979); also see article (E).

6.9. C. Y. Prescott, W. B. Atwood, R. L. A. Cottrell, *et al.*, *Phys. Lett.*, **77B**(3), 347 (1978) [article (F)].

6.10. M. Holder, J. Knobloch, J. May, *et al.*, *Phys. Lett.*, **71B**(1), 222 (1977) [article (G)].

6.11. M. K. Gaillard, B. Lee, and J. Rosner, *Rev. Mod. Phys.*, **47**, 227 (1975).

6.12. B. Aubert *et al.*, in *Proceedings of the 17th International Conference on High Energy Physics, London England, 1974*, J. R. Smith, ed. (Rutherford High Energy Lab., Didcot, Berkshire, England, 1975); and in *Neutrinos–1974*, C. Baltay, ed. (AIP Conference Proceedings No. 22, AIP, New York, 1974), p. 201.

6.13. A. Benvenuti *et al.*, *Phys. Rev. Lett.*, **34**, 419 (1975).

6.14. E. Cazzoli *et al.*, *Phys. Rev. Lett.*, **34**, 1125 (1975).

6.15. J. von Krogh *et al.*, *Phys. Rev. Lett.*, **36**, 710 (1976).

6.16. Additional evidence based on four dimuon events was given by B. C. Barish *et al.*, Colloq. Int. CNRS 245, 131 (1975).

6.17. H. Deden *et al.*, *Phys. Lett*, **58B**, 361 (1975); also J. Blietschau *et al.*, *Phys. Lett.*, **60B**, 207 (1976).

6.18. G. Goldhaber *et al.*, *Phys. Rev. Lett.*, **37**, 255 (1976).

Observation of Single-Pion Production by a Weak Neutral Current*

S. J. Barish, Y. Cho. M. Derrick, L. G. Hyman, J. Rest, P. Schreiner,
R. Singer, R. P. Smith, and H. Yuta
Argonne National Laboratory, Argonne, Illinois 60439

and

D. Koetke
Concordia Teachers College, River Forest, Illinois 60305

and

V. E. Barnes, D. D. Carmony, and A. F. Garfinkel
Purdue University, Lafayette, Indiana 47907
(Received 28 May 1974)

In exposures of the Argonne National Laboratory 12-ft bubble chamber filled with hydrogen and deuterium to a neutrino beam, we have observed events consisting of (1) a single π^+ meson originating in the liquid, and (2) a proton with an e^+e^- pair pointing to it. Only a small fraction of these events can be ascribed to known reactions such as $np \to nn\pi^+$ and $np \to np\pi^0$. The remaining events, which correspond to a signal of about 4.5 standard deviations, we ascribe to the reactions $\nu p \to \nu n \pi^+$ and $\nu p \pi^0$.

In the conventional current-current theory of the weak interaction, only charge-changing lepton bilinear combinations ($\nu_\mu \mu^-$), etc., are used and, although the theory is not renormalizable, it accounts in a quantitative way for almost all experimental facts. Direct evidence for neutral-current combinations such as (ν, ν) is difficult to obtain, and it is only recently that experiments searching for the existence of the inclusive reactions $\nu N \to \nu + \text{anything}$ and $\bar{\nu} N \to \bar{\nu} + \text{anything}$ have reported positive results.[1] In strangeness-changing decays, no evidence has been seen for neutral currents and very stringent limits have been set. Formulations of weak interaction theory that are renormalizable have been proposed[2] and require the existence of neutral currents or heavy leptons. Thus, the question of the existence and properties of the neutral-current interaction is of paramount importance.

Our experiment measures single-pion production through the reactions[3]

$$\nu_\mu p \to \nu_\mu n \pi^+, \tag{1}$$

$$\nu_\mu p \to \nu_\mu p \pi^0. \tag{2}$$

Three separate exposures of the Argonne National Laboratory 12-ft bubble chamber were used: the first with a hydrogen filling and the other two with deuterium fillings. Each exposure consisted of about 300 000 pictures and had about 4×10^{17} protons incident on the primary production target. An analysis of single-pion production in the

charged-current reaction

$$\nu p \to \mu^- p \pi^+ \tag{3}$$

and a brief description of the experimental arrangement have been published.[4]

The signal for Reaction (1) is a single π^+ meson originating in the chamber fiducial volume of 11.1 m³. Ten events were uniquely identified as π^+ mesons because the single positive track stopped in the chamber and decayed via the chain $\pi^+ \to \mu^+ \to e^+$. Four events were identified because the track scattered giving a unique kinematic fit to $\pi^+ p$ elastic scattering. In addition, four leaving π^+ tracks were uniquely identified on the basis of energy loss. If interpreted as protons, they were overstopped by 5 or more standard deviations, and they are inconsistent with incoming π^-.[5]

The signal for Reaction (2) is a proton track originating in the chamber fiducial volume with a converted γ ray (e^+e^- pair) pointing to the origin of the proton track. We find thirteen events of this topology which we refer to as one-prong $+\gamma$. Because our experiment is done near threshold, 99% of the protons from the $\nu p \to \mu^- p \pi^+$ events have the azimuth and dip of the proton track within 60° of the neutrino beam direction, and we apply this selection to the proton track of the one-prong $+\gamma$ events. In addition, 88% of protons from Reaction (3) have momentum < 1 GeV/c, so we apply this cut. Eight one-prong $+\gamma$ events satisfy these criteria.

Since the Reactions (1) and (2) are not constrained, we cannot prove that any individual
event originates from these reactions and not
from such processes as

$$np \rightarrow nn\pi^+, \tag{4}$$

$$np \rightarrow np\pi^0. \tag{5}$$

We measure the neutron background using events
of the reaction[6]

$$np \rightarrow pp\pi^-, \tag{6}$$

which are observed as three-prong events and
are identified by a one-constraint kinematic fit.
Since the final states $nn\pi^+(\pi^0)$ and $pp\pi^-(\pi^0)$ are
charge symmetric, the number of events fitting
the $pp\pi^-$ hypothesis is a direct measure of the
number of π^+ mesons coming from the $nn\pi^-$ final
state.[6]

In order to scale our observed number of np
$\rightarrow pp\pi^-$ events to measure the background from
Reactions (4) and (5) within our selection criteria,
a number of correction factors are necessary.
These corrections were measured in a separate
experiment in which the bubble chamber was exposed to a 0- to 3-GeV/c neutron beam. This exposure well simulates the neutron background we
observe in the neutrino exposure.

Figure 1 shows the pion momentum spectrum

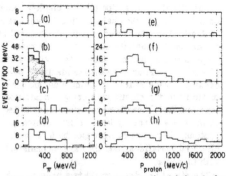

FIG. 1. Pion momentum spectrum for (a) the single-
π^+-meson sample, eighteen events; (b) $\nu p \rightarrow \mu^- p\pi^+$,
133 events; (c) $np \rightarrow pp\pi^-$ in the neutrino film, twelve
events; and (d) $np \rightarrow pp\pi^-$ in the neutron film, 51 events.
The shaded events in (b) represent the $\nu p \rightarrow \mu^- p\pi^+$
events after applying our detection efficiency on the
single π^+ sample. Proton momentum spectrum for (e)
the one-prong +γ sample, nine events; (f) $\nu p \rightarrow \mu^- p\pi^+$,
102 events; (g) $np \rightarrow pp\pi^-$ in the neutrino film, fourteen
entries; and (h) $np \rightarrow pp\pi^-$ in the neutron film, 102 entries.

for (a) the single-π^+ meson sample, (b) the $\mu^- p\pi^+$
events, (c) the $pp\pi^-$ events in the ν film, and
(d) the $pp\pi^-$ events observed in the auxiliary neutron experiment. As seen in Fig. 1(b), only 12%
of the charged-current events have pion momentum ≥ 400 MeV/c; in addition, our detection efficiency for single π^+ is low above this momentum
so we select as a neutral-current signal π^+ mesons with momentum below 400 MeV/c. Furthermore, we find that 90% of the background neutrons which give rise to Reaction (6) enter through
the top of the chamber (where the shielding is
weakest because of the camera ports), and produce a π^- which also goes downwards. Hence,
we choose events with the pion traveling upwards
in the chamber. These selections give a neutron-
induced background from Reaction (4) of 0.55
± 0.55 events.

In order to measure the background from photoproduced mesons $\gamma p \rightarrow nn^+$, etc., we use the
e^+e^- pairs found in a scan of 55 000 pictures. Almost all of these pairs are close to cosmic-ray
tracks as the photon must be have been produced by bremsstrahlung in the magnet iron above the chamber.
By rejecting all events having the vertex within
20 cm of any cosmic ray, we drastically reduce
this background.[5] After this selection, the measured photoproduced background to Reaction (1)
is 0.33 ± 0.10 events.

For those events where the π^+ leaves the cham-
ber, a possible background is $\bar{\nu} p \rightarrow \mu^+ n$. Using
our $\bar{\nu}$ beam contamination, we calculate a background of 0.04 ± 0.04 events from this reaction.
Finally, K_L^0-induced background from, for ex-
ample, $K_L^0 \rightarrow \Lambda \pi^+$ has been measured and is found
to be negligible.

We now discuss the background contributions
to the one-prong +γ events. Figure 1(e) shows
the proton momentum distribution for these after
the angle selections discussed previously. Figures 1(f)-1(h) show the proton momentum spec-
trum from the $\nu p \rightarrow \mu^- p\pi^+$ events and the $np \rightarrow pp\pi^-$
events from the ν film and the neutron film, re-
spectively. Applying the same selections to the
$pp\pi^-$ events as to the one-prong +γ events, we
are left with five $pp\pi^-$ events. We have measured the empirical ratio of one-prong +γ events
to $pp\pi^-$ events to be 0.19 ± 0.04 in our neutron experiment, so after some small corrections, the
neutron background to Reaction (2) is 0.92 ± 0.40
events.

The photoproduced background $\gamma p \rightarrow p\pi^0$ was
measured as described above and is only 0.02
± 0.01 events. The probability of a converted γ

ray accidentally pointing to a one-prong event was determined to be 0.11 ± 0.11 events. The K_L^0-induced background was again found to be negligible. Finally, for the one-prong $+\gamma$ events, there is a background source consisting of an incoming π^- that undergoes charge exchange with the subsequent conversion of one γ ray from the π^0. We have measured this background by analyzing $\pi^- p \to \pi^- p$ scatters of all π^- that enter the chamber and whose tracks could not be distinguished from a leaving proton. By scaling from the known elastic and charge-exchange cross sections, this background was measured to be 0.52 ± 0.23 events.

Table I gives a summary of the events found. With the selections mentioned, we observe with no other corrections seven single-π^+ events and seven one-prong $+\gamma$ events with a total measured background of 2.49 ± 0.73 events. Folding the Poisson-distributed fluctuations of the background with the uncertainty of 0.73 in the measured background level, the probability of seeing fourteen or more events when 2.49 are expected is 1.2×10^{-5} which corresponds to a 4.3 standard deviation effect. We conclude then that neutral currents exist and are responsible for the exclusive reactions (1) and (2).

In order to compare to the measured cross sec-

tion for $\nu p \to \mu^- p \pi^+$, we must treat the π^+ and proton tracks in the $\mu^- p \pi^+$ events in the same way as the neutral-current candidates and know the overall detection efficiencies for the $\nu n \pi^+$ and $\nu p \pi^0$ final states. These are 75% and 6%, respectively. The number of $\mu^- p \pi^+$ events on the sample of film is about 100. Applying the corrections, we obtain the ratios[7]

$$R_0 = \frac{\sigma(\nu p \to \nu p \pi^0)}{\sigma(\nu p \to \mu^- p \pi^+)} = 0.51 \pm 0.25, \quad p_p < 1 \text{ GeV}/c,$$

and

$$R_+ = \frac{\sigma(\nu p \to \nu n \pi^+)}{\sigma(\nu p \to \mu^- p \pi^+)} = 0.17 \pm 0.08,$$

$$p_{\pi^+} < 400 \text{ MeV}/c,$$

with $R_0 + R_+ = 0.68 \pm 0.28$. The value of $R_0 + R_+$ is somewhat larger than, but not inconsistent with, the value predicted in the Salam-Weinberg theory,[8] which gives $0.15 < R_0 + R_+ < 0.44$ for our spectrum. Our measured ratio

$$\frac{\nu p \to \nu p \pi^0}{\nu p \to \nu n \pi^+} = 3.1 \pm 2.1$$

suggests isospin-$\frac{1}{2}$ dominance of the final state, which does not support the suggestion of Sakurai[9] that the neutral current may be an isoscalar, but the errors on this ratio are too large to rule this out.

Finally, we note that all other experiments on neutral currents have problems of muon identification and/or neutron background that must be estimated using a Monte Carlo calculation. The events of the present experiment consist of a single positive track and so the absence of a μ^- particle is evident. In addition, we make direct measurements of all background contributions.

We wish to thank the staff of the zero-gradient synchrotron and the bubble chamber operators for fine cooperation in obtaining the pictures. Our scanners, under Hope Chafee, have responded well to a difficult and tedious job. J. Campbell, R. Engelmann, A. Mann, U. Mehtani, and B. Musgrave gave help in the earlier stages of the experiment.

TABLE I. Event and background summary. For the H_2 exposure, all three π^+ modes have been analyzed; for the D_2 exposure, somewhat different fractions of the film have been processed for the three modes.

Events	H_2	D_2	Sum	Cosmic cut	π^+ dip cut
π^+ Events					
Stop	2	8	10	7	6
Scatter	3	1	4	3	1
Leave	3	1	4	3	0
One-prong					
$+\gamma$ events	3	5	8	7	7
Total			26	20	14

Background	π^+	One-prong $+\gamma$
Neutron	0.55 ± 0.55	0.92 ± 0.40
γ induced	0.33 ± 0.10	0.02 ± 0.01
$\nu p \to \mu^+ n$	0.04 ± 0.04	$\cdots$
$\pi^- p \to \pi^0 n$	$\cdots$	0.52 ± 0.23
Accidental γ pointing	$\cdots$	0.11 ± 0.11
Totals	0.92 ± 0.56	1.57 ± 0.47
Total background		2.49 ± 0.73

*Work supported by the U. S. Atomic Energy Commission.

[1]F. Hasert et al., Phys. Lett. 46B, 138 (1973); A. Benvenuti et al., Phys. Rev. Lett. 32, 800 (1974); B. Aubert et al., Phys. Rev. Lett. 32, 1454 (1974).
[2]A. Salam and J. C. Ward, Phys. Lett. 13, 168 (1964); S. Weinberg, Phys. Rev. Lett. 19, 1264 (1967).

[3]We are really measuring the sum of the cross sections $\nu p \to \nu n \pi^+ l \pi^0$ and $\nu p \to \nu p \pi^0 l \pi^0$, $l \geq 0$, but since our neutrino spectrum peaks at 500 MeV/c and is down by an order of magnitude by 1500 MeV/c, we expect the contribution of the final states with additional π^0's to be very small.

[4]J. Campbell *et al.*, Phys. Rev. Lett. <u>30</u>, 335 (1973).

[5]More details of the experiment are given by S. J. Barish, Argonne National Laboratory Report No. ANL/ HEP 7418 (unpublished).

[6]Y. Cho *et al.*, in Proceedings of the Sixteenth International Conference on High Energy Physics, The University of Chicago and National Accelerator Laboratory, 1972 (unpublished), paper 473.

[7]In doing this we are implicitly assuming that the characteristics of our neutral- and charged-current events are the same. This is true on the Salam-Weinberg model but may not be true in general. For the charged-current events, we measure the ratio $N(\nu p \to \mu^- p \pi^+ \pi^0)/N(\nu p \to \mu^- p \pi^+) = 0.1 \pm 0.05$ and, therefore, we reduce the observed $\nu N \pi/\mu^- p \pi^+$ ratios by 10%. In addition, for the one-prong + γ events, a small contribution from the reaction $\nu d \to \nu n \pi^0(p_s)$ has been subtracted.

[8]S. Adler, private communication.

[9]J. Sakurai, Phys. Rev. D <u>9</u>, 250 (1974).

Observation of Elastic Neutrino-Proton Scattering*

D. Cline, A. Entenberg, W. Kozanecki, A. K. Mann, D. D. Reeder, C. Rubbia, J. Strait,
L. Sulak, and H. H. Williams

*Department of Physics, Harvard University, Cambridge, Massachusetts 02138, and
Department of Physics, University of Pennsylvania, Philadelphia, Pennsylvania 19174, and
Department of Physics, University of Wisconsin, Madison, Wisconsin 53706*
(Received 13 April 1976)

We have observed thirty events of the process $\nu p \to \nu p$ with a background expectation of seven events. The neutral-current to charged-current ratio $\sigma(\nu p \to \nu p)/\sigma(\nu n \to \mu^- p)$ is measured to be 0.17 ± 0.05 for $0.3 < q^2 < 0.9$ $(GeV/c)^2$ where $-q^2$ is the square of the four-momentum transfer to the proton.

Because of its simplicity, one of the most interesting weak neutral-current reactions is the elastic scattering of neutrinos by protons. Previous searches[1] for this reaction have been hampered by high neutron background, poor pion-proton separation, and/or low statistics. The addition of shielding does not necessarily eliminate the neutron-background problem because of the presence of ν-induced neutrons in equilibrium with neutrinos. In this experiment the problem is significantly alleviated by using a detector of such large size that ν-induced neutrons can be absorbed or detected through their interactions in the outer regions of the detector.

The experiment was performed at Brookhaven National Laboratory in a "wide-band" horn-focused neutrino beam. The target-detector [Fig. 1(a)] consists of twelve calorimeter modules containing a total of 33 tons of liquid scintillator.[2] Each module [Fig. 1(b)] is segmented into sixteen cells which are viewed at each end by phototubes. For an energy deposition greater than 3 MeV in a given cell, precise timing and the energy depositions are recorded for each tube. This information determines the position of the source of the energy deposition along the cylinder to ± 10 cm and its timing to ± 0.5 nsec.

The front half of the detector utilizes a close-packed geometry to be fully sensitive to neutrons and charged particles entering from the sides, top, and bottom. For example, a neutron passing through the detector would signal itself by colliding with protons in several, separated cells. The last half of the detector has four large drift chambers[3] interspersed among the calorimeters.

Each chamber contains two x planes and two y planes so that the angle as well as position of any tracks exiting from a module may be determined. We have measured a single-gap efficiency of 98% for particles with angles of up to 60° relative to the beam direction. The entire apparatus is housed in a blockhouse of 1.5-m-thick heavy concrete to shield against neutrons. A 2.4-m × 3.5-m liquid scintillation counter upstream of the first calorimeter is used to veto charged particles.

The calorimeters and drift chambers are continuously calibrated by accepting beam-associated muons along with neutrino-induced triggers; in addition, vertical cosmic-ray events, recorded between machine bursts, monitor the pulse height and timing of each phototube in the system.

To estimate the cosmic-ray background, the detector is activated between beam bursts for a period of time equal to the duration of the beam

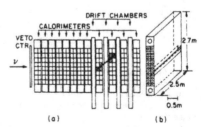

FIG. 1. (a) Side view of the apparatus showing a typical recoil proton event. (b) Diagram of a single calorimeter module.

spill. None of the events recorded during this gate satisfy the criteria imposed on neutrino-induced candidates.

The initial selection criterion for neutral-current events is containment, i.e., that neither the veto counter nor any cell in the farthest upstream calorimeter fires, and no energy deposition occurs within 40 cm horizontally or 46 cm vertically of the edge of any module nor in the most downstream 20 cm of the detector. From a data sample representing 1.8×10^{18} protons incident on the ν target, events were selected which contained a single track that originated and stopped in the scintillator, and had an energy greater than 150 MeV.

The measured range, the pattern of energy depositions, and the total energy identify the particle. Since the range and energy loss for protons typically differ from that for pions by more than a factor of 2, the particle identification is relatively unambiguous. For example, shown in Fig. 2(a) is a plot of range versus energy for a sample of events selected as protons. A more quantitative discrimination is achieved by comparison of the calculated and observed energy deposition in each cell, an example of which is shown in the inset of Fig. 2(a). The results of fits to the proton and pion hypotheses are presented in Fig. 2(b). The χ^2 is on the average greater than the degrees of freedom because of the Landau distribution and calibration errors. A clear separation, however, occurs for all but 15% of the events which deposited ≥ 150 MeV of energy. Since we observe comparable numbers of proton and pion events, this ambiguity does not modify the total number of proton candidates.

Because of the rf structure of the proton beam, the neutrino beam consists of twelve bunches of 35-nsec duration (full width at half-maximum) occurring every 220 nsec. Figure 3(a) shows a plot of event time (modulo 220 nsec) for the selected proton events. The distribution is identical to that for $\nu n - \mu^- p$ events, except for a few events which occur later than 30 nsec (which are excluded from the final sample). We conclude that at most 1.5 of the in-time events could have been induced by slow neutrons originating far upstream.

We have included in this analysis only those events with energy T greater than 150 MeV. An inspection of all candidates suggests that those with $T < 150$ MeV should be excluded for two reasons: (1) The proton-pion separation becomes increasingly ambiguous, and (2) nearly all triggers which occur outside the rf structure have $T < 150$ MeV.

Additional evidence supporting the view that the proton candidates are neutrino induced is presented in Figs. 3(b) and 3(c), which display the vertex distributions for the proton events along the length of the detector and in the plane normal to the incident beam; within statistical error both distributions are uniform, showing no attenuation in the front or away from the edges. The distribution of angles of the recoil protons show

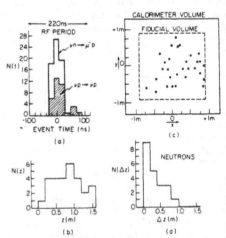

FIG. 3. (a) Distribution of event times for single protons (shaded) compared to that for a sample of charged current elastic events. (b) Distribution of p vertices along the beam direction, and (c) in the plane normal to the beam direction. (d) Distribution of the distance between successive interactions for neutron events.

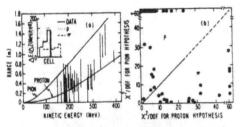

FIG. 2. (a) Plot of range versus energy for the selected p events (several events have been omitted for clarity). Each line indicates a measured range of the particle as determined by the number of cells that fire. Shown in the inset is the observed and calculated energy loss for a proton event. (b) Results of fits of the calculated to the observed energy depositions for proton and pion hypotheses.

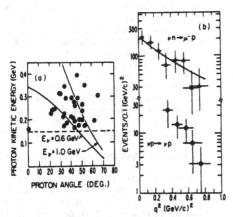

FIG. 4. (a) Kinetic energy versus angle for p events. The uncertainty in the energy is $\pm 15\%$. The absence of events below 150 MeV is due to the energy cut. Kinematic curves for $\nu p \to \nu p$ are shown for two neutrino energies. (b) Observed, corrected, q^2 distributions for $\nu n \to \mu^- p$ and $\nu p \to \nu p$. The solid line is the calculated q^2 distribution for $m_A = 0.95$ GeV/c^2, $m_V = 0.84$ GeV/c^2. The horizontal error bars represent the uncertainty in the energy measurement.

no angular asymmetries, either up-down or left-right. In contrast, neutron-induced events, selected by requiring totally contained events with a second separate energy deposition, indicate a neutron interaction length of 40 cm [see Fig. 3(c)]. We conclude that the background from in-time neutrons, produced either far upstream or in nearby masses, is small.

Additional confirmation that the events arise from ν elastic scattering is obtained from the angle-energy correlation of the protons. The neutrino flux is peaked at 1 GeV: Approximately 60% of the $\nu n - \mu^- p$ events observed correspond to ν energies between 0.6 and 1.7 GeV. Shown in Fig. 4(a) is a plot of kinetic energy versus angle for the $\nu p \to \nu p$ events. The majority of these events lie in the region expected for νp elastic scattering for incident neutrinos of this energy range. The energy distribution for $\nu p \to \nu p$ events, when corrected for geometric acceptance, agrees with that for $\nu n - \mu^- p$ events. Protons from $np \to np$ reactions induced by neutrons produced in nearby masses upstream would populate predominantly the region of the plot corresponding to $E_\nu < 0.6$ GeV. From the T-θ distribution and the timing distribution, we estimate that a total of at most three of the events could be neutron induced.

Background from the ν-induced reactions $\nu n \to \nu \pi^- p$ and $\nu n \to \mu^- p$ occurs if the second prong is contained in the vertex cell. From an extrapolation of the rate of two-prong events as the length of the second prong approaches the cell size, we estimate a background of at most two events. By a similar technique we estimate that the background from three- or more-prong events is negligible. The reaction $\nu p \to \nu p \pi^0$ in which the π^0 is not detected constitutes another source of background. The observed number of $p \pi^0$ events[4] in the sample and the calculated π^0 detection efficiency (75%) imply at most two events with an unseen π^0.

Charged currents of the type $\nu n \to \mu^- p$ have been observed in the same sample as the $\nu p \to \nu p$ events. For $q^2 > 0.3$ (GeV/$c)^2$, the selection requirements for $\nu n \to \mu^- p$ and $\nu p \to \nu p$ are identical except for the presence of a muon; for $q^2 < 0.3$ (GeV/$c)^2$, $\nu n \to \mu^- p$ candidates were primarily identified by their muon signature. The q^2 distributions of the two samples of events, corrected for geometric acceptance[5] (calculated both analytically and by Monte Carlo techniques), are presented in Fig. 4(b). The observed shape of the q^2 distribution for $\nu n \to \mu^- p$ and the measured ratio of $\nu n \to \mu^- p$ to the total cross section, 0.25 ± 0.03, are consistent with previous measurements.[6]

In conclusion, then, we have observed a total of thirty events that possess the required characteristics of neutrino-proton[7] elastic scattering; the background is estimated to be at most three events from $np \to np$, two events from $\nu n \to \mu^- p$ and $\nu n \to \nu \pi^- p$, and two events from $\nu p \to \nu p \pi^0$. After subtraction, the number of $\nu p \to \nu p$ events for $0.3 < q^2 < 0.9$ (GeV/$c)^2$, when compared with the number of quasi-elastic events in the same q^2 interval, yields a cross section ratio $\sigma(\nu p \to \nu p)/\sigma(\nu n \to \mu^- p)$ of 0.17 ± 0.05 where the error is statistical. This value is consistent with most of the broken-gauge-symmetry models involving the weak neutral current,[8] and with previous measurements.[1] A larger ν sample and a $\bar{\nu}$ exposure are currently being analyzed to provide a more quantitative test of the model.

We thank R. Rau, H. Foelsche, W. D. Walker, A. Pendzick, and the staff of the Brookhaven National Laboratory as well as the staffs of the High Energy Physics Laboratories at Harvard University, University of Pennsylvania, and University of Wisconsin for their immeasurable support during this work. A. Conners, D. DiBitonto, G. Gollin, J. Horstkotte, J. LoSecco, M. Merlin,

J. Rich, W. Smith, and M. Yudis aided in the set-up and initial running of the experiment.

*Work supported in part by the U. S. Energy Research and Development Administration.

[1]D. C. Cundy *et al.*, Phys. Lett. **31B**, 478 (1970); P. Schreiner *et al.*, in *Proceedings of the Seventeenth International Conference on High Energy Physics, London, England, 1974*, edited by J. R. Smith (Rutherford High Energy Laboratory, Didcot, Berkshire, England, 1974), p. IV-123; C. Y. Pang *et al.*, unpublished.

[2]L. Sulak, in *Proceedings of the Calorimeter Workshop*, edited by M. Atac (Fermi National Accelerator Laboratory, Batavia, Ill., 1975), p. 155.

[3]D. C. Cheng *et al.*, Nucl. Instrum. Methods **117**, 157 (1974).

[4]If we assume that the number of π^+ and π^- events are equal, the observed number of π^0/π^- neutral current events is consistent with the Gargamelle measurement on Freon. See G. H. Bertrand Cosemans *et al.*, Phys. Lett. **61B**, 207 (1976).

[5]The geometric acceptance for $E_\nu = 1$ GeV and for tracks which go through at least one drift chamber is approximately 0.2, 0.45, and 0.33 for $q^2 = 0.35$, 0.55, and 0.75 $(\text{GeV}/c)^2$, respectively.

[6]The value, 0.25 ± 0.03, that we determine for the ratio of the total quasielastic cross section to the total cross section, integrated over the Brookhaven National Laboratory neutrino energy distribution, is consistent with that obtained from data taken with the Brookhaven National Laboratory 7-ft bubble chamber: M. J. Murtagh, private communication; W. A. Mann *et al.*, Phys. Rev. Lett. **31**, 844 (1973).

[7]Events from the reaction $\nu n \to \nu n$ would appear in our sample if np charge exchange occurs in the target nucleus or if it occurs after the neutron leaves the nucleus provided that the de-excitation of the target goes undetected (i.e., < 3 MeV of energy is deposited in the scintillator). We are currently evaluating the probability of these processes.

[8]S. Weinberg, Phys. Rev. D **5**, 1412 (1972); J. J. Sakurai and L. F. Urrutia, Phys. Rev. D **11**, 159 (1975); A. De Rújula, H. Georgi, and S. L. Glashow, Phys. Rev. D **12**, 3589 (1975); J. T. Gruenwald *et al.*, Bull. Am. Phys. Soc. **21**, 17 (1976); H. Fritzsch, M. Gell-Mann, and P. Minkowski, Phys. Lett. **59B**, 256 (1975); F. Wilczek, A. Zee, R. L. Kingsley, and S. B. Treiman, Phys. Rev. D **12**, 2768 (1975); M. Barnett, Phys. Rev. Lett. **34**, 41 (1975), and Phys. Rev. D **11**, 3246 (1975); S. Pakvasa, W. A. Simmons, and S. F. Tuan, Phys. Rev. Lett. **35**, 703 (1975).

Measurement of Muon-Neutrino and -Antineutrino Scattering off Electrons

H. Faissner, H. G. Fasold,[a] E. Frenzel, T. Hansl,[b] D. Hoffmann, K. Maull,[c] E. Radermacher,
H. Reithler, and H. de Witt

III. Physikalisches Institut, Technische Hochschule, Aachen, Germany

and

M. Baldo-Ceolin, F. Bobisut,[d] H. Huzita, M. Loreti, G. Puglierin, I. Scotoni,[e] and M. Vascon

*Istituto di Fisica Galileo Galilei dell'Università di Padova, Padova, Italy, and
Istituto Nazionale di Fisica Nucleare, Sezione di Padova, Italy*

(Received 6 April 1978)

Muon-neutrino and -antineutrino scattering off electrons was detected in a 19-ton Al spark chamber, exposed to the wide-band ν ($\bar{\nu}$) beam from the CERN proton synchrotron. The background was determined experimentally. 11 (10) genuine ν_μ- ($\bar{\nu}_\mu$-) e scattering events were found. The respective cross sections are $(1.1 \pm 0.6) \times 10^{-42}$ (E_ν/GeV) cm^2 and $(2.2 \pm 1.0) \times 10^{-42}$($E_\nu$/GeV) cm^2. The analysis excludes a pure $V-A$ interaction, and makes a pure V or A theory improbable. The data agree well with the Salam-Weinberg model and $\sin^2\theta_W = 0.35 \pm 0.08$.

Measuring muon-neutrino–electron scattering

$$\nu_\mu + e \to \nu_\mu + e, \tag{1}$$

$$\bar{\nu}_\mu + e \to \bar{\nu}_\mu + e, \tag{$\bar{1}$}$$

is one of the key experiments for testing unified gauge theories of weak and electromagnetic interactions.[1] These reactions are mediated by the neutral weak leptonic current only,[2] in contrast to $\bar{\nu}_e e$ scattering,[3] which would also proceed by the usual charged-current $V-A$ coupling.[4] The scattering of muon antineutrinos off electrons was discovered in the heavy-liquid bubble chamber Gargamelle.[5] The final result[6] is based on three $\bar{\nu}_\mu e$ candidates. The present experiment measured both $\bar{\nu}_\mu e$ scattering.[7,8]

The ν_μ- ($\bar{\nu}_\mu$-) e scattering cross section is,[9] in a general V,A framework, and for neutrino energies $E_\nu \gg m_e$,

$$d\sigma/dy = \sigma_e[A + B(1-y)^2], \tag{2}$$

$$d\bar{\sigma}/dy = \sigma_e[A(1-y)^2 + B], \tag{$\bar{2}$}$$

with $y = E_e/E_\nu$, and $\sigma_e = 17.2 \times 10^{-42}(E_\nu/\text{GeV})$ cm^2. The total cross sections are

$$\sigma = \sigma_e(A + B/3), \tag{3}$$

$$\bar{\sigma} = \sigma_e(A/3 + B). \tag{$\bar{3}$}$$

Their ratio $r_e = \bar{\sigma}/\sigma$ and also the average energy transfers, $\langle y \rangle$ and $\langle \bar{y} \rangle$, depend on the ratio A/B only. The singular case of a parity-conserving V (or A) electron current implies $A = B$, and thus equality of all ν_μ and $\bar{\nu}_\mu$ data. The minimal gauge model[10] has only one free parameter, $x_w = \sin^2\theta_W$,

which fixes $A = (-\frac{1}{2} + x_w)^2$, and $B = x_w^2$.

The apparatus was exposed to the neutrino beam, derived from the CERN proton synchrotron (PS). A high-intensity ($\approx 10^{13}$ protons per pulse), 26-GeV proton beam was steered onto an external target. Focusing of the emitted secondaries resulted in a fairly pure ν_μ ($\bar{\nu}_\mu$) beam, with 0.2% (0.3%) wrong-type ν_μ, and 0.4% ν_e (0.3% $\bar{\nu}_e$) contamination. The ν ($\bar{\nu}$) spectrum has its maximum energy at 1.5 (1.4) GeV, and an average of 2.2 (2.0) GeV. Absolute neutrino fluxes are known with an average uncertainty near 10%.

The optical spark chamber[7,8] consisted of 1-cm-thick aluminum plates, with an effective mass of 19 tons. The chamber was fired every PS radiation pulse. Two views were photographed, with a stereo angle of 90°. Since the average radiation length is only 22 cm, electrons in the GeV range are easily identified by the electromagnetic shower they induce. But also γ rays initiate such showers, and their pattern cannot be distinguished from an electron-induced shower. The shower energy was measured by spark counting. An absolute calibration has been obtained from exposing part of the chamber to a monochromatic electron beam. In the energy range 0.1–2 GeV the electron energy E_e corresponding to N observed sparks is well described by $E_e = 6.5N/(1-0.0025N)$. MeV. The resolution in energy and projected angle were found to be $\Delta E_e/E_e \approx 22\%$ and $\Delta\theta_{\text{proj}} \approx 12/\sqrt{E_e}$ mrad (E_e in GeV), respectively.

In selecting candidates for neutrino-induced recoil electrons two topological requirements were made: (a) The spark pattern must show multipli-

cation (≥ 2 double sparks, on the average, per
radiation length); (b) the shower must not be associated with any other structure in the picture.

The final selection is based on the νe kinematics. For high energies it may be summarized by
the following relation between electron angle θ_e,
energy E_e, and energy transfer $y = E_e/E_\nu$:

$$E_e \theta_e^2 = 2m_e(1-y). \qquad (4)$$

Hence $\theta_e \lesssim 2°$, for $E_e = 1$ GeV. According to Eqs.
(2) and ($\bar{2}$), and for the average ν energies stated,
one expects $\langle E_e \rangle$ to lie between 0.5 and 1.1 GeV.
This suggests the acceptance criteria (c) $\theta_e < 5°$,
(d) 0.2 GeV $< E_e < 2$ GeV. Moreover, the measurable left-hand side of Eq. (5) is essentially a determination of the target particle's mass. Usual-
ly we required (e) $\frac{1}{2}E_e\theta_e^2 < m_e$, i.e. $y > 0$, but, considering the finite angular resolution, sometimes
the weaker condition (e') $\frac{1}{2}E_e\theta_e^2 < 2m_e$. An additional cut could be imposed on the reconstructed
neutrino energy: (f) E_ν $(E_{\bar{\nu}}) > 0.8$ GeV.

With criteria (a)–(d) and (e'), 32 (17) candidates
for νe scattering were found in 842 000 (812 000)
pictures taken in the ν ($\bar{\nu}$) runs. Within statistics
they are uniformly distributed over the fiducial
volume. This shows that they are not primarily
induced by gammas, or neutrons, coming from
the outside. Scatter plots of their distribution in
θ^2 and E are given in Figs. 1(a) and 1($\bar{a}$).

Extended studies[6-8,11] have clarified that the
background stems mainly from muonless neutral
pions without visible recoil, and with one decay
gamma lost. In general this means

$$^{(\bar{\nu})}_\mu + n \to ^{(\bar{\nu})}_\mu + n + \pi^0, \quad \pi^0 \to \gamma(+\gamma), \qquad (5)$$

where the target had been a neutron. The probability for one of these gammas to satisfy our acceptance conditions (a) through (f) was computed
by a Monte Carlo technique. The π^0 distributions
in energy and angle were measured for the ν_μ
($\bar{\nu}_\mu$) exposure from a sample of 321 (101) "naked"
π^0's with both gammas seen. Thus, the calculation proper was reduced to evaluating decay kinematics, γ conversion, and geometry. It yielded
both the absolute number of 1γ events, and their
distribution in energy and angle. Thus the 1γ
background could be assessed either directly, or
by extrapolating the number of single-shower
events observed above the hyperbola in Figs.
1(a) and 1($\bar{a}$) into the physical νe region below.
In addition, the γ background was determined
through an extrapolation of the large-angle ($\lesssim 10°$)
single-γ distribution into the νe region. The γ's
were found distributed practically constant over

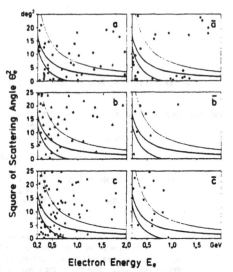

FIG. 1. Scatter plots in energy and square of angle
for ν ($\bar{\nu}$) exposures and (a),($\bar{a}$) isolated single showers;
(b),($\bar{b}$) single γ showers, associated with proton recoil(s);
and (c),($\bar{c}$) γ showers associated with muons (and protons). (The curves mark the borders of the acceptance
regions used; from top to bottom, $y = -1$, $y = 0$; E_ν
= 0.8 GeV.)

θ^2, as one would expect from π^0 decay.

An independent check on the background was
provided by neutral-current π^0 production, analogous to (5), but off a *proton*. Thus for either exposure (ν or $\bar{\nu}$), one further control sample was
collected, namely single showers associated with
a visible proton recoil and fulfilling the acceptance criteria (1), (c), and (d). These p-associ-
ated single gammas give the one-gamma background directly, provided they fall inside the accepted region [Figs. 1(b) and 1($\bar{b}$)]. The number
of the p events outside can again be extrapolated
to the inside by the Monte Carlo simulation. The
effective ratio of p to n events was derived from
samples of two-gamma events and, independently,
from single-gamma events in the large-angle region ($5° < \theta < 30°$). An analogous procedure was
undertaken with single gamma rays from charged-
current–produced π^0's, Figs. 1(c) and 1($\bar{c}$). These
reactions can be used to calibrate the background,
provided their dynamics is sufficiently close to
that of Reaction (5). This was checked by com-

TABLE I. $\nu_\mu e$ and $\bar\nu_\mu e$ cross sections, and parameters from the maximum-likelihood fit. The quoted confidence level is the probability that the measured number of background events does not account for the observed number of candidates.

Kinematical cuts	$E_e\theta_e^2 \le 4\,m_e$ 0.2 < E_e < 2 GeV		$E_e\theta_e^2 \le 4\,m_e$ 0.6 < E_e < 2 GeV		$E_e\theta_e^2 \le 2\,m_e$ 0.2 < E_e < 2 GeV E_ν > 0.8 GeV	
Exposure	ν	$\bar\nu$	ν	$\bar\nu$	ν	$\bar\nu$
No. of candidates	32	17	11	8	11	7
Total no. of background events	20.5±2.0	7.4±1.0	3.9±0.5	1.7±0.3	5.4±0.5	2.0±0.3
Confidence of ν_μ, $\bar\nu_\mu$-e scattering to exist	>99%	>99.9%	>99.9%	>99.9%	>99%	>99.9%
Cross sections σ, $\bar\sigma$, in units of 10^{-42}cm$^2 E_\nu$/GeV	1.1±0.6	2.2±1.0	1.0±0.7	2.3±1.1	0.8±0.6	2.4±0.9
(V-A)-constant A	$0.022^{+0.034}_{-0.022}$		$0.016^{+0.037}_{-0.016}$		$0.0^{+0.03}_{-0.00}$	
(V+A)-constant B	0.122±0.055		0.13±0.06		0.14±0.05	
Probability for V-A	<0.3%		<0.3%		<0.1%	
Probability for pure V or A	<15%		<15%		<5%	
$\sin^2\theta_W$	0.35±0.08		0.36±0.08		0.38±0.07	

paring the single-gamma densities of Figs. 1(b) and 1($\bar{b}$), 1(c) and 1($\bar{c}$), with those derived from the Monte Carlo simulation based on the "naked" π^0's. Within statistics, they did agree.

Consequently, the various estimations of the background are consistent, too. Their average is given in Table I and Fig. 2, for different sets of acceptance criteria. The small error of the background numbers reflects the fact that its estimation is based on a large control sample. Systematic uncertainties are added in quadrature. The lower background for antineutrinos can be understood from the difference in flux and from the neutral-current-induced π^0 cross-section ratio $\bar\sigma/\sigma = 0.49 \pm 0.12$.[11] Contributions from inverse beta decay are small: 0.13 (0.04) events. The ν_e flux was checked by observing $e\not{p}$ events. Scattering of ν_e ($\bar\nu_e$) off electrons is expected to contribute 0.47 (0.09) events, assuming the standard minimal model. Furthermore, genuine neutral-current single-gamma production: $\nu N \rightarrow \nu\gamma N$, would have been seen in our 1γ test samples, but it has been shown to be small.[12]

The parameters A and B, and hence the total cross sections (3) and ($\bar{3}$), were evaluated by optimizing a maximum-likelihood function for ν and $\bar\nu$ together. The probability assigned to a candidate of a given energy was the ratio of the number of $\nu_\mu e$ ($\bar\nu_\mu e$) events expected plus the background determined, at this energy, to the total number of candidates expected in the complete energy range. This was related to the observed number by a Poisson distribution. The results of this maximum-likelihood calculation are given in Table I for the full sample, and also for two subsamples chosen in order to increase the signal-to-noise ratio, and to demonstrate the stability of the results.

From Table I we select representative total cross sections for $\nu_\mu e$ and $\bar\nu_\mu e$ scattering: $\sigma = (1.1 \pm 0.6) \times 10^{-42}(E_\nu$/GeV) cm^2 and $\bar\sigma = (2.2 \pm 1.0) \times 10^{-42}(E_\nu$/GeV) cm^2, with a ratio $r_e = 2.0^{+3.0}_{-1.0}$. The stated uncertainties include an estimate of the systematic errors. The results of the fit are corroborated by the observation that the average electron energy for $\bar\nu_\mu e$ events is larger than for $\nu_\mu e$ events.[8] Following Table I, we make the following conclusions: (1) $\bar\nu_\mu e$ and $\nu_\mu e$ scattering have been observed; the signal is established with more than 99% confidence. (2) Any $V - A$ interaction is excluded at > 99% C.L. (3) Pure V or pure A currents are improbable at about 90% C.L. (4) The results indicate dominant $V + A$, with some $V - A$ mixed in; an estimate of their ratio is $a = A/B = (15^{+25}_{-15})\%$. (5) The data are in good agreement with the Salam-Wein-

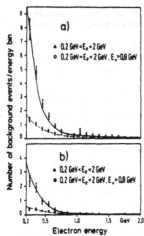

FIG. 2. Mean value of the experimentally determined background, for different acceptance criteria, as a function of shower energy. Cuts in γ were $y > -1$ for triangles, and $y > 0$ for circles. The curves are fits to the data. (a) neutrino; (b) antineutrino.

berg model with

$$\sin^2\theta_W = 0.35 \pm 0.08.$$

The implications of these findings for other gauge models are discussed elsewhere.[13]

We are grateful to CERN for unusual technical support. We thank our technicians and scanners for their devoted work. This experiment has been supported by the German Bundesministerium für Forschung und Technologie.

(a)Now with Ruhrgas Aktiengesellschaft, Essen, Germany.

(h)Presently at CERN, Geneva, Switzerland.

(c)Now with Vereinigte Elektrizitätswerke, Dortmund, Germany.

(d)Now visitor at the University of Pennsylvania, Philadelphia, Penn. 19104.

(e)Now at Istituto di Fisica, Università di Trento, Trento, Italy.

[1]See M. Gourdin, in *Proceedings of the International Neutrino Conference, Aachen, West Germany, 1976*, edited by H. Faissner, H. Reithler, and P. Zerwas (Vieweg, Braunschweig, 1977), p. 234.

[2]To first order, and to the extent that ν_μ carries a conserved muon quantum number.

[3]F. Reines, H. S. Gurr, and H. W. Sobel, Ref. 1, p. 217, and Phys. Rev. Lett. 37, 315 (1976).

[4]R. P. Feynman and M. Gell-Mann, Phys. Rev. 109, 193 (1958); E. C. G. Sudarshan and R. E. Marshak, Phys. Rev. 109, 1860 (1958).

[5]F. J. Hasert *et al.*, Phys. Lett. 46B, 121 (1973).

[6]J. Blietschau *et al.*, Nucl. Phys. B114, 189 (1976).

[7]H. Faissner *et al.*, Ref. 1, p. 223.

[8]H. Faissner *et al.*, in *Proceedings of the International Neutrino Conference, Baksan, 1977*, edited by M. A. Markov *et al.*, (Nauka, Moscow, 1978), Vol. 2, p. 142; H. Reithler, in *Proceedings of the International Symposium on Lepton and Photon Interactions at High Energies*, edited by F. Gutbrod (DESY, Hamburg, 1977), p. 343.

[8]G. 't Hooft, Phys. Lett. 37B, 195 (1971).

[10]A. Salam and J. C. Ward, Phys. Lett. 13, 168 (1964); A. Salam, in *Elementary Particle Theory*, edited by N. Svartholm (Almqvist and Wiksell, Stockholm, 1968), p. 367; S. Weinberg, Phys. Rev. Lett. 19, 1264 (1967), and 27, 1688 (1971), and Phys. Rev. D 5, 1412, 1962 (1972).

[11]H. Faissner *et al.*, Phys. Lett. 68B, 377 (1977).

[12]E. L. Choban and V. M. Shekhter, Yad. Fiz. 26, 1064 (1977) [Sov. J. Nucl. Phys. (to be published)].

[13]H. Faissner, in Proceedings of the Ben Lee Memorial Conference on Parity Nonconservation, Weak Neutral Currents and Gauge Theories, Batavia, Illinois, 1977 (to be published).

Evidence for Parity Nonconservation in the Weak Neutral Current*

A. Benvenuti, D. Cline, F. Messing,† W. Ford, R. Imlay,‡ T. Y. Ling, A. K. Mann,
D. D. Reeder, C. Rubbia, R. Stefanski, L. Sulak, and P. Wanderer§
*Department of Physics, Harvard University, Cambridge, Massachusetts 02138, and
Department of Physics, University of Pennsylvania, Philadelphia, Pennsylvania 19174, and
Department of Physics, University of Wisconsin, Madison, Wisconsin 53706, and
Fermi National Accelerator Laboratory, Batavia, Illinois 60510*
(Received 3 May 1976)

Measurements of R^ν and $R^{\bar\nu}$, the ratios of neutral current to charged current ν and $\bar\nu$ cross sections, yield neutral current rates for ν and $\bar\nu$ that are consistent with a pure $V-A$ interaction but 3 standard deviations from pure V or pure A, indicating the presence of parity nonconservation in the weak neutral current.

The existence of a weak neutral-current interaction has been inferred from the discovery of ν and $\bar\nu$ inelastic collisions with nucleons that have no final-state muons.[1] This Letter reports the most recent results of our study of the nature of the weak neutral current. The 3500 events reported here are greater in number and were produced by neutrino beams of substantially higher purity than those utilized previously by us.[2] Furthermore, the data were acquired during different running conditions, making possible important tests for systematic errors. We have extracted from the data values of R^ν and $R^{\bar\nu}$, the ratios of neutral- to charged-current cross sections for ν and $\bar\nu$, which are used to obtain values of the neutral-current cross sections. The observed inequality of the ν and $\bar\nu$ neutral-current cross sections leads directly to the conclusion that a parity-nonconserving component is present in that current.

The apparatus, which is described in greater detail elsewhere,[3] is shown in Fig. 1(a). The spectra of the incident ν and $\bar\nu$ beams are shown in Fig. 1(b). Neutrino interactions which produced a hadronic cascade with energy $E_H > 4$ GeV triggered the apparatus with an efficiency greater than 99%. Counter A and the first seven calorimeters were in anticoincidence. Pulse-height information from the last eight calorimeter modules yielded the value of E_H. In addition, the wide-gap spark chambers (WGSC) were photographed in two ± 7.5° stereo views and in a 90° stereo view. Muons were identified by their presence in detectors 1 or 2 after passing through the iron hadron absorbers shown in Fig. 1(a).

Detector 1 consists of a 3.6-m × 3.6-m scintillation counter together with a 2.8-m × 2.8-m WGSC; detector 2 is solely a 2.8-m × 2.8-m WGSC. The muon identifiers have (1) high acceptance ϵ_μ for charged-current events (CC), (2) a small prob-

ability ϵ_p that hadrons associated with neutral-current (NC) interactions will punch through the absorbers to simulate muons, and (3) redundancy. In the experiments reported here, ϵ_μ is very near unity and therefore the determination of the NC/CC ratios is only weakly dependent on the details of the calculation of ϵ_μ. That calculation takes into account the spatial and energy distribution of the ν beams, muon-ionization energy loss and multiple scattering, resolution in the measurement of E_H, and the properties of the CC events as determined from these same data and reported elsewhere.[4] Values of ϵ_μ and other characteristics of the data are given in Table I. The dependence of ϵ_p on E_H and longitudinal position was obtained directly from measurements

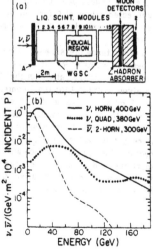

FIG. 1. (a) Sketch of the apparatus. (b) Neutrino spectra used in these experiments.

TABLE I. Relevant parameters for the four experiments reported here. The numbers of events have been corrected for muon-detection efficiency and hadron punchthrough. The parameter α, obtained from the measured CC events, indicates the composition of the beams.

Beam type	$\alpha \equiv \dfrac{N_{\mu^-}}{N_{\mu^-}+N_{\mu^+}}$	Number of events		Muon detection efficiency ϵ_μ		Hadron filter thickness (cm)	Hadron punchthrough probability ϵ_p	
		Neutral	Charged	Det. 1	Det. 2		Det. 1	Det. 2
Single horn	0.94	266	857	0.83	a	35	0.24	~0
1974 quadru-pole triplet	0.83	158	599	0.84	a	35	0.20	~0
1975 quadru-pole triplet	0.86	300	1042	0.95	0.84	70	0.12	~0
Double horn with plug	0.27	69	198	0.98	0.84	70	0.07	~0

[a]Detector 2 had a different geometry for these experiments.

made on the CC data.

Candidates for both charged- and neutral-current interactions were selected from a scan of the film. The scanning efficiency was greater than 95% and was the same for NC and CC events because the scanning criteria involved only properties of event vertices. The vertex was measured in three views and its location checked for consistency with the electronic information. The small fiducial region shown in Fig. 1(a) was chosen to insure high muon-detection efficiency and good shower containment, and to reject neutron backgrounds. As with previous data,[2] no significant evidence for neutron contamination was found.

The values of R, after correction for muon efficiency and hadron punchthrough, were found to be independent of the transverse and longitudinal position of the interaction. Figure 2 illustrates this for the longitudinal coordinate. The agreement between the corrected values of R for the detectors shown in Fig. 2(b) suggests that the corrections are well understood. Comparison of the raw and corrected values of R for detector 1 shows the small magnitude of the net correction for this detector.

For each of the three measurements with nearly pure ν beams a value of R^ν for $E_H > 4$ GeV was obtained by correcting the ratio R of the NC and CC events given in Table I for the $\bar\nu$ content of the beams. Similarly, a value of $R^{\bar\nu}$ for $E_H > 4$ GeV was extracted from a measurement with a relatively pure $\bar\nu$ beam. These results are given with their respective values of $\langle E_{\nu,\bar\nu}\rangle$ in Table II. The corrections for ν_e contamination and single strange-particle production in the charged-current channel are not important relative to the statistical limitations of these data.

We previously reported the values $R^\nu = 0.11 \pm 0.05$ and $R^{\bar\nu} = 0.32 \pm 0.09$,[2] also for $E_H > 4$ GeV. We believe that the difference between the earlier value of R^ν and the values given in Table II underscores the difficulty of the method used to extract R^ν and $R^{\bar\nu}$ from the first data samples which were either statistically limited or obtained from a beam with a significant admixture of $\bar\nu$ ($\alpha \approx 0.62$).

To determine the space-time structure of the weak neutral current from these data, it is necessary to obtain $\sigma_N{}^{\bar\nu}/\sigma_N{}^\nu$ from the relation $\sigma_N{}^\nu = (R^{\bar\nu}/R^\nu)(\sigma_c{}^{\bar\nu}/\sigma_c{}^\nu)$, because the ratio of the

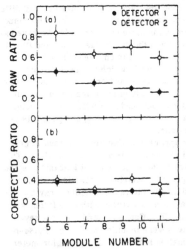

FIG. 2. (a) Raw ratio of the number of muonless events to the number with detected muons, versus module number as defined in Fig. 1(a). (b) R, the raw ratio corrected for muon detection inefficiencies and hadron punchthrough, versus module number.

TABLE II. The values of R^{ν} or $R^{\bar{\nu}}$ for $E_H > 4$ GeV.

Beam type	R^{ν}	$R^{\bar{\nu}}$	Comment
Single horn	0.31 ± 0.06		Pure ν beam, $\langle E_\nu \rangle = 53$ GeV.
1974 quadru- pole triplet	0.24 ± 0.06		Mixed beam, $\langle E_\nu \rangle = 78$ GeV.
1975 quadru- pole triplet	0.29 ± 0.04		Mixed beam, $\langle E_\nu \rangle = 85$ GeV.
Double horn with plug		0.39 ± 0.10	Pure $\bar{\nu}$ beam, $\langle E_{\bar{\nu}} \rangle = 41$ GeV.

charged-current cross sections $\sigma_c^{\bar{\nu}} \sigma_c^{\nu}$ is changing with energy.[4] To obtain $\sigma_N^{\bar{\nu}} \sigma_N^{\nu}$, however, the effect of the experimental requirement $E_H > 4$ GeV must first be considered. If the neutral current has the same form as the charged current, exclusion of events with $E_H < 4$ GeV has no effect on the measured values of R^{ν} and $R^{\bar{\nu}}$. More generally, for any linear combination of V and A, $R^{\nu}(E_H > 0) \geqslant R^{\nu}(E_H > 4)$, while $R^{\bar{\nu}}(E_H > 0) \leqslant R^{\bar{\nu}}(E_H > 4)$, so that $\sigma_N^{\bar{\nu}}/\sigma_N^{\nu} \leqslant [\sigma_N^{\bar{\nu}}/\sigma_N^{\nu}]_{V-A}$. We find directly the numerical upper limit on $\sigma_N^{\bar{\nu}}/\sigma_N^{\nu}$ at $\langle E_{\nu, \bar{\nu}} \rangle = 41$ from

$$\sigma_N^{\bar{\nu}}/\sigma_N^{\nu} \leqslant (R^{\bar{\nu}}/R^{\nu})(\sigma_c^{\bar{\nu}}/\sigma_c^{\nu})$$
$$= [(0.39 \pm 0.10)/(0.29 \pm 0.04)](0.45 \pm 0.08)$$
$$= 0.61 \pm 0.25, \tag{1}$$

where R^{ν} and $R^{\bar{\nu}}$ are obtained from Table II, and $\sigma_c^{\bar{\nu}}/\sigma_c^{\nu}$ at 41 GeV is taken from Ref. 4. Note that the values of R^{ν} in Table II are approximately constant over the average energy interval from 53 to 85 GeV. This largely justifies the extrapolation of constant R^{ν} to 41 GeV, and implies, within experimental error, a linear rise with energy of the total neutral-current cross section for neutrinos.[5]

We can obtain $\sigma_N^{\bar{\nu}}/\sigma_N^{\nu}$, rather than its upper limit, by correcting for unobserved events with $E_H < 4$ GeV according to various forms of the neutral-current interaction. The high energy of the neutrino beams, particularly the 1975 quadrupole triplet beam, gives rise to few events with $E_H < 4$ GeV, so that for any combination of V and A the difference between $R^{\nu}(E_H > 0)$ and the measured values of $R^{\nu}(E_H > 4)$ is negligible. The ratio $R^{\bar{\nu}}$ is more sensitive to the exact form of the interaction because of the lower $\langle E_{\bar{\nu}} \rangle$. Thus for $V - A$, $R^{\bar{\nu}}(E_H > 0) = R^T(E_H > 4)$, while for pure V or A, $R^{\bar{\nu}}(E_H > 0) = 0.71 R^{\bar{\nu}}(E_H > 4)$; and for $V + A$, $R^{\bar{\nu}}(E_H > 0) = 0.63 R^{\bar{\nu}}(E_H > 4)$. We compare in Table III the corrected, experimental values of $\sigma_N^{\bar{\nu}}/\sigma_N^{\nu}$ with the expected values of this ratio for several admixtures of V and A. It is clear that $V + A$ is ruled out. Furthermore, the experimental value for pure V or A is 3 standard deviations away from the value expected for either of those pure forms, while $V - A$ is within 1 standard deviation of the expected value.

We obtain the best fit for the form of the weak neutral current by using the general forms of the $y \equiv E_H/E_\nu$ distribution expected for any V, A combination:

$$d\sigma_N^{\nu}/dy = a + b(1-y)^2, \quad d\sigma_N^{\bar{\nu}}/dy = b + a(1-y)^2.$$

TABLE III. The measured values of $\sigma_N^{\bar{\nu}}/\sigma_N^{\nu}$, after correction for the loss of events with $E_H < 4$ GeV according to the form of the weak neutral current in the first column. The corresponding values of $\sigma_N^{\bar{\nu}}/\sigma_N^{\nu}$, expected from theory, are given in column three. An antiquark contribution of 5% has been assumed.

Form of the weak neutral current	$\sigma_N^{\bar{\nu}}/\sigma_N^{\nu}$ Corrected experimental value	Expected value
$V - A$	0.61 ± 0.25	0.38
V or A	0.40 ± 0.17	1.00
$V \cdot A$	0.37 ± 0.16	2.65

For a $V - A$ interaction, $a \approx 1$ and $b \approx 0$, while a V or A interaction has $a = b = \frac{1}{2}$. The best fit is obtained by varying a and b until the expected value of $\sigma_N^{\bar{\nu}}/\sigma_N^{\nu}$ agrees with the corrected experimental value. For the best fit, $\sigma_N^{\bar{\nu}}/\sigma_N^{\nu} = 0.48 \pm 0.20$, $a = 0.85$, and $b = 0.15$. These results are confirmed by the measured, essentially uniform dependences of R^{ν} and $R^{\bar{\nu}}$ on E_H which do not, however, sensitively discriminate among the possible forms of the neutral current.

In summary, measurements of neutral-current and charged-current inelastic scattering of ν and $\bar{\nu}$ rule out $V + A$, and are incompatible with a pure V or pure A form for the weak neutral current. The experimental results require a significant parity-nonconserving component in the weak neutral current, and are consistent with $V - A$.

We thank Mr. W. Grant and Mrs. A. Black for their diligent scanning efforts. One of us (P.W.) thanks the Brookhaven National Laboratory for support during the last phase of this work.

*Work supported in part by the U. S. Energy Research and Development Administration.

†Now at Cornell University, Ithaca, N. Y. 14853. Submitted in partial fulfillment of the Ph.D. requirements at the University of Pennsylvania.

‡Now at Rutgers University, New Brunswick, N. J. 08903.

§Now at Brookhaven National Laboratory, Upton, N. Y. 11973.

[1]G. Myatt, in *Proceedings of the Sixth International Symposium on Electron and Photon Interactions at High Energies, Bonn, 1973*, edited by H. Rollnik and W. Pfeil (North-Holland, Amsterdam, 1974), p. 389; F. J. Hasert *et al.*, Phys. Lett. 46B, 138 (1973); A. Benvenuti *et al.*, Phys. Rev. Lett. 32, 800 (1974).

[2]B. Aubert *et al.*, Phys. Rev. Lett. 32, 1454, 1457 (1974).

[3]A. Benvenuti *et al.*, Nucl. Instrum. Methods 125, 447, 457 (1975).

[4]A. Benvenuti *et al.*, Phys. Rev. Lett. 37, 189 (1976).

[5]No account has been taken of a possible isoscalar component of the neutral current which is no $V-A$.

Observation of parity nonconservation in atomic transitions

L. M. Barkov and M. S. Zolotorev

Nuclear Physics Institute, Siberian Division, USSR Academy of Sciences
(Submitted 14 February 1978)
Pis'ma Zh. Eksp. Teor. Fiz. 27, No. 6, 379–383 (20 March 1978)

Parity nonconservation in atomic transitions has been observed. The rotation of the plane of polarization of light was measured on the components of the hyperfine splitting of the 6477 Å line of bismuth.

PACS numbers: 35.10.Fk, 32.30.Jc, 11.30.Er

Search for effects of parity conservation in atomic transitions was undertaken in our institute in 1974, prompted by a discussion with I. B. Khriplovich. The first to call attention to the possible existence of such effects was Ya. B. Zel'dovich in 1959,[1] and these effects have since been discussed many times.[2-6] Similar studies with a different procedure are being carried out at Oxford and in Seattle.[7-12]

Figure 1 shows a block diagram of our setup for the measurement of the plane of polarization of light passing through bismuth vapor. The measurements were performed on the 6477 Å line corresponding to the magnetic-dipole transition. The light source was a model-375 Spectra Physics dye laser. An element was introduced into the laser to permit operation in the single-frequency lasing regime and scanning the laser wavelength at a frequency of 1 kHz. The radiation power in the single-frequency regime was 15 mW.

Light modulated at the emission frequency passed through a polarizer, a cell with bismuth vapor, and an analyzer. The polarizer and the analyzer were situated inside the vacuum volume of the cell. To protect them against sputtering and to maintain a

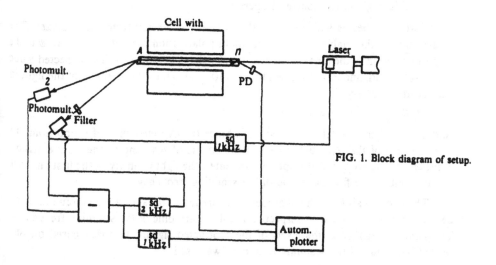

FIG. 1. Block diagram of setup.

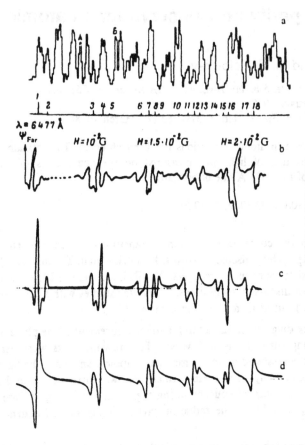

FIG. 2. a) Absorption spectrum of bismuth vapor; b) experimentally measured Faraday rotation on the hyperfine structure line; c) theoretical curve for the Faraday rotation[13]; d) rotation of plane of polarization calculated in accordance with the Weinberg-Salam model.

constant vapor pressure, helium was used as a buffer gas. The axes of the polarizer and the analyzer made an angle θ_0. The mounting for the polarizer was designed to be able to vary this angle in the course of operation.

Two light beams with orthogonal polarizations emerge from the analyzer. The angle between the analyzer axis and the light polarization is $\theta = \theta_0 + \psi$, where ψ is the angle of rotation of the plane of polarization by the bismuth vapor. The expected value was $\psi \sim 10^{-7}$ rad, and θ_0 was chosen to be $\sim 10^{-3}$ rad. Thus, the intensity of the light in one of the channels is

$$I \approx I_0 \theta^2 \approx I_0 \theta_0^2 (1 - 2\psi/\theta_0),$$

and in the other channel it is $I_0 \cos^2\theta \approx I_0$. Since the expected effect if proportional to the real part of the refractive index, in the course of scanning of the laser emission wavelength relative to the absorption line center, the light intensity in the first channel should contain the first harmonic of the scanning frequency.

The two signals from the photodetectors are fed to the subtraction circuit and are detected synchronously with respect to the first harmonic of the scanning frequency. To maintain equality of the levels of the subtracted signals, feedback based on the second harmonic of the scanning frequency was used.

TABLE I.

Line	$F \rightarrow F'$	$\psi_{exp} \cdot 10^8$	$\psi_{theor} \cdot 10^8$
1	6 – 7	-11.8 ± 5.5	-12.2
3	6 – 6	-4.7 ± 2.2	-3.1
7	5 – 5	-3.6 ± 4.0	-4.2
10	6 – 4	0.0 ± 1.8	$+0.1$
12	4 – 4	-11.3 ± 3.2	-4.2
A	–	$+11.9 \pm 11.9$	$+0.1$
B	–	$+6.6 \pm 2.8$	$+0.1$

[1]The calculated curves shown in Fig. 2 were kindly provided by V. I. Novikov and O. P. Sushkov.

The feedback with respect to the first harmonic in the signal of the channel with the higher intensity of the light was used to maintain the correct position of the scanner relative to the bismuth-vapor absorption line.

The cell with the bismuth vapor is located inside a double magnetic screen. The average magnetic field along the light beam axis is 2×10^{-5} G. The cell can operate at temperatures up to 1500 K. At these temperatures, the partial pressures of the atomic and molecular bismuth vapor are approximately equal and the vibrational-rotational spectrum of the molecule is superimposed on the absorption line of the hyperfine structure of the investigated atomic line.

Figure 2a shows the spectrum of the absorption line of the bismuth vapor. The intervals in the histogram correspond to the distance between the neighboring longitudinal modes of the laser resonator and is equal 400 MHz. Measurements of the Faraday rotation of the plane of polarization of the light, shown in Fig. 2b, and their comparison with the theoretical calculations[13] (see Fig. 2c)[1] have made it possible to identify uniquely the hyperfine structure of the 6477 Å line of atomic bismuth. The results of these measurements agree also with the results obtained in Oxford (P. Sandars, private communication).

Measurements of the rotation of the plane of polarization were carried out on the lines 1, 3, 7, 10, 12, A, and B. The results of the measurements are shown in Table I. The measurements were made at a total bismuth-vapor pressure 24 Torr and an effective cell length on the order of 30 cm. The scanning amplitude corresponded to one or two Doppler widths. The values of the rotation angle given in the table were defined as $\partial\psi/\partial\omega \Delta\omega$, where $\Delta\omega > 0$ is the scanning amplitude. We used the standard optics defi-

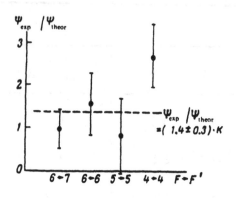

FIG. 3. Ratio of the experimentally measured rotational angles of the plane of polarization to those calculated by the Weinberg-Salam model.

nition, whereby positive rotation is clockwise as seen by an observer looking at the source. The theoretical values given in the table for the rotation angle of the plane of polarization of the light, for the Weinberg model, are based on the results of [14].

As seen from the table, rotation of the polarization plane is observed for all the working lines 1, 3, 7, and 12. The average rotation angle for these lines is $\bar{\psi}_{exp}$ $=(-6.7\pm1.6)\times10^{-8}$ rad, whereas the average rotation angle on the control lines 10, A, and B is $(+2.1\pm1.5)\times10^{-8}$ rad. Figure 3 shows the ratio of the experimentally measured angle of rotation of the plane of polarization of the light to the value predicted for the Weinberg–Salam model. The mean value of this ratio is

$$\overline{\psi_{exp}/\psi_{W-S}} = (+1.4\pm0.3)\,k\,.$$

The factor k was introduced because of inexact knowledge of the bismuth vapor, and also because of some uncertainty in the relative positions of the resonator modes and the absorption-line contours. According to our estimates, this factor lies in the interval from 0.5 to 1.5.

Figure 4 shows a comparison of the results of the present work at $k=1$ with the

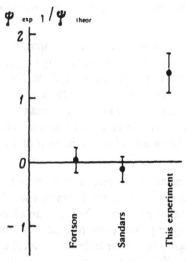

FIG. 4. Comparison of the results of the present paper (at $k=1$) with the results of [11,12].

results obtained in Seattle and Oxford. The results testify to the existence of parity nonconservation in atomic transitions and does not contradict the predictions of the Weinberg–Salam model.

The authors are sincerely grateful to the late G. I. Budker for support, to A. N. Skrinskiǐ, V. A. Sidorov, and I. I. Gurevich for constant interest in the work, to I. B. Khriplovich for numerous discussions of all the stages of the work, to V. P. Cherepanov, E. A. Kuper, and A. A. Litvinov for constructing and adjusting the electronic circuitry, and to V. M. Khoreev, V. S. Mel'nikov, and I. F. Legostaev for help with the design and constructure of the apparatus.

[1]Ya. B. Zel'dovich, Zh. Eksp. Teor. Fiz. 36, 964 (1959) [Sov. Phys. JETP 9, 681 (1959)].
[2]M. S. Bouchiat and C. C. Bouchiat, Phys. Lett. B 48, 111 (1974).
[3]A. N. Moskalev, Pis'ma Zh. Eksp. Teor. Fiz. 19, 229 (1975) [JETP Lett. 19, 141 (1975)].
[4]I. B. Khriplovich, Pis'ma Zh. Eksp. Teor. Fiz. 20, 686 (1974) [JETP Lett. 20, 315 (1974)].
[5]P. G. H. Sandars, Atomic Physics IV, Plenum Press, N.Y., 1975.
[6]D. C. Soreide and E. N. Fortson, Bull. Am. Phys. Soc. 20, 491 (1975).
[7]D. C. Soreide, D. E. Roberts, E. G. Lindahl, L. L. Lewis, G. R. Apperson, and E. N. Fortson, Phys. Rev. Lett. 36, 352 (1976).
[8]E. N. Fortson, Invited talk at the Fifth Intern. Conf. on Atomic Physics, Berkeley, 1976.
[9]P. E. Baird, M. W. Brimicombe, G. J. Roberts, P. G. H. Sandars, and D. N. Stacey, Invited Talk at Fifth Intern. Conf. on Atomic Physics, Berkeley, 1976.
[10]P. E. Baird et al., E. N. Fortson et al., Nature (London) 264, 528 (1976).
[11]L. L. Lewis, J. H. Hollister, D. C. Soreide, E. G. Lindahl, and E. N. Fortson, Phys. Rev. Lett. 39, 795 (1977).
[12]P. E. Baird, M. W. Brimicombe, R. G. Hunt, G. J. Roberts, P. G. H. Sandars, and D. N. Stacey, Phys. Rev. Lett. 39, 798 (1977).
[13]V. N. Novikov, O. P. Sushkov, and I. B. Khriplovich, Opt. Spektrosk. 43, 621 (1977) [Opt. Spectrosc. (USSSR) 43, 370 (1977)].
[14]V. N. Novikov, O. P. Sushkov, and I. B. Khiplovich, Zh. Eksp. Teor. Fiz. 71, 1665 (1976) [Sov. Phys. JETP 44, 872 (1976)].

PARITY NON-CONSERVATION IN INELASTIC ELECTRON SCATTERING ☆

C.Y. PRESCOTT, W.B. ATWOOD, R.L.A. COTTRELL, H. DeSTAEBLER, Edward L. GARWIN,
A. GONIDEC [1], R.H. MILLER, L.S. ROCHESTER, T. SATO [2], D.J. SHERDEN, C.K. SINCLAIR,
S. STEIN and R.E. TAYLOR
Stanford Linear Accelerator Center, Stanford University, Stanford, CA 94305, USA

J.E. CLENDENIN, V.W. HUGHES, N. SASAO [3] and K.P. SCHÜLER
Yale University, New Haven, CT 06520, USA

M.G. BORGHINI
CERN, Geneva, Switzerland

K. LÜBELSMEYER
Technische Hochschule Aachen, Aachen, West Germany

and

W. JENTSCHKE
II. Institut für Experimentalphysik, Universität Hamburg, Hamburg, West Germany

Received 14 July 1978

We have measured parity violating asymmetries in the inelastic scattering of longitudinally polarized electrons from deuterium and hydrogen. For deuterium near $Q^2 = 1.6$ (GeV/c)^2 the asymmetry is $(-9.5 \times 10^{-5})Q^2$ with statistical and systematic uncertainties each about 10%.

We have observed a parity non-conserving asymmetry in the inelastic scattering of longitudinally polarized electrons from an unpolarized deuterium target. In this experiment a polarized electron beam of energy between 16.2 and 22.2 GeV was incident upon a liquid deuterium target. Inelastically scattered electrons from the reaction

$$e(\text{polarized}) + d \rightarrow e' + X, \qquad (1)$$

were momentum analyzed in a magnetic spectrometer at 4° and detected in a counter system instrumented to measure the electron flux, rather than to count individual scattered electrons. The momentum transfer, Q^2, to the recoiling hadronic system varied between 1 and 1.9 (GeV/c)^2 (see table 1).

Parity violating effects may arise from the interference between the weak and electromagnetic amplitudes. Calculations of the expected effects in deep inelastic experiments have been reported by several authors [1–7], and asymmetries at the level of $10^{-4} Q^2$ are predicted for the kinematics of our experiment. Previous experiments with muons [8] and electrons [9,10] have not achieved sufficient accuracy to observe such small effects. This same interference of amplitudes may also give rise to measurable effects in

☆ Work supported by the Dept. of Energy.

[1] Permanent address: Annecy (LAPP), 74019 Annecy-le-Vieux, France.

[2] Permanent address: National Laboratory for High Energy Physics, Tsukuba, Japan.

[3] Present address: Department of Physics, Kyoto University, Kyoto, Japan.

Table 1
Kinematic conditions at which data were taken. The average Q^2 and y values were calculated for the shower counter using a Monte Carlo program.

Beam energy E_0 (GeV)	$g - 2$ precession angle θ_{prec} (rad)	Spectrometer setting E' (GeV)	Kinematic quantities averaged over spectrometer	
			Q^2 (GeV/c)^2	y
16.18	5.0π	12.5	1.05	0.18
17.80	5.5π	13.5	1.25	0.19
19.42	6.0π	14.5	1.46	0.21
22.20	6.9π	17.0	1.91	0.21

atomic spectra; experiments on transitions in the spectrum of bismuth have already been reported [11–13].

Of crucial importance to this experiment was the development of an intense source of longitudinally polarized electrons. The source consisted of a gallium arsenide crystal mounted in a structure similar to a regular SLAC gun with the GaAs replacing the usual thermionic cathode. The polarized electrons were produced by optical pumping with circularly polarized photons between the valence and conduction bands in the GaAs, which had been treated to assure a surface with negative electron affinity [14,15]. The light source was a dye laser operated at 710 nm and pulsed to match the linac (1.5 μs pulses at 120 pulses per second). Linearly polarized light from the laser was converted to circularly polarized light by a Pockels cell, a crystal with birefringence proportional to the applied electric field. The plane of polarization of the light incident on the Pockels cell could be varied by rotating a calcite prism. Reversing the sign of the high voltage pulse driving the Pockels cell reversed the helicity of the photons which in turn reversed the helicity of the electrons. This reversal was done randomly on a pulse to pulse basis. The rapid reversals minimized the effects of drifts in the experiment, and the randomization avoided changing the helicity synchronously with periodic changes in experimental parameters. Pulsed beam currents of several hundred milliamperes were achieved, with intensity fluctuations of a few percent.

The longitudinally polarized electrons were accelerated with negligible depolarization as confirmed by earlier tests [16] [+1]. Both the sign and the magnitude of the polarization of the beam at the target were mea-

sured periodically by observing the asymmetry in Møller (elastic electron–electron) scattering from a magnetized iron foil [16]. The polarization, $|P_e|$, averaged 0.37. Each measurement had a statistical error less than 0.01; we estimate an overall systematic uncertainty of 0.02. The beam intensity at the target varied between 1 and 4×10^{11} electrons per pulse.

A schematic of the apparatus is shown in fig. 1. The target was a 30 cm cell of liquid deuterium. The spectrometer consisted of a dipole magnet, followed by a single quadrupole and a second dipole. The scattering angle was 4° and the momentum setting was about 20% below the beam energy (see table 1 for the kinematic settings). The acceptance was ±7.4 mrad in scattering angle, ±16.6 mrad in azimuth and about ±30% in momentum, as determined from a Monte Carlo model of the spectrometer.

Two separate electron detectors intercepted electrons analyzed by the spectrometer. The first was a nitrogen-filled Cerenkov counter operated at atmospheric pressure. The second was a lead-glass shower counter with a thickness of nine radiation lengths (the TA counter). Approximately 1000 scattered electrons per pulse entered the counters.

The high rates were handled by integrating the outputs of each phototube rather than by counting individual particles. For each pulse, i, the integrated output of each phototube, N_i, was divided by the integrated beam intensity (charge), Q_i, to form the yield for that pulse, $Y_i = N_i/Q_i$. For the distributions of the Y_i we verified experimentally that the (charge weighted) means of the distributions, $\langle Y \rangle$, were independent of Q, within errors of about ±0.3%, and that the (charge weighted) standard deviations, ΔY, consistent with the statistical fluctuations expected from the number of scattered electrons per pulse. For a run with n beam pulses the statistical uncertainty on $|Y|$ was given by $\Delta Y/\sqrt{n}$.

As a check on our procedures we measured the asymmetry for a series of runs using the unpolarized beam from the regular SLAC gun for which the asymmetry should be zero. For a given run the experimental asymmetry was given by:

$$A_{exp} = [\langle Y(+) \rangle - \langle Y(-) \rangle] / [\langle Y(+) \rangle + \langle Y(-) \rangle], \quad (2)$$

[+1] The present experiment used the same target as ref. [16], but used a different spectrometer and detectors.

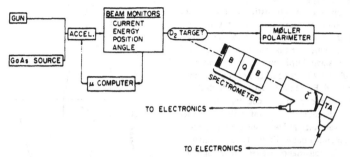

Fig. 1. Schematic layout of the experiment. Electrons from the GaAs source or the regular gun are accelerated by the linac. After momentum analysis in the beam transport system the beam passes through a liquid deuterium target. Particles scattered at 4° are analyzed in the spectrometer (bend-quad-bend) and detected in two separate counters (a gas Cerenkov counter, and a lead-glass shower counter). A beam monitoring system and a polarization analyzer are only indicated, but they provide important information in the experiment.

where + and − were assigned by the same random number generator that determined the sign of the voltage applied to the Pockels cell. For the shower counter we obtained a value of $(-2.5 \pm 2.2) \times 10^{-5}$ for A_{exp} divided by 0.37, the average value of $|P_e|$ for polarized beams from the GaAs source. The individual values were distributed about zero consistent with the calculated statistical errors. We conclude that asymmetries can be measured in this apparatus to a level of about 10^{-5}.

The same procedures were next applied to a similar series of runs using polarized beams. The helicity of the electrons coming from the source depended on the orientation of the linearly polarizing prism as well as on the sign of the voltage on the Pockels cell. Rotation of the plane of polarization by rotating the calcite prism through an angle ϕ_p caused the net electron helicity to vary as $\cos(2\phi_p)$. We chose three operating conditions:

   (a) prism orientation at 0°, producing + (−) helicity electrons for + (−) Pockels cell voltage;

   (b) prism orientation at 45°, producing unpolarized electrons for either sign of Pockels cell voltage; and

   (c) prism orientation at 90°, producing − (+) helicity electrons for + (−) Pockels cell voltage.

Positive helicity indicates that the spin is parallel to the direction of motion. As the prism is rotated by 90°, A_{exp} should change sign since it is defined only with respect to the sign of the voltage on the Pockels cell. We may define a physics asymmetry, A, whose sign depends on the helicity of the beam at the target

$$A_{exp} = |P_e| A \cos(2\phi_p), \qquad (3)$$

where ϕ_p is the angle of orientation of the calcite prism. Fig. 2 shows the results at 19.4 GeV for $A_{exp}/|P_e|$. For the 45° point we used a value of 0.37 for $|P_e|$. These data are in satisfactory agreement with expecta-

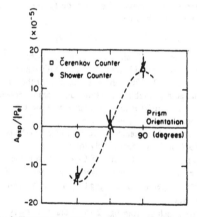

Fig. 2. The experimental asymmetry shows the expected variation (dashed line) as the beam helicity changes due to the change in orientation of the calcite prism. The data are for 19.4 GeV and deuterium. Since the same scattered particles strike both counters, they are not statistically independent. No systematic errors are shown. No corrections have been made for helicity dependent differences in beam parameters.

tions, and serve to separate effects due to the helicity of the beam from possible systematic effects associated with the reversal of the Pockels cell voltage. Only statistical errors are shown. The results at $45°$ are consistent with zero and indicate that other sources of error in A_{exp} must be small. Furthermore, the asymmetries measured at $0°$ and $90°$ are equal and opposite, within errors, as expected. Fig. 2 shows data from both the Cerenkov counter and the shower counter. Although these two separate counters were not statistically independent, they were analyzed with independent electronics and responded quite differently to potential backgrounds. The consistency between these counters serves as a check that such backgrounds are small.

At 19.4 GeV with the prism at $0°$ the helicity at the target was positive for positive Pockels cell voltage. However, this helicity depended on beam energy, owing to the $g - 2$ precession of the spin in the transport magnets which deflected the beam through $24.5°$ before reaching the target. Because of the anomalous magnetic moment of the electron, the electron spin direction precessed relative to the momentum direction by an angle

$$\theta_{prec} = \frac{E_0}{m_e c^2}\frac{g-2}{2}\theta_{bend} = \frac{E_0\,(\text{GeV})}{3.237}\,\pi\text{ rad},\quad(4)$$

where m_e is the mass and g the gyromagnetic ratio of the electron. Thus we expect

$$A_{exp} = |P_e|\,A\cos[(E_0\,(\text{GeV})/3.237)\pi]\,,\quad(5)$$

where the signs of values of A_{exp} for the prism at $90°$ have been reversed before combining with values for the prism at $0°$. Fig. 3 shows the results for the kinematic points in table 1 as a function of beam energy. At each point Q^2 is different. Since we expect A to be proportional to Q^2, we divide A_{exp} by Q^2 [*2]. Fig. 3 also shows the expected curve normalized to the point at 19.4 GeV. The data clearly follow the $g - 2$ modulation of the helicity. At 17.8 GeV the spin is transverse; any effects from transverse components of the spin are expected to be negligible, in agreement with our data.

We conclude from figs. 2 and 3 that the observed asymmetries are due to electron helicity. Nevertheless,

---

[*2] This fact is true in all models. It arises because the electromagnetic amplitude has a $1/Q^2$ dependence, giving an asymmetry proportional to Q^2.

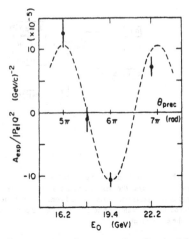

Fig. 3. The experimental asymmetry shows the expected variation (dashed line) as the beam helicity changes as a function of beam energy due to the $g - 2$ precession in the beam transport system. The data are for the shower counter and the deuterium target. No systematic errors are shown. No corrections have been made for helicity dependent differences in beam parameters.

it is essential to search for and set limits on asymmetries due to effects other than helicity. Systematic effects due to slow drifts in phototube gains, magnet currents, etc., were minimized by the rapid, random reversals of polarization, and had negligible effects on A_{exp}. Effects due to random fluctuations in the beam parameters were small compared to the 3% pulse to pulse fluctuations due to counting statistics in the detectors. This was verified experimentally by measuring A_{exp} with unpolarized beams from the regular SLAC gun, and also by generating "fake" asymmetries using pulses of the same helicity from the polarized data runs themselves.

A more serious source of potential error came from small systematic differences between the beam parameters for the two helicities. Small changes in position, angle, current or energy of the beam can influence the measured yields. If these changes are correlated with reversals of the beam helicity, they may cause apparent parity violating asymmetries. Using an extensive beam monitoring system based on microwave cavities, measurements were made for each beam pulse of the average energy and position [17]. Angles were determined from cavities 50 m apart. The beam charge was determined using the standard toroid monitors [18]. The resolutions per pulse were about 10 μm in position, 0.3 μrad in angle, 0.01% in energy, and 0.02% in beam intensity. A microcomputer driven feedback system used position and energy signals to stabilize the average beam position, angle, and energy. Using the measured pulse to pulse beam information together with the measured sensitivities of the yield to each of the beam parameters, we made corrections to the asymmetries for helicity dependent differences in beam parameters. For these corrections, we have assigned a systematic error equal to the correction itself. The most significant imbalance was less than one part per million in E_0 which contributed -0.26×10^{-5} to A/Q^2.

We combine the values of A/Q^2 from the shower counter for the two highest energy points to obtain

$$A/Q^2 = (-9.5 \pm 1.6) \times 10^{-5} \, (\text{GeV}/c)^{-2} \, (\text{deuterium}). \tag{6}$$

We do not include the point at 16.2 GeV because it contains fairly strong elastic and resonance contributions. The sign implies a greater yield from electrons with spin antiparallel to momentum. For this combined point the average value of $y = 1 - E'/E_0$ is 0.21 and the average value of Q^2 is 1.6 $(\text{GeV}/c)^2$. The quoted error, based on preliminary analysis, is derived from a statistical error of $\pm 0.86 \times 10^{-5}$ added linearly to estimated systematic uncertainties of 5% in the value of $|P_e|$, and of 3.3% from asymmetries in beam parameters. We determined experimentally that the π^- background contributed less than 0.1×10^{-5} to A/Q^2. The result in eq. (6) includes normalization corrections of 2% for the π^- background, and 3% for radiative corrections.

Any observation of non-conservation of parity in interactions involving electrons adds new information on the nature of neutral currents and gauge theories. Certain classes of gauge theory models predict no observable parity violations in experiments such as ours. Among these are those left—right symmetric models in which the difference between neutral current neutrino and anti-neutrino scattering cross sections is explained as a consequence of the handedness of the neutrino and anti-neutrino, while the underlying dynamics are parity conserving. Such models are incompatible with the results presented here.

The simplest gauge theories are based on the gauge

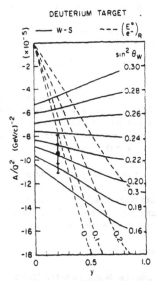

Fig. 4. Comparison of our result for deuterium with two SU(2) X U(1) predictions using the simple quark-parton model for nucleons. The outer error bars correspond to the error quoted in the text (eq. (6)). The inner error bars correspond to the statistical error. The y-dependence of A/Q^2 for various values of $\sin^2\theta_W$ is shown for two models: Weinberg–Salam (solid lines) and the hybrid model (dashed line).

group SU(2) X U(1). Within this framework the original Weinberg–Salam (W–S) model makes specific weak isospin assignments: the left-handed electron and quarks are in doublets, the right-handed electron and quarks are singlets [19]. Other assignments are possible, however. In particular, the "hybrid" or "mixed" model that assigns the right-handed electron to a doublet and the right-handed quarks to singlets has not been ruled out by neutrino experiments.

To make specific predictions for parity violation in inelastic electron scattering, it is necessary to have a model for the nucleon, and the customary one is the simple quark-parton model. The predicted asymmetries depend on the kinematic variable y as well as on the weak isospin assignments and on $\sin^2\theta_W$, where θ_W is the Weinberg angle. Fig. 4 compares our result for two SU(2) X U(1) models. The simplest model (W–S) is in good agreement with our measurement for $\sin^2\theta_W = 0.20 \pm 0.03$ which is consistent with the values obtained in neutrino experiments. The hybrid model is consistent with our data only for values of $\sin^2\theta_W$ $\lesssim 0.1$.

We took a limited amount of data at 19.4 GeV using a liquid hydrogen target with the result

$$A/Q^2 = (-9.7 \pm 2.7) \times 10^{-5} \, (\text{GeV}/c)^{-2} \text{ (hydrogen)}, \tag{7}$$

where the error contains both statistical and systematic uncertainties. A proton target provides a different mix of quarks and is expected to give a slightly smaller asymmetry than deuterium [7]. Our results are not inconsistent with this expectation.

It is a pleasure to acknowledge the support we received from many people at SLAC. In particular we would like to thank M.J. Browne, G.J. Collet, R.L. Eisele, Z.D. Farkas, H.A. Hogg, C.A. Logg and H.L. Martin for especially significant contributions.

References

[1] A. Love et al., Nucl. Phys. B49 (1972) 513.
[2] E. Derman, Phys. Rev. D7 (1973) 2755.
[3] W.W. Wilson, Phys. Rev. D10 (1974) 218.
[4] S.M. Berman and J.R. Primack, Phys. Rev. D9 (1974) 2171; D10 (1974) 3895 (erratum).
[5] M.A.B. Beg and G. Feinberg, Phys. Rev. Lett. 33 (1974) 606.
[6] S.M. Bilenkii et al., Sov. J. Nucl. Phys. 21 (1975) 657.
[7] R.N. Cahn and F.J. Gilman, Phys. Rev. D17 (1978) 1313; further references to the theory may be found in this reference.
[8] Y.B. Bushnin et al., Sov. J. Nucl. Phys. 24 (1976) 279.
[9] M.J. Alguard et al., Phys. Rev. Lett. 37 (1976) 1258, 1261; 41 (1978) 70.
[10] W.B. Atwood et al., SLAC preprint SLAC-PUB-2123 (1978).
[11] L.L. Lewis et al., Phys. Rev. Lett. 39 (1977) 795.
[12] P.E.G. Baird et al., Phys. Rev. Lett. 39 (1977) 798.
[13] L.M. Barkov and M.S. Zolotorev, Zh. Eskp. Teor. Fiz. Pis'ma 26 (1978) 379.
[14] E.L. Garwin, D.T. Pierce and H.C. Siegmann, Swiss Physical Society Meeting (1974), Helv. Phys. Acta 47 (1974) 393 (abstract only); the full paper is available as SLAC-PUB-1576 (1975) (unpublished).
[15] D.T. Pierce et al., Phys. Lett. 51A (1975) 465; Appl. Phys. Lett. 26 (1975) 670.
[16] P.S. Cooper et al., Phys. Rev. Lett. 34 (1975) 1589.
[17] Z.D. Farkas et al., SLAC-PUB-1823 (1976).
[18] R.S. Larsen and D. Horelick, in: Proc. Symp. on Beam intensity measurement, DNPL/R1, Daresbury Nuclear Physics Laboratory (1968); their contribution is available as SLAC-PUB-398.
[19] S. Weinberg, Phys. Rev. Lett. 19 (1967) 1264; A. Salam, in: Elementary particle theory: relativistic groups and analyticity, Nobel Symp. No. 8, ed. N. Svartholm (Almqvist and Wiksell, Stockholm, 1968) p. 367.

MEASUREMENT OF THE NEUTRAL TO CHARGED CURRENT
CROSS SECTION RATIO IN NEUTRINO AND ANTINEUTRINO INTERACTIONS

M. HOLDER, J. KNOBLOCH, J. MAY, H.P. PAAR, P. PALAZZI, F. RANJARD, D. SCHLATTER,
J. STEINBERGER, H. SUTER, H. WAHL, S. WHITAKER and E.G.H. WILLIAMS

CERN, Geneva, Switzerland

F. EISELE, C. GEWENIGER, K. KLEINKNECHT, G. SPAHN and H.-J. WILLUTZKI

Institut für Physik * der Universität, Dortmund, Germany

W. DORTH, F. DYDAK, V. HEPP, K. TITTEL and J. WOTSCHACK

Institut für Hochenergiephysik * der Universität, Heidelberg, Germany

A. BERTHELOT, P. BLOCH, B. DEVAUX, M. GRIMM, J. MAILLARD,
B. PEYAUD, J. RANDER, A. SAVOY-NAVARRO and R. TURLAY

D.Ph.P.E., CEN-Saclay, France

F.L. NAVARRIA

Istituto di Fisica dell'Università, Bologna, Italy

Received 17 August 1977

We report on the analysis of inclusive neutral current events produced in neutrino and antineutrino narrow band beams. We find for incident neutrino energies in the range $12 - 200$ GeV and for hadron energies above 12 GeV a neutral to charged current cross-section ratio of $R_\nu = 0.293 \pm 0.010$ for incident neutrinos, and $R_{\bar\nu} = 0.35 \pm 0.03$ for antineutrinos. These ratios are consistent with the Weinberg-Salam model, with $\sin^2\theta_W = 0.24 \pm 0.02$.

Since the discovery of the semi-leptonic neutral current interaction, $\nu(\bar\nu) + N \to \nu(\bar\nu) + \text{hadrons}$ [1], interest has been concentrated on precise measurements of its strength and structure. We have measured inclusive neutral current (NC) interactions on iron nuclei induced by neutrinos and antineutrinos. In this letter we present results on the ratios of neutral to charged current (CC) inclusive cross-sections for neutrinos (R_ν) and antineutrinos $(R_{\bar\nu})$.

The experiment has been performed in the narrow-band neutrino beam at the CERN SPS. The incident neutrinos are produced by the decay of charge selected pions and kaons in a parallel beam of well defined momentum, 200 ± 9 (rms) GeV/c. Fig. 1 shows the primary energy spectrum for neutrinos (fig. 1a) and antineutrinos (fig. 1b).

* Supported by Bundesministerium für Forschung und Technologie.

The detector [2] combines the function of target, hadron calorimeter, and muon spectrometer integrally in 19 similar modules. Each module consists of multiple iron plates 3.75 m in diameter, with a total thickness of 75 cm of iron and a weight of 65 tons per module. The hadron shower is measured by means of scintillator sheets divided into eight horizontal strips inserted between the iron plates. The first seven modules are equipped with scintillators placed every 5 cm of iron, the next eight sample every 15 cm, and the last four every 75 cm. The vertical position of the showers is determined by weighting coordinates of the scintillator strips with their energy response. The horizontal shower position is calculated from the ratio of the pulseheights at the right and left scintillator edge, making use of the light absorption in the scintillator sheets (absorption length ~2 m). The radial distance R of each event is thus determined with a resolution of $\sigma = 22$ cm. The trigger relevant to this analysis re-

Table I

Data reduction for the ratio of neutral to charged current inclusive cross-sections for $E_H > 12$ GeV

	Neutrinos	Antineutrinos
NC candidates	10770 ± 104	3314 ± 58
Cosmic-ray background	−59 ± 7	−119 ± 10
WBB background	−286 ± 126	−646 ± 116
CC background	−1493 ± 64	−235 ± 49
NC with $L < 16$ and $l. > L_{cut}$	+150	+43
K_{e3} correction	−1008	−154
NC signal	8074 ± 156	2203 ± 130
CC candidates	26509 ± 163	6483 ± 81
WBB background	−239 ± 117	−323 ± 83
CC extrapolation	+1467 ± 64	+253 ± 42
NC with $L > L_{cut}$	−134	−35
CC signal	27603 ± 211	6378 ± 123
NC/CC (error only statistical)	0.293 ± 0.006	0.346 ± 0.021
NC/CC (final result, systematic error included)	0.293 ± 0.010	0.35 ± 0.03

to muons escaping at the side.

A NC candidate is defined by the requirement: $16 < L < L_{cut}$ (cm of iron). The lower cut at 16 cm excludes most of the cosmic ray background, L_{cut} is chosen as small as possible while maintaining an efficiency of ~99%. A study of event length distributions, such as are given in fig. 2 but in bins of hadron-shower energy, leads to the parameterisation $L_{cut} = 75 + 38 \times \ln E_H$ (cm of iron) with E_H in GeV. The NC event losses due to the lower and upper cuts in L are together of the order of 1%.

The CC background in the NC signal is determined on the basis of the observed events in a CC monitor region, defined by $311 < L < 511$ cm of iron.

To obtain the number of genuine NC and CC events, the following corrections to the number of candidates must be applied (see also table 1):

1. Cosmic ray background: It is determined from the event rate out of time with the 23 μsec long beam spill. The difference of the electronics dead-time for events inside and outside the spill is accounted for. The contamination by cosmic ray muons is significant only for hadron energies below 15 GeV and is determined with good statistical and systematic accuracy.

2. Wide-band beam (WBB) background: In the region between the proton target and the beginning of the beam-forming elements, hadrons of both signs and with the full energy spectrum are present. Their decays give rise to a flux of both neutrinos and antineutrinos. A subtraction of the events induced by the WBB flux is necessary because of the very different energy spectrum, peaking at low energies. It is determined in special runs where the momentum-defining collimator, located halfway downstream between the beam-forming elements, was closed. Thus the hadrons which normally produce the neutrinos are stopped there. The main flux of the WBB background does not point to our detector, however this sort of background constitutes a major source of statistical and systematic error in our analysis. A fraction of approximately 5% of the running time has been devoted to closed collimator studies. The events observed in these runs are scaled by the ratio of the collimator open and collimator closed fluxes. They are globally subtracted in the NC signal, CC signal and CC monitor region, respectively. The subtraction is of the order of 3% for neutrinos, and 20% for antineutrinos.

3. CC background: The CC events with an event length $l. < L_{cut}$ must be subtracted from the NC candidates, and added to the CC candidates. Their number is determined by a Monte Carlo extrapolation from the measured number of events in the CC monitor region. For this extrapolation the distribution of the scaling variable $y = E_H/E_\nu$ is assumed to be $1 + \alpha_{CC}(1 - y)^2$ for CC neutrino events, and $\alpha_{CC} + (1 - y)^2$ for CC antineutrino events. In the quark parton model $\alpha_{CC} = \bar{q}/q$ measures the contribution of antiquarks. We have chosen $\alpha_{CC} = 0.1$ independent of the neutrino energy [3]. However, since essentially only CC events with y near 1 constitute the background in the NC signal, this extrapolation is insensitive to the actual value of α_{CC}. The small contribution of events where the interaction takes place near the edge of the detector and the muon escapes, is taken into account by the Monte Carlo simulation. The systematic uncertainties in the extrapolation are estimated to be ~5% for neutrinos and ~8% for antineutrinos.

4. K_{e3} background: Electron neutrinos produce CC and NC interactions in the detector, and all of them simulate NC interactions, because the electromagnetic cascade generated by the final state electron

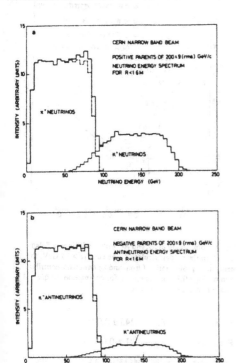

Fig. 1. Monte Carlo simulation of the incident neutrino energy spectrum for neutrinos (a) and antineutrinos (b), with a cut in the radial distance $R < 1.6$ m.

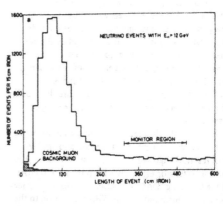

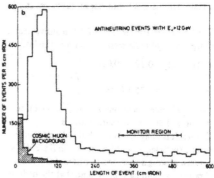

Fig. 2. Event length distribution for neutrinos (a) and antineutrinos (b). The data are not corrected.

quires an energy deposition of more than 3 GeV in the detector.

Events with a hadron-shower of at least 12 GeV are selected; this is well above the hardware trigger threshold. The vertex position z_v along the beam direction is chosen in the interval $35 < z_v < 825$ cm of iron. The lower cut excludes possible neutron induced showers in the front part of the spectrometer although we have no evidence for such an effect. The upper cut allows for a sufficient penetration depth in the rear part of the spectrometer to be used for the separation of NC and CC events. The overall length of iron is 1425 cm. To minimize problems of side leakage, the radial distance R is required to be less than 1.6 m.

For each event, an event length L is calculated from the scintillator information and is equal to the thickness of iron between the vertex and the end of the event, measured parallel to the axis of the detector. In fig. 2 the distribution of L is shown for neutrinos (fig. 2a) and antineutrinos (fig. 2b). We see a strong signal of NC events for an event length less than 200 cm of iron. The region above 200 cm is rather flat and consists of CC events with a short muon track. It is emphasized that in our analysis no muon reconstruction is required. Also, by virtue of the large diameter of our set-up, the background of CC events in the NC signal is mostly due to stopping muons rather than due is contained in the hadron shower. The observed NC/CC ratio is therefore affected. The K_{e3} decay is the only important source of electron neutrinos. This background is suppressed by the relative branching ratio of the K_{e3} and $K_{\mu2}$ decays, and by the lower energy of the neutrinos produced in three body decays. The background is further reduced by the admixture of neutrinos arising from pion decay. Since the relative amount of kaons and pions in the parent beam is measured to be $K^+/\pi^+ = 0.17$ and $K^-/\pi^- = 0.05$, the K_{e3} correction is straightforward. It amounts to ~10% for neutrinos and ~5% for antineutrinos, respectively.

In table 1 the data reduction and the results on R_ν and $R_{\bar\nu}$ are summarized. With the exception of the final result, the quoted errors are statistical only. The quoted systematic error arises from the uncertainties in the CC background subtraction and the K_{e3} correction. The contribution of the WBB subtraction to the systematic error is small with respect to its statistical error.

The final result for the ratio of the inclusive neutral to charged current cross sections on iron nuclei is:

$$R_\nu = (NC/CC)_\nu = 0.293 \pm 0.010 \left.\begin{array}{c}\\\\\end{array}\right\} \quad E_H > 12 \text{ GeV} .$$
$$R_{\bar\nu} = (NC/CC)_{\bar\nu} = 0.35 \pm 0.03$$

The average incident neutrino energy is 110 GeV for neutrinos, and 90 GeV for antineutrinos.

Our result is in agreement with the recent results of other experiments [4–6], as shown in fig. 3. However, when making a direct comparison of the measured ratios it should be kept in mind that the hadron energy cut-offs and the neutrino energy spectra are quite different. In particular, the cut-off correction is large for the Gargamelle values [4]. Since the cut-off correction is small for the high energy experiments, we prefer to compare our result, obtained with a cut-off, to the uncorrected values of the two other high-energy experiments [5,6], and to the corrected value of the low energy Gargamelle experiment [4].

Our results may be compared with the prediction of the Weinberg-Salam model. For this purpose, the 12 GeV cut in the hadron energy, the small deviation of iron from an isoscalar target, and antiquark contribution have been incorporated into the model [7] (see also fig. 3). The Weinberg mixing angle can be determined from R_ν and $R_{\bar\nu}$ separately, with the results:

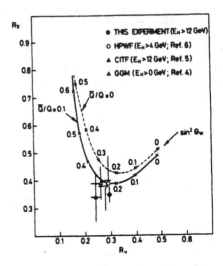

Fig. 3. Comparison of R_ν and $R_{\bar\nu}$ with the Weinberg-Salam model. In the model calculation, the cut $E_H > 12$ GeV, the neutron to proton ratio of iron, and an antiquark contribution of $\bar q/q = 0.1$ (solid line) and, for comparison, of $\bar q/q = 0$ (dashed line) are incorporated.

From R_ν : $\quad \sin^2\theta_w = 0.243 \pm 0.021$

From $R_{\bar\nu}$: $\quad \sin^2\theta_w = 0.21 \pm 0.09$.

The quoted error includes variations of $\bar q/q$ in the range $0.05 < \bar q/q < 0.20$. Both values are in good agreement, thus supporting the Weinberg-Salam model. Combining both results, we get

From R_ν and $R_{\bar\nu}$: $\quad \sin^2\theta_w = 0.24 \pm 0.02$.

We thank most sincerely our many technical collaborators and the members of the SPS staff for the operation of the accelerator.

References

[1] F.J. Hasert et al., Phys. Lett. 46B (1973) 121; 46B (1973) 138.
[2] M. Holder et al., A detector for high-energy neutrino interactions, submitted to Nuclear Instrum. Methods.
[3] M. Holder et al., Is there a high-y anomaly in antineutrino interactions? submitted to Phys. Rev. Lett.

[4] J. Blietschau et al., Nucl. Phys. B118 (1977) 218.

[5] B.C. Barish et al., Proc. Int. Neutrino Conf. Aachen, 1976 (F. Vieweg & Sohn, Verlagsgesellschaft mbH, Braunschweig, 1977) p. 289.

[6] T.Y. Ling, Proc. Int. Neutrino Conf. Aachen, 1976 (F. Vieweg & Sohn, Verlagsgesellschaft mbH, Braunschweig, 1977) p. 296;
P. Wanderer et al., Measurement of the neutral current interactions, preprint HWPF-77/1, submitted to Phys. Rev. D.

[7] L.M. Sehgal, Nucl. Phys. B65 (1973) 141.
The antiquark contribution is specified by

$$\int x(\bar{u} + \bar{d})\ dx/\int x(u + d)\ dx = 0.07\ ,$$

$$\int 2 \times \bar{s}\ dx/\int x(u + d)\ dx = 0.03\ ,$$

$$\bar{c} = 0\ ,$$

and $s = \bar{s}$, $c = \bar{c}$ is assumed.

CHAPTER 7
ON TO THE *W* AND *Z* PARTICLES

The Electroweak Force Rises (1976 - 1983)

7.1. THE STANDARD MODEL AND THE IVB

The concept of the IVB dates back to the late 1930s and '40s (see Fig. 7.1), and it was the subject of intense speculation and experimental search starting in the 1950s (see Chapters 1 and 2 and the references therein). By the end of the '50s, the limit on the IVB mass was of order 2 GeV. The earliest neutrino experiments at BNL, and later at CERN, put slightly better limits [7.1]. The 200-GeV machine at FNAL was constructed largely to discover the IVB. However, no one had any idea about the mass of the IVB and, in principle, it could have been at the unitary limit of the electroweak scale of ~ 300 GeV [7.1]. However, there was already an explosion of predictions of different types of IVBs, as shown in Fig. 7.1.

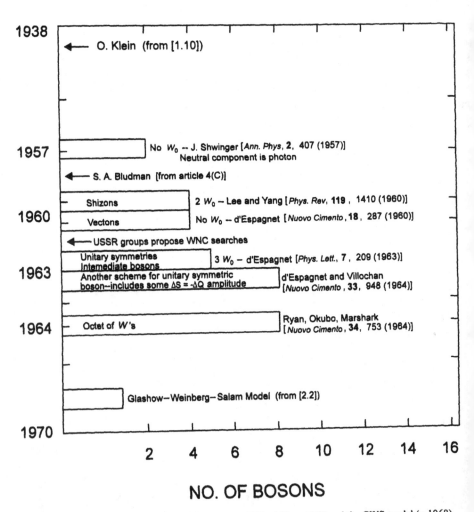

Figure 7.1. The history of various suggestions for the IVBs before 1964 and the GWS model (~ 1968).

While the IVB was not directly observed, the phenomenality of weak interactions strongly pointed to such an object. Thus, by 1970, there was a feeble limit on the W mass, no evidence for WNC, strong constraints on FCNC processes, and general excitement about the new high-energy machines at Batavia and CERN. No one could guess the great revolution that lay three years ahead. (The rise of the standard model is illustrated in Fig. 1.1.)

The early theory and experiments on weak interactions did not distinguish between the various space–time operators like tensor or axial vector, *etc.* In 1957, with the discovery of parity violation in the weak interaction, a simple picture of the weak interaction emerged that selected the V and V-A interaction. [See article 2(A) by C-S. Wu.] At that time, the known or suspected particles could be classified into doublets such as

$$\begin{pmatrix} \nu \\ e \end{pmatrix} \begin{pmatrix} \nu' \\ \mu \end{pmatrix} \begin{pmatrix} p \\ n \end{pmatrix} . \tag{7.1}$$

It was not known whether these neutrinos were the same or different, although there was strong speculation that two types of neutrinos existed. (We now know that there is a ν_e and a ν_μ.) At the start, there was not a clear attempt to incorporate the newly discovered strange particle into the picture.

The multiplets (7.1) could account for much of the known physics:

$$\bar{\nu}_e + p \to e^+ + n \ ,$$

$$n \to p + e^- + \nu \ ,$$

$$\mu^- \to e + \nu + \bar{\nu} \ .$$

Each of the doublets was coupled to an equal mixture of V and A couplings.

In the early 1960s, following a suggestion by M. Schwartz and B. Pontecorvo, a second neutrino was discovered at BNL [see reprinted article 2(E) by Danby and the references in this chapter]. The lepton doublet structure was thus

$$\begin{pmatrix} \nu_e \\ e^- \end{pmatrix} \begin{pmatrix} \nu_\mu \\ \mu^- \end{pmatrix} .$$

In these same neutrino experiments, there were attempts to observe a very low-mass IVB (it was actually an intermediate V-A boson). The limits on the IVB mass were then in the 1 to 2-GeV range. However, since no adequate theory existed that provided a mass estimate (beyond the unitary limit argument, which gave ~ 300 GeV as an upper limit) [7.1], these null searches were considered very important.

Let us look at the remarkable developments of the period of 1973–1978 and following from the point of view of the whole story of neutral currents, as displayed in Figs. 5.2 and 6.3. These set the stage and the components of the standard model of today.

The measurements of $\sin^2\theta_W$ were carried out with amazing rapidity during the period of 1974–1978, as discussed in Chapter 6 and shown in Fig. 6.3. By 1978, a fairly small error was obtained and established the standard model at the time. At this point, the estimated mass of the W and Z particles was known, and this provided a motivation for the $\bar{p}p$ collider and the subsequent discovery of the W and Z.

7.2. THE DEVELOPMENT OF THE $\bar{p}p$ COLLIDER

Colliding beam machines were developed in the 1950s, largely following a paper by Don Kerst *et al.* [7.2], which is certainly a seminal work in this field. In these storage rings, intense beams of protons or electrons and positrons were made to collide at a special interaction region where a particle detector was placed. After this, low-energy e^+e^- and e^-e^- colliding-beam machines were developed at Stanford, Novosibirsk, and Frascati. In the early 1970s, CERN built the intersecting storage ring for proton–proton collisions. The combined experience with the e^+e^- and pp machines made it possible to propose higher energy pp machines by the mid-1970s; in addition, superconducting bending magnets were starting to be developed.

In 1976, an idea was put forward by RMC [C. Rubbia (CERN and Harvard), P. McIntyre (Harvard, currently at Texas A&M), and this author] to convert either the CERN or FNAL synchrotron to an antiproton–proton ($\bar{p}p$) collider in order to discover the W and Z particles [7.3]. (This was not the first suggestion of a $\bar{p}p$ collider, as can be seen in the RMC references.) In 1983, the W and Z were discovered at CERN, the pinnacle of the rise of the standard model! For completeness, we show (Table 7.1) the parameters of the $\bar{p}p$ machine proposed by RMC in 1973 [7.3]. The actual CERN $\bar{p}p$-collider parameters that were used to discover the W were very close to these values. [The article (A) is reprinted here.] Figure 7.2 shows the scheme outlined by RMC.

During the late 1970s, both the FNAL and CERN laboratories perfected the art of beam cooling. [See the *Physics Today* article (B) by Cline and Rubbia.] The CERN $\bar{p}p$ collider, the S$\bar{p}p$S, is shown schematically in Fig. 7.3. However, the FNAL $\bar{p}p$ program, which was started in 1976, was delayed, mainly because of the resignation of R. R. Wilson, director of the laboratory. The next FNAL director, Leon Lederman, decided to wait for the superconducting tevatron machine to initiate $\bar{p}p$ collider work.

A strong $\bar{p}p$ collider program was in progress at CERN in 1982. Two large experimental detectors, UA1 and UA2, were being commissioned to search largely for W and Z. Figure 7.4 shows a schematic of the UA1 experiment. This detector

Table 7.1. Original parameters for a $\bar{p}p$ collider.[*]

Main Ring (Fermilab)

Beam momentum	250 (400) GeV/c
Equivalent laboratory energy for $\bar{p}p$	133 (341) TeV
Accelerating and bunching frequency	53.14 Mc/s
Harmonic number	1113
RF peak voltage/turn	3.3×10^6 V
Residual gas pressure	$< 0.5 \times 10^{-7}$ Torr
Beta functions at interaction point	3.5 m
Momentum compaction at interaction point	~ 0 m
Invariant emittances ($N_p = 10^{12}$):	
Longitudinal	3 eV–s
Transverse	$50\ \pi\ 10^{-6}$ rad-m
Bunch length	2.3 m
Design luminosity	5×10^{29} (8×10^{29}) cm^{-2} s^{-1}

Antiproton Source (Stochastic Cooling)

Nominal stored $\bar{p}$ momentum	3.4 GeV/c
Circumference of ring	100 m
Momentum acceptance	0.02
Betatron acceptances	$100\ \pi\ 10^{-6}$ rad-m
Bandwidth of momentum stochastic cooling	400 Mc/s
Maximum stochastic accelerating RF voltage	3000 V
Bandwidth of betatron stochastic cooling	200 Mc/s
Final invariant emittances ($N_{\bar{p}} = 3.10^{10}$):	
Longitudinal	0.5 eV–s
Transverse	$10\ \pi\ 10^{-16}$ rad-m

[*](From [7.3].) The actual values of the machine parameters for the S$\bar{p}p$S were close to these values.

was, in many ways, the prototype of the 4π detector for hadron colliders, except that it employed a dipole magnetic field, whereas a solenoidal magnetic field has been chosen for subsequent detectors.

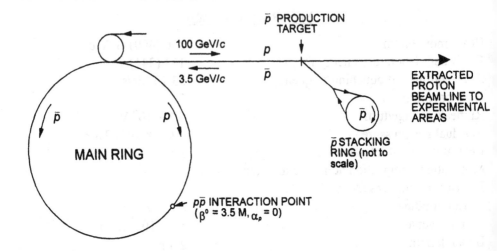

Figure 7.2. General layout of the $\bar{p}p$ colliding scheme proposed by RMC in 1976. Protons (100 GeV/c) are periodically extracted in short bursts and produce 3.5-GeV/c antiprotons, which are accumulated and cooled in the small stacking ring. Then $\bar{p}$'s are reinjected in an RF bucket of the main ring and accelerated to top energy. They collide head on against a bunch filled with protons of equal energy and rotating in the opposite direction.

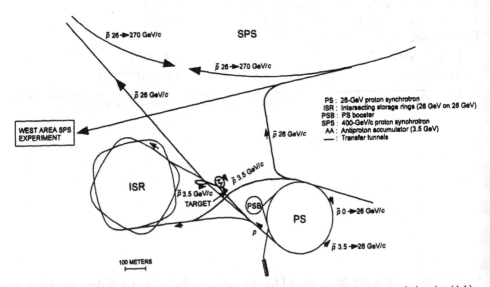

Figure 7.3. Schematic of the CERN $\bar{p}p$ collider, including the $\bar{p}$-beam cooling and accumulation ring (AA).

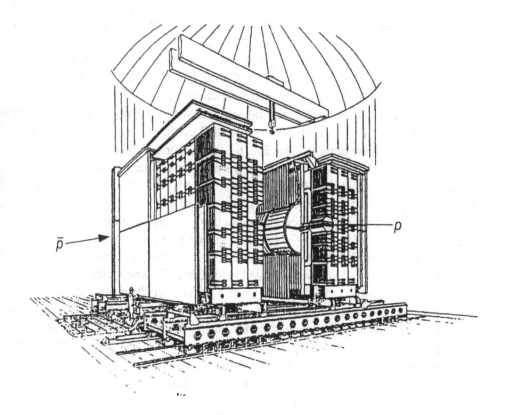

Figure 7.4. The UA1 detector at CERN in 1983. The central region is the drift chamber, followed by an electromagnetic and hadronic calorimeter and a muon identifier on the outside.

7.3. THE *W* AND *Z* BOSON DISCOVERY AND PROPERTIES

7.3.1.   The Estimated Mass of the *W* and *Z* in 1976

After the discovery of WNCs in 1973, and the clarification of the parameters and that
the GWS model provided a good description of the data, it was natural to turn to the
search for the *Z* (and *W*) boson implied by these theories and by the earlier studies
of weak interactions. During 1973–1975, the discovery of WNCs was consolidated,
and the value of $\sin^2\theta_W$ was being determined, as we discussed in Chapters 5 and
6. The GWS model was so successful that it provided a direct prediction for m_W and
m_Z once ρ and $\sin^2\theta_W$ were determined using $m_Z^2 = m_W^2/\cos^2\theta_W$ and
$m_W = \overline{v}\alpha/\sqrt{2}\,G_F\sin^2\theta_W$.

During this period, m_W and m_Z were estimated to be in the range of 40 to 50
GeV. [See the reprinted paper (A) of RMC to obtain some feeling for the general
belief in the GWS model and the estimated parameters in late 1975 and early 1976.]
During this period, there were several suggestions and one approved project
(ISABELLE at BNL) to search for the *W* and *Z* particles. In 1976, the first low
values of $\sin^2\theta_W$ were obtained (see Fig. 6.3) that indicated that the *W* mass could
be well above 60 GeV, which caused a serious problem for the *W* discovery potential
of the ISABELLE machine. It was realized by RMC that the FNAL and CERN
machines might be converted to $\bar{p}p$ colliders (using a single ring) if intense and
cooled $\bar{p}$'s could be made available, and that the $\bar{p}p$ energy would be high enough
to discover even an 80- or 90-GeV IVB. At that point in time, the ISABELLE
colliding proton–proton machine proposed for BNL seemed the most likely place to
search for the *W* and *Z*. However, as stated before, in 1976, a new idea was proposed
by RMC to convert the FNAL or CERN synchrotrons to colliders
(antiproton–proton) with the addition of an antiproton source. (As one might guess,
this suggestion was just met with widespread disbelief and even with some humor.)
The key idea was to prove that beam cooling worked, and that the estimated
luminosity needed to detect the *W* and *Z* was correct.

Two schemes for cooling $\bar{p}$'s had been invented: (1) electron cooling by
G. Budker and (2) stochastic cooling by S. Van der Meer [see paper (A) for
references]. Both schemes were invoked by RMC to provide an adequate source of
$\bar{p}$'s. Initially, the plan to convert FNAL to a $\bar{p}p$ collider was rejected (but later
approved), and the focus of attention fell on CERN for this effort. By 1981, in one
of the most incredible technical advances made in science, a powerful source of
cooled $\bar{p}$'s was constructed at CERN, and $\bar{p}p$ collisions were observed. This tour de
force of technology and science was first reported by the CERN group in late 1981
[7.4]. [The paper is reproduced here (C) to show the reader the magnitude of this
remarkable accomplishment.] Within two years, *W* and *Z* particles were discovered
at CERN in the UA1 and UA2 experiments [7.3]–[7.8] [all discovery papers are
reproduced here (C)–(G) so the reader can experience this event]. [During this

period, two large detectors had been approved for the S$\bar{p}p$S colliders, UA1 and UA2 (Underground Areas 1 and 2).]

In 1982, a large-statistics $\bar{p}p$-collider run was made at CERN, and by the end of 1982 the first evidence for the *W* was recorded [7.5]. By 1983, the *Z* was discovered [7.7], thus completing the search for the origin of the weak force that dates back to the early part of this century.

7.3.2. The Discovery of the *W* and *Z* and Early Properties of these Particles

During the Rome $\bar{p}p$ conference in early 1983, the *W* discovery was announced by the UA1 group, and the UA2 group quickly observed *W*'s as well. Later in the spring, the UA1 group observed several examples of

$$Z^0 \rightarrow e^+e^- \quad ,$$
$$Z^0 \rightarrow \mu^+\mu^- \quad .$$

A short time later, *Z*'s were also observed by UA2.

The observed parameters of the *W* and *Z* were almost exactly those expected from the precise values of ρ and $\sin^2\theta_W$ measured at that time and, thus, this was an enormous triumph for the GWS standard model of electroweak interactions (see Fig. 7.5).

Once the *W* and *Z* were discovered and the masses known within a few percent of the current values, the average properties were quickly measured by UA1 and UA2 and later by the CDF detector at the tevatron (the finally approved Fermilab $\bar{p}p$ collider). Special e^+e^- colliders had been planned since the early 1980s at SLAC and CERN. The SLAC machine invoked a new principle of linear colliders but still managed to detect the first Z^0 produced by an e^+e^- collision at the SLC.

In late 1988, the LEP I e^+e^- collider at CERN was turned on and immediately regained the world record of Z^0 production. The only major discovery so far from these studies of the Z^0 (other than the precision parameters and the *t* quark, which will be discussed in Chapter 8) was the measurement of the number of light neutrinos (families) through the process

$$Z^0 \rightarrow \nu_i\bar{\nu}_i \quad .$$

Both the SLC and LEP observed that only three families of neutrinos were contributing to the Z^0 width in neutrinos.

During 1989-1992, the Z^0 parameters were measured mainly by using the four large detectors at LEP (ALEPH, DELPHI, OPAL, and L3). We will describe these parameters in detail in Chapter 8 when we discuss the radiative corrections to the *W* and *Z* particles.

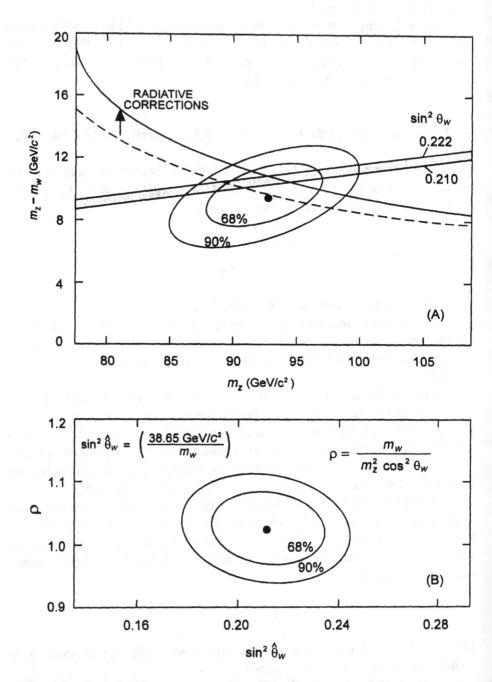

Figure 7.5. Early result on the electroweak parameter in the 1980s. (A) Fit result in the m_W versus m_Z plane. The error curves reflect the uncertainty in the energy scale at the 68% and 90% CLs. The broken curve shows the standard model prediction for $\rho = 1$ as a function of the IVB masses. The shaded band shows the expectation from low-energy neutrino-induced charged- and neutral-current measurements. (B) ρ versus $\sin^2\theta_W$ determined from the measurements of m_W and m_Z. The 68% and 90% CLs are shown.

The combined measurements of UA1, UA2, and CDF groups in 1987 established the basic properties of the *W* bosons. During 1989–1990, the improved UA2 detector and the CDF detector determined the *W* properties more precisely; all observations fit the GWS and standard model to date.

In 1984, Carlo Rubbia and Simon Van der Meer received the Nobel Prize for the *W* discovery. The author and others accompanied them to Stockholm for the award. The period of 1973 (WNC discovery) to 1983 (*W*, *Z* discovery) is certainly one of the more exciting and intense in all of modern physics.

REFERENCES

7.1. For example, see the article by T. D. Lee, *Physics Today* (April 1972), p. 23 for a discussion.

7.2. D. Kerst *et al.*, *Phys. Rev.*, **102**, 590 (1956).

7.3. C. Rubbia, P. McIntyre, and D. Cline, in *Proceedings of the International Neutrino Conference* (Aachen, 1976), H. Faissner, H. Reithler, and P. Zerwas, eds. (Vieweg, Braunschweig, 1977), p. 683 [article (A)].

7.4. Staff of the CERN proton–antiproton project, *Phys. Lett.*, **107B**(1), 306 (1981) [article (C)].

7.5. UA1 Collaboration, G. Arnison *et al.*, *Phys. Lett.*, **122B**(1), 103 (1983) [article (D)].

7.6. UA2 Collaboration, G. Banner *et al.*, *Phys. Lett.*, **122B**(5,6), 476 (1983) [article (E)].

7.7. UA1 Collaboration, G. Arnison *et al.*, *Phys. Lett.*, **126B**(5), 398 (1983) [article (F)].

7.8. UA2 Collaboration, P. Bagnaia *et al.*, *Phys. Lett.*, **129B**(1,2), 130 (1983) [article (G)].

PRODUCING MASSIVE NEUTRAL INTERMEDIATE VECTOR BOSONS WITH EXISTING ACCELERATORS*)

C. Rubbia and P. McIntyre

Department of Physics, Harvard University, Cambridge, Massachusetts 02138
and

D. Cline

Department of Physics, University of Wisconsin, Madison, Wisconsin 53706

Presented by C. Rubbia

Abstract: We outline a scheme of searching for the massive weak-boson ($M = 50 - 200$ GeV/c^2). An antiproton source is added either to the Fermilab or the CERN SPS machines to transform a conventional 400 GeV accelerator into a p$\bar{\text{p}}$ colliding beam facility with 800 GeV in the center of mass ($E_{eq} = 320,000$ GeV). Reliable estimates of production cross sections along with a high luminosity make the scheme feasible.

The past ten years have seen remarkable progress in the understanding of weak interactions. First there is the experimental discovery of $\Delta S = 0$ weak neutral currents [1], which when contrasted with the previous limits on $\Delta S = 1$ neutral current decay processes [2] leads to the suggestion of additional hadronic quantum numbers in nature [3]. Strong evidence now exists for new hadronic quantum numbers that are manifested either directly [4, 5] or indirectly [6]. The experimental discoveries are complemented by the theoretical progress of unified gauge theories [7, 8]. These developments lead to the expectation that very massive intermediate vector bosons ($50 - 100$ GeV/c^2) may exist in nature [7, 8]. The search for these massive bosons require three separate elements to be successful: a reliable physical mechanism for production, very high center of mass energies, and an unambiguous experimental signature to observe them. In this note we outline a scheme which satisfies these requirements and that could be carried out with a relatively modest program at existing proton accelerators.

We first turn to the production process. We concentrate on neutral bosons because of the extremely simple experimental signature and because production is largely dominated by a single production resonant pole in the particle-antiparticle cross section. The best production reaction would of course be:

$$e^+ + e^- \rightarrow W^0 \begin{array}{l} \nearrow e^+ + e^- \\ \searrow \mu^+ + \mu^- \\ \rightarrow \text{hadrons} \end{array}$$

$$\tag{1}$$

where a sharp resonance peak is expected for $2E_{e^+} = 2E_{e^-} = M$. In the Breit-Wigner approximation near its maximum we get:

$$\sigma(e^+ e^- \rightarrow W^0) \simeq \frac{3}{4}\pi\lambda^2 \frac{\Gamma_i \Gamma}{(2E - M)^2 + \frac{\Gamma^2}{4}} \tag{2}$$

---
*) Since no manuscript was submitted, the Editors decided to reproduce a reprint of the authors', circulated in March 1976.
 – Some references were updated.

Work supported in part by the U.S. Energy Research and Development Administration.

where Γ_i, Γ are the partial width to the initial $e^+ e^-$ state and the total width, respectively. The decay widths into $e^+ e^-$ (and $\mu^+ \mu^-$) pairs can be calculated in the first order of the semi-weak coupling constant: $\Gamma_{e^+e^\pm} \cong \Gamma_{\mu^+\mu^-} = 1.5 \times 10^{-7} M_W^3$ (GeV). For M = 100 GeV, $\Gamma_{e^+e^-} \simeq 150$ MeV, which is surprisingly large. The total width is related to the above quantity by the branching ratio $B_{e^+e^-} = \Gamma_{e^+e^-}/\Gamma$ which is unknown. Crude guesses based on quark models suggest $B_{e^+e^-} \simeq 1/10$, giving $\Gamma = 1.5$ GeV or $\Gamma/2E = 1.5$ % for M = 100 GeV/c². At the peak of the resonance, $\sigma(e^+ e^- \rightarrow W^0, 2E = M) = 3\pi\lambda^2 B_i \simeq 2 \cdot 10^{-31}$ cm². Neutrino experiments [9] have found that $M_{W^\pm} > 20$ GeV/c². Therefore, if $M_{W^\pm} \sim M_{W^\circ}$, the neutral intermediate boson is out of reach of existing $e^+ e^-$ storage rings.

A more realistic production process is the one initiated by proton-antiproton collisions:

$$p + \bar{p} \rightarrow W^0 + \text{(hadrons)}$$

which, according to the quark (parton) picture, proceeds by a reaction analog to (1), except that now incoming e^+ and e^- are replaced with q and $\bar{q}$. Strong support to the idea that W's are directly coupled to spin 1/2 point-like constituents comes from neutrino experiments [10] and from semi-leptonic hadron decays [11]. Furthermore neutrino experiments provide the necessary structure functions and have set limits [9] (> 20 GeV) on any nonlocality in the parton form factor. The main difference with respect to $e^+ e^-$ is that now the kinematics is largely smeared out by the internal motion of q's and $\bar{q}$'s. The average center of mass energy squared of the $q - \bar{q}$ collision is roughly [12]:

$$\langle S_{q\bar{q}}\rangle \sim S \langle x_q\rangle_p \langle x_{\bar{q}}\rangle_{\bar{p}}$$

where S is the center of mass energy squared of the $\bar{p}p$ system and $\langle x_q\rangle_p$ ($\langle x_{\bar{q}}\rangle_{\bar{p}}$) is the mean fractional momentum of q's($\bar{q}$'s) in the proton (antiproton). From the neutrino measurements [9] and $\langle x_q\rangle_p = \langle x_{\bar{q}}\rangle_{\bar{p}}$ we find $\langle S_{q\bar{q}}\rangle \sim 0.04$ S. For M = 100 GeV/c² this suggests $S \geq 2 \times 10^5$ GeV² or $\sqrt{S} \geq 450$ GeV. The production cross section can be evaluated by folding the (narrow) resonance (2) over the q and $\bar{q}$ momentum distributions:

$$\sigma(q\bar{q} \rightarrow W^0 \rightarrow \mu^+\mu^-) = 3\pi\lambda^2 \frac{\Gamma_{q\bar{q}}}{\Gamma} \cdot \frac{\Gamma_{\mu\mu}}{\Gamma} \cdot \frac{dN}{dE}(E = M) \cdot 2\Gamma \qquad (3)$$

where dN/dE is the probability (per unit of energy) of finding a $q\bar{q}$ collision with center of mass energy E, and the other symbols have the same meaning as in (2). Note that $\Gamma_{q\bar{q}} \simeq O(1)$ is a model-dependent parameter. The resultant cross section is $\sigma(p\bar{p} \rightarrow W^0 + \text{hadrons} \rightarrow \mu^+ + \mu^- + \text{hadrons}) \simeq 6\pi\lambda^2 \cdot (\Gamma_{q\bar{q}}/\Gamma) \cdot dN/dE (E = M) \cdot \Gamma_{\mu\mu} \cong 10^{-32}$ cm². The numerical value is given for M = 100 GeV/c², $\sqrt{S}$ = 500 GeV and $\Gamma_{q\bar{q}}/\Gamma = 1/2$. This derivation of the cross section exposes the basic simplicity of the assumptions and gives the order of magnitude of the expected cross section. More sophisticated calculations give similar results [12]. We note that calculations of $W^\pm$ production in proton-proton collisions are very uncertain in contrast to the present one due due to the apparent small antiparton content in the nucleon and the unknown distributions of this component [13].

We turn now to the question of the experimental observation. The cleanest experimental signature for the program outlined here is:

$$\bar{p} + p \rightarrow W^0 + \text{hadrons}$$
$$\quad\quad\quad \hookrightarrow \mu^+ + \mu^-$$

with the observation of a peak in the $\mu^+ \mu^-$ invariant mass spectrum with the cross section of equation (3). A modest magnetized iron detector system is adequate to detect the high energy decay muons ($P_\mu \sim 50$ GeV) in the center of mass system. Electromagnetic production of $\mu^+ \mu^-$ pairs is expected to be suppressed by a factor of $\approx (\alpha^2/G^2 M_W^4)$. Note that a similar suppression is expected to hold for any hadronic vector meson. Note also that the production and decay of charged vector bosons is more problematic since the decay sequence

$$\bar{p} + p \rightarrow W^+ + X$$
$$\hookrightarrow \mu^+ + \nu_\mu$$

leads to one muon and a missing neutrino which is difficult if not impossible to detect. In many previous discussions it has been assumed that the W^+ would be produced with very little transverse momentum with respect to the incident beam direction and therefore the transverse momentum of the decaying μ would exhibit a sharp peak at $p_{\mu\perp} \sim M_W/2$ [14]. Present evidence in case of the production of massive strongly interacting vector bosons (i.e., J/Ψ) indicate that the parent is produced at relatively large $p_\perp$ and therefore the Jacobian peak is largely smeared out [15]. There is no obvious reason why the production of massive intermediate vector bosons should not follow the same behavior [16]. Without a sharp structure in the $p_{\mu\perp}$ distribution, a crucial experimental signature for the W^+ is absent.

We now briefly outline the scheme of transforming an existing proton accelerator into high luminosity $p\bar{p}$ colliding beams [17] using standard vacuum ($p \simeq 10^{-7}$ Torr) and the separate function magnet system. The main elements are (1) an extracted proton beam to produce an intense source of antiprotons at 3.5 GeV/c, and (2) a small ring of magnets and quadrupoles that guides and accumulates the $\bar{p}$ beam, (3) a suitable mechanism for damping the transverse and longitudinal phase spaces of the $\bar{p}$ beam (either electron cooling [18] or stochastic cooling [19]), (4) an R. F. system that bunches the protons in the main ring and in the cooling ring, (5) transport of the "cooled" R. F. bunched $\bar{p}$ beam back to the main ring for injection and acceleration. A long straight section of the main ring is used as $p\bar{p}$ interaction region. A schematic drawing of these elements for the FNAL accelerator is presented in Fig. 1. The main parameters of the scheme are summarized in Table 1.

Fig. 1

General layout of the $p\bar{p}$ colliding scheme. Protons (100 GeV/c) are periodically extracted in short bursts and produce 3.5 GeV/c antiprotons which are accumulated and cooled in the small stacking ring. Then $\bar{p}$'s are reinjected in an R.F. bucket of the main ring and accelerated to top energy. They collide head-on against a bunch filled with protons of equal energy and rotating in the opposite direction.

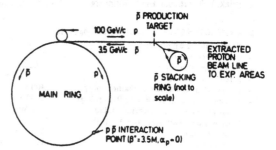

Table 1. List of Parameters

1. MAIN RING (Fermilab)

– Beam momentum	250 (400) GeV/c
– Equivalent laboratory energy for (p$\bar{p}$)	133 (341) TeV
– Accelerating and bunching frequency	53.14 Mc/s
– Harmonic number	1113
– R. F. peak voltage/turn	3.3×10^6 Volt
– Residual gas pressure	$< 0.5 \times 10^{-7}$ Torr
– Beta functions at interaction point	3.5 m
– Momentum compaction at int. point	~ 0 m
– Invariant emittances ($N_p = 10^{12}$)	
– longitudinal	3 eV s
– transverse	50×10^{-6} rad m
– Bunch length	2.3 m
– Design luminosity	5×10^{29} (8×10^{29}) cm^{-2} s^{-1}

The luminosity for two bunches colliding head-on is estimated using the relation

$$L = N_p N_{\bar{p}} \, \Phi/a$$

where N_p and $N_{\bar{p}}$ are the number of protons and antiprotons circulating in the machine, respectively, Φ is the revolution frequency and a is the effective area of interaction of the two beams. N_p is taken as 10^{12} protons in one R. F. bunch. The value of N_p is limited by the maximum allowed beam-beam tune shift ($N_p = 10^{12}$ for $\Delta\nu = 0.01$). We have verified the longitudinal stability of the bunch, the phase area growth due to R. F. noise, the transverse wall instability, the headtail effect and non-linear resonances, including those arising from beam-beam interactions. None of these effects appears to be important [20]. We note that $N_p = 10^{12}$ corresponds to $i_{av} = 10$ mA and $i_{peak} = 25$ A for $\ell_{bunch} = 2.5$ m and that the Brookhaven AGS currently accelerates twelve bunches of similar characteristics.

The production of antiprotons at 3.5 GeV is done with protons from the same accelerator and with an overall efficiency $\bar{p}/p \simeq 4 \times 10^{-6}$. In order to reach $N_{\bar{p}} = 3 \times 10^{10}$ we need 750 pulses with 10^{13} ppp. About 10 seconds must elapse between pulses in order to clear away the freshly injected antiprotons [21]. Therefore the formation of $\bar{p}$'s would take of the order of few hours.

In order to make the beam as small as possible one can reduce the value of the betatron function in the collision point ($\beta_V \simeq \beta_h = 3.5$ m) and make the momentum compaction factor close to zero [22]. Then for standard beam emittances [23] and $E_p = E_{\bar{p}} = 250$ GeV we calculate $L = 5 \times 10^{29}$ cm^{-2} s^{-1} for $N_{\bar{p}} = 3 \times 10^{10}$. In order to observe one event/hour at our estimated cross section we require a luminosity of 3×10^{28} cm^{-2} s^{-1}. If the more pessimistic cross section of 10^{-33} cm^2 is used, a luminosity of 3×10^{29} cm^{-2} s^{-1} is needed which is still appreciably less than the calculated value. Finally, the half-life of the luminosity due to beam-gas scattering is about 24 hours for an average residual pressure of 0.5×10^{-7} Torr.

We would like to acknowledge Drs. T. Collins, R. Herb, S. Glashow, E. Picasso, G. Petrucci, N. Ramsey, L. Sulak, L. Thorndahl, and S. Weinberg for helpful discussions and suggestions.

2. ANTIPROTON SOURCE (Stochastic Cooling [21])

- Nominal stored $\bar{p}$ momentum   3.5 GeV/c
- Circumference of ring   100 m
- Momentum acceptance   0.02
- Betatron acceptances $100 \, \pi \, 10^{-6}$ rad m
- Bandwidth of momentum stochastic cooling   400 Mc/s
- Maximum stochastic accelerating R. F. voltage   3000 V
- Bandwidth of betatron stochastic cooling   200 Mc/s
- Final invariant emittances ($N_{\bar{p}} = 3 \times 10^{10}$)
   - longitudinal   0.5 eV s
   - transverse $10 \, \pi \, 10^{-6}$ rad m

References

[1] F. J. Hasert et al., Phys. Letters 46B, 121 and 138 (1973),
     A. Benvenuti et al., Phys. Rev. Letters 32, 800 (1974).

[2] U. Camerini, D. Cline, W. Fry and W. Powell, Phys. Rev. Letters 13, 318 (1964),
     M. Bott-Bodenhausen et al., Phys. Letters 24B, 194 (1967).

[3] S. L. Glashow, J. Iliopoulos and L. Maiani, Phys. Rev. D2, 1285 (1970).

[4] B. Aubert et al., "Experimental Observation of $\mu^+ \mu^-$ Pairs Produced by Very High Energy Neutrinos", in Proceedings of the Seventeenth International Conference on High Energy Physics, London, 1974 and in Neutrinos - 1974. AIP Conference Proceedings No. 22, edited by C. Baltay (American Institute of Physics, New York, 1974), p. 201.
     A. Benvenuti et al., Phys. Rev. Letters 34, 419 (1975), ibid. 34, 597 (1975).

[5] J. Blietschau et al., Phys. Letters 60B, 207 (1976).
     J. von Krogh et al., Phys. Rev. Letters 36, 710 (1976).

|6| J. J. Aubert *et al*., Phys. Rev. Letters **33**, 1404 (1974).
J. E. Augustin *et al*., Phys. Rev. Letters **33**, 1406 (1974).

|7| S. Weinberg, Phys. Rev. Letters **19**, 1264 (1967).

|8| A. Salam in Elementary Particle Physics (edited by N. Svortholm, Almquist and Wiksells, Stockholm, 1968), p. 367.

|9| A. Benvenuti *et al*., "Test of Locality of the Weak Interaction in High Energy Neutrino Collisions" to be submitted to Phys. Rev. Letters (March 1976).

|10| See for instance A. de Rujula: Quark Tasting with Neutrinos, *Proceedings 1976 Coral Gables Conference*, Miami, Florida, January 1976.

|11| L. M. Chounet, J. M. Gaillard and M. K. Gaillard, Physics Reports **4C**, 5 (1972).
M. Roos, Phys. Letters, **36B**, 130 (1971).

|12| S. D. Drell, T. M. Yan, Phys. Rev. Letters **25**, 316 (1970),
S. Pakvasa, D. Parashar, and S. F. Tuan, Phys. Rev. **D10**, 2124 (1974),
S. M. Berman, J. D. Bjorken, J. B. Kogut, Phys. Rev. **D4**, 3388 (1971),
G. Altarelli *et al*., Nucl. Phys. **B92**, 413 (1975),
and R. B. Palmer, E. A. Paschos, N. Samios and L. L. Wang, BNL preprint 20634.

|13| A recent estimate of the antiparton content of the nucleon has been obtained using antineutrino scattering data below 30 GeV and is reported by A. Benvenuti *et al*., Phys. Rev. Letters **36**, 1478 (1976); **37**, 189 (1976).

|14| Y. Yamaguchi, Nuovo Cim. **43**, 193 (1966);
L. Lederman and B. Pope, Phys. Rev. Letters **27**, 765 (1971).

|15| B. Knapp *et al*., Phys. Rev. Letters **34**, 1044 (1975),
Y. M. Antipov, *et al*., IHEP 75–125, Serpukhov (1976),
F. W. Büsser *et al*., Phys. Letters **56B**, 482 (1975),
K. J. Anderson *et al*., Phys. Rev. Letters **36**, 237 (1976) and **37**, 799 (1976),
D. C. Hom *et al*., Phys. Rev. Letters **36**, 1236 (1976).

|16| F. Halzen, private communication.

|17| There are also various schemes for producing at modest cost a pp colliding beam machine of sufficient energy to produce the intermediate vector bosons. One such scheme is to collide the Fermilab main ring with the projected energy doubler ring (Cline, Richter, and Rubbia, private communication with R. R. Wilson, 1975); another scheme is to collide the Fermilab main ring with a small ring of 25 GeV protons (J. K. Walker *et al*., proposal submitted to Fermilab, 1976).

|18| G. I. Budker, Atomic Energy **22**, 346 (1967).
G. I. Budker, Ya. S. Derbenev, N. S. Dikansky, V. I. Kudelainen, I. N. Meshkov, V. V. Parkhomchuk, D. V. Pestrikov, B. N. Sukhina, A. N. Skrinskiy, Experiments on Electron Cooling, paper presented at the National Conference, Washington, March 1975.

|19| L. van der Meer, CERN-ISR-PO/72–31 August 1972 (unpublished).
P. Braham *et al*., NIM **125**, 156 (1975).
L. Thorndahl, CERN-ISR-RF/75 (unpublished).

|20| See, for a complete discussion, M. Month, in *Proceedings IX. International Conference on High Energy Accelerators*, SLAC, May 1974, page 593.

|21| L. Thorndahl, CERN-ISR, RF/75 and C. Rubbia, Report in preparation.

|22| An electron target for NAL, in *Proceedings 1973 NAL Summer Study*, Vol. 2, page 21.

|23| T. Collins, private communication.

Antiproton–proton colliders and Intermediate Bosons

Phase-space cooling makes p̄p colliders practical;
CERN and Fermilab plan to use them to find and measure
the properties of the Intermediate Vector Bosons.

David Cline and Carlo Rubbia

The development of particle accelerators in the 1920–30's was strongly influenced by the availability of intense sources of ions as well as the ability to shape magnetic fields and to produce high electric fields. Later, electron synchrotrons required intense electron sources. Still later the development of electron–positron colliding-beam machines required positron sources. At present CERN and Fermilab are developing intense sources of antiprotons and related beam-cooling techniques needed to make intense sources. We expect these antiproton sources to influence the development of future particle accelerators and storage rings just as profoundly as ion, electron and positron sources have in the past. One immediate use for the antiprotons is the creation of high-energy antiproton-proton colliders.

CERN is building an antiproton–proton collider that will have 270 GeV in each beam. This collider is scheduled to be operating in summer 1981. The Antiproton Accumulator at CERN (figure 1) cooled its first proton beam early last month. Fermilab plans to build an antiproton–proton collider that will have 1000 GeV in each beam. The energy available with these colliders is expected to be for the first time large enough to cross the threshold for creation of intermediate vector bosons, expected to have a mass in the range 80–90 GeV.

David Cline is at Fermilab and is a professor of physics at the University of Wisconsin. Carlo Rubbia is at CERN and is a professor of physics at Harvard University.

The development of intense sources of antiprotons will undoubtedly provide for other experiments as well. One can search for bound states of baryons and antibaryons (baryonium). At CERN a storage ring—the Low-Energy Antiproton Ring—is being developed. This ring will use the intense source of antiprotons and provide for antiproton collisions with resting protons. Other areas of research may be affected; for example, it will be possible to accelerate antiprotons in a high-energy synchrotron to 400–1000 GeV, extract the beam and allow it to strike a target, as is routinely done for proton synchrotrons.

To estimate the rate of production of intermediate vector bosons, we must calculate the luminosity of an antiproton-proton collider. The luminosity is given by

$$L = \frac{N_{\bar{p}}(f\gamma)(\Delta\nu)_{max}}{r_p\beta^*} \text{ cm}^{-2}\text{ sec}^{-1}$$

where β^* is related to the size of the beams at the interaction point, $N_{\bar{p}}$ is the total number of antiprotons, $(\Delta\nu)_{max}$ is the maximum tune shift due to beam-beam interaction, r_p is the classical proton radius and $f\gamma$ is the revolution frequency f, times the Lorentz factor for the beam. The rate of interactions is given by $L\sigma$ where σ is the cross section for the collision. The tune shift due to beam–beam interactions is given by

$$(\Delta\nu)_{max} = \frac{r_p N_{\bar{p}}}{n_b \epsilon}$$

where n_b is the number of bunches in the machine and ϵ is the invariant emittance of the beam.

Table 1 compares the various values of these parameters for the antiproton-proton colliders under construction or being discussed. To obtain a high luminosity, the total number of antiprotons produced by the source must be greater than 10^{11}. The collection of such a large number of antiprotons poses an extremely difficult technical problem.

Consider the yield of antiprotons from a beam of high-energy protons striking a target:

$$\frac{N_{\bar{p}}}{N_p} = \frac{1}{\sigma_0} E \frac{d^3\sigma}{dp^3}(p\Delta p)\Delta\Omega(\epsilon_1)(\epsilon_2)$$

where ϵ_1 is the factor for absorption in the target, $\Delta\Omega$ the collection solid angle, ϵ_2 is the target efficiency, p the momentum of the proton beam and $Ed^3\sigma/dp^3$ the invariant production cross section.

The invariant antiproton production cross section has been measured at different incident energies (figure 2). Above 100 GeV the increase in cross section is small. The optimum yield occurs for antiprotons produced at rest in the center of mass of the collision. A reasonable estimated cross section is

$$E\frac{d^3\sigma}{dp^3} \sim 1 \text{ mb/GeV}^2$$

and for the case of a feasible storage ring to collect the antiprotons with acceptance $\epsilon_x \sim \epsilon_y \sim 100$ mm-mr and momentum acceptance $\Delta p/p = \pm 2\%$, the ratio of antiprotons to protons interacting in the target

$$N_{\bar{p}}/N_p = 9\times10^{-6}$$

For $N_p = 4\times10^{12}$/sec at CERN or Fermilab, this gives 4×10^7 p̄/second and

greater than 5×10^{11} p̄/day.

Thus the accumulation of large numbers of antiprotons is feasible provided a method of storing the large number of individual pulses is devised. Because high-energy protons are required, the number of locations at which intense sources of antiprotons can be produced is limited at present.

The accumulation of a large number of antiprotons requires phase-space compression because after a few injections of antiprotons, nearly any conceivable storage ring will have its phase space completely filled. In addition the transverse temperature $T_\perp$ must be reduced. The average transverse temperature of antiprotons produced by proton beams is

$$kT_\perp = \tfrac{1}{2}mv_\perp{}^2 + \frac{(\Delta p_\perp)^2}{2m_p}$$

$$\langle \Delta p_\perp \rangle \sim 300 \text{ MeV}/c$$

and thus

$$T_\perp \simeq 5 \times 10^6 \text{ eV}$$

The transverse temperature accepted by a high-energy storage ring is

$$T_\perp \simeq 1.2 \times 10^4 \text{ eV}$$

Two phase-space cooling techniques to reduce the transverse temperature of a beam have been developed—electron cooling[1] and stochastic cooling.[2] The use of these techniques is fundamental to the collection of antiprotons.

The general scheme for antiproton-proton colliders is to focus high-energy protons on a target; the antiprotons produced are then transported to a storage ring (cooling ring), which provides for cooling of the transverse and longitudinal temperatures of the p̄ beams and also provides for storage of the accumulated antiprotons. Once greater than 10^{11} p̄ are collected, they are accelerated and injected along with protons into a high-energy storage ring for antiproton–proton collisions.

Phase-space cooling

The beams of particles stored in accelerators are largely subject to Liouville's theorem because they are under the influence of conservative forces. Simply put, Liouville's theorem states: "Under the action of a force that can be derived from a Hamiltonian, the motion of a group of particles is such that the local density of the representative points in phase space remains everywhere constant." However, it is

Antiproton Accumulator at CERN cooled its first proton beam early in July. After cooling, the antiprotons are accelerated, first to 26 GeV/c, then to 270 GeV/c. They then collide with a 270-GeV/c proton beam.   Figure 1

possible to introduce "non-Liouvillian" processes into a beam, and this is what is meant by "beam cooling." Table 2 lists the kinds of Liouvillian and non-Liouvillian forces that we know of. Stochastic cooling is not included in the table because it results from fluctuations.

The beam can be described by a temperature and entropy as well. When two beams are brought together the laws of thermodynamic events can be applied; this is the basic concept for electron cooling.

The suggestion of electron cooling came from Gersh I. Budker many years ago, and it is schematically shown[1] in figure 3. Suppose that a "hot" proton or antiproton beam circulates in a storage ring. The temperature is meant to be related to the average residual energy in the frame of reference of the ideal particle of the equilibrium orbit. Suppose that "cold"

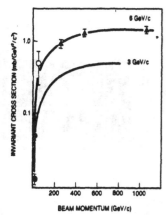

Antiproton production as a function of incident proton energy. Above 1000 GeV the increase in cross section is small.   Figure 2

electrons are put in "thermal" contact with the protons. To ensure a thermal contact with no additional "friction," electrons must have the same average (vector) velocity as the heavy particles throughout the cooling section. This means $\langle v_e \rangle = \langle v_p \rangle$ or equivalently

$$\langle E_e \rangle = \frac{m_e}{m_p} \langle E_p \rangle \simeq \frac{1}{1830} \langle E_p \rangle$$

The "heat" is then removed by the electrons, which are either produced cold enough or radiate their acquired heat away by synchrotron radiation.[2]

Another more recent suggestion, called stochastic cooling, has been put forward by Simon Van der Meer.[3] The principle is shown schematically in figure 4. It consists of picking up the fluctuation noise due to the passage of particles across a pick-up. A particle that passes through the pick-up induces a short pulse in it. The electrical delay in the system is such that after having traversed a high-gain, wide-band amplifier this pulse arrives at the kicker together with the particle. The

kicker is designed to correct for the deviation of the particle from the equilibrium orbit. At the same time, other particles also produce pulses. They are not infinitely narrow, due to the finite system bandwidth; so some of them will also influence the particle under consideration. The mean effect of this noise will be zero if the system does not transmit the dc component. It will, however, lead to an increase in the rms deviation of the beam. Because the pulse duration is quite short compared to the revolution time and because different particles have different revolution times, each particle will be influenced by a small and continuously changing factor contributed by the other particle. The quasi-random effect has been analyzed quantitatively, and it was found that the blow-up is similar to the one due to purely random kicks. That is, the mean square spread is proportional to time and to the square of the electronic gain. The proportionality factor depends on the amount by which particles overtake each other at each turn. Therefore, because the damping factor on the particle is linearly proportional to time and electronic gain, there is always a sufficiently low gain at which the "cooling" action overtakes the "heating" action due to random noise.

Electron cooling

Let us consider in more detail the idealized case of a massive particle (such as a proton or antiproton) slowing down due to the longitudinal component of the momentum transfer from collisions with electrons. We consider first the collision between an electron of initial velocity v_e and a particle of initial velocity v_p. We shall also assume that v_e and v_p have the same order of magnitude and that they are both much smaller than c. Then in the center of mass of the collision, the particle $(m_p \gg m_e)$ is essentially at rest,

$$v_{cm} = \frac{m_e v_e + m_p v_p}{m_e + m_p} \simeq v_p$$

and we can describe the scattering as classic Rutherford scattering of the electron on a fixed potential.

We can easily evaluate the longitudinal momentum transfer to the particle per unit of time. After integrating over all momentum transfers and electron velocities, we get the final expression for the drag force F:

$$F = \frac{2\pi e^4 L n_e}{m_e} \int d\mathbf{K}_e f(v_e) \frac{v_e - v_p}{|v_e - v_p|^3}$$

where L, the Coulomb logarithm, $= \log q^2_{max}/q^2_{min} \simeq 20$; e, m_e are the charge and mass of the electron respectively and $f(v_e)$ is the electron velocity distribution.

The simple two-component plasma relaxation picture is used to obtain some general guidelines for the case $|v_p| \gg |v_e|$. There are both theoretical and experimental reasons to expect that this picture gives a reasonable description of at least the (initial) part of the process.

A simplified formula for the cooling time can be derived for the electron velocity distribution of a δ-function $(v_p \gg v_e)$, equivalent to the point-charge approximation of the electrostatic analog. The drag force becomes:

$$F = - \frac{2\pi e^4 L n_e}{m_e} \frac{v_p}{|v_p|^3}$$

The formula for the cooling time τ in the laboratory frame is

$$\tau = \frac{1}{6\pi} \frac{e\beta\gamma^2}{r_p r_e jL\eta} \left[\frac{p^*}{m_p} \right]^3$$

where β, γ are the usual relativistic factors of the average electron speed; j is the electron current density; η is the fraction of the time the particle is traversing the electron beam and p^* is the effective initial particle momentum in the electron frame. For pure longitudinal cooling $p^* = \Delta p/\gamma$, where Δp is the deviation from the momentum of one of the equilibrium particles with the orbit. Note that the cooling time increases very rapidly with p^*; thus electron cooling is really only effective for very low-energy antiprotons. (However at very high energy a special circumstance makes electron cooling effective again.[2])

Stochastic cooling

Liouville's theorem predicts that electron oscillations cannot be damped by the use of electromagnetic fields deflecting the particles. (See table 2.) However, this theorem is based on statistics and is only strictly valid either for an infinite number of particles, or for a finite number if no information is available about the position in the phase plane of the individual particles. Clearly, if each particle could be separately observed and a correction

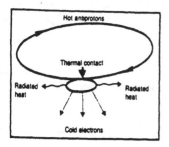

Cooling of antiprotons with cold electron beam occurs when electrons have the same average velocity as the antiprotons. The electrons remove the "heat."   Figure 3

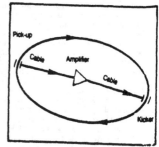

Stochastic cooling. The kicker corrects for the deviation of a particle from the equilibrium orbit, and beam cooling occurs.   Figure 4

applied to its orbit, the oscillations could be suppressed. It is well known that coherent betatron oscillations (where the beam behaves like a single particle) can be damped by means of pickup-deflector feedback systems. In the same way, the statistical fluctuations of the average beam position, caused by the finite number of particles, can be detected with pickup electrodes and a corresponding correction applied. In other words, the small fraction of the oscillations that happens to be coherent at any time due to the statistical fluctuations can be damped.

A special trick, the notch filter, has been invented at CERN to increase the damping of the beam considerably.[4] In the notch filter method, information regarding a particle's momentum is obtained through its relationship with its revolution frequency. A filter system in the pickup-kicker chain appropriately conditions signals to accelerate or decelerate particles toward a specific rotation frequency (that is, momentum). A useful filter element for this purpose is a shorted transmission line whose length corresponds to half the rotation period. Such an element exhibits "zeros" in its input impedance at all harmonics of the rotation frequency. The resultant transfer functions of such an element, when used in a voltage-divider configuration, appears as a series of notches, hence the term "notch filter."

Cooling measurements

Over the past few years beam cooling measurements have been made at Novosibirsk,[5] CERN[6,7] and Fermilab, and the theory of both stochastic and electron cooling has improved.[4,8] The pioneering Novosibirsk measurements demonstrated electron cooling for the first time. Recent measurements at CERN fully confirm the Novosibirsk results.[7] An electron cooling experiment is also being carried out at Fermilab,[9] where construction of a special cooling ring started in 1977. Recently both the momentum spread and transverse beam dimensions were stochastically cooled in this ring by a Lawrence Berkeley Lab group.

Very important cooling results have been obtained from the Initial Cooling Experiment ring at CERN,[7] which was specifically constructed to study both stochastic and electron cooling. The observed stochastic cooling of the beam momentum is compared with the theory of stochastic cooling in figure 5. The excellent agreement between theory and measurements gives confidence that the antiproton collection rings at CERN and Fermilab will work.

The characteristics of the phase-space cooling techniques are quite different and complementary. Stochastic

cooling is relatively independent of the beam energy but less effective at high beam intensity. Also the cooling of transverse beam dimensions is expected to be very slow. Unlike stochastic cooling, electron cooling mainly is effective for low-energy beams, and the cooling time is relatively independent of the beam intensity. Transverse beam cooling is very fast and effective.

CERN and Fermilab colliders

The CERN antiproton–proton collider[10] (figure 6) now under construction accepts a fixed antiproton beam energy—3.5 GeV/c. These are produced with a momentum spread of 0.7×10^{-2} by 27 GeV/c protons from the Proton Synchrotron. The antiprotons are transferred to a very large aperture storage ring—the Antiproton Accumulator—cooled rapidly (a few seconds) in momentum space and slowly in transverse phase space by stochastic cooling. (The Antiproton Accumulator, which was just finished, only two years after it was approved, has properties very similar to those made in our 1976 proposal[11] to convert existing synchrotrons into colliders.) After $1-5 \times 10^{11}$ $\bar{p}$ are collected, they are transferred into the Proton Synchrotron, accelerated to 26 GeV/c, bunched and injected into the Super Proton Synchrotron. Meanwhile protons are injected into the SPS at 26 GeV/c. Both the proton beam and the antiproton beam are accelerated to 270 GeV/c each and then collide at Section F in the SPS, where a large experimental detector will be placed. Early in July the Antiproton Accumulator cooled its first proton beam, and by mid-July antiproton injection was expected.

The Fermilab antiproton–proton collider[12] (figure 7) uses two sequences: precooling to reduce the initial phase

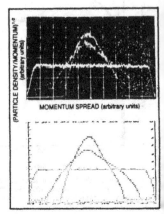

MOMENTUM SPREAD (arbitrary units)

Stochastic cooling of momentum as a function of time in the Initial Cooling Experiment at CERN. Experimental results (top) show square root of particle density in momentum space increases as the beam is cooled. Theoretical calculation (bottom) agrees well with the experiment.   Figure 5

space and freezing to produce very cool beams for storage.[13] The Main Ring will accelerate 1.8×10^{13} protons to 80 GeV, extract and aim them at an antiproton production target. The antiprotons are then collected in a large-aperture Precooler ring, roughly the same size as the present Booster ring. The antiprotons, at 4.5 GeV with a transverse emittance of 4.8 mm-mrad in each plane and momentum spread of ± 2%, are stochastically momentum cooled by a factor of about a hundred in several seconds. This occurs in three or four steps; each time cooling is followed by some deceleration. Then the cooled beam is decelerated to 200 MeV,

---

Table 1: Parameters for Three Antiproton–Proton Colliders

	γ	$f\gamma$	β^*(m)	Δ_{max}	L (cm^{-2} sec^{-1})	N_p
CERN SPS $\bar{p}p$ Collider	2.7×10^2	1.3×10^7	1.5	2×10^{-3}	10^{30}	6×10^{11}
Fermilab Tevatron $\bar{p}p$ Collider	10^3	5×10^7	1.5	2×10^{-3}	4×10^{30}	5×10^{11}
5–10 TeV $\bar{p}p$ Collider at CERN or Fermilab		10^8	2	5×10^{-3}	10^{32}	5×10^{12}

Table 2: Liouville's Theorem Applied to Particle Beams

Liouvillian	Non-Liouvillian
External fields	*Dissipative forces*
Time periodic	Synchrotron radiation
Constant magnetic or electric fields	dE/dX in thin foils
Long-range forces	*Single-particle collisions or decay*
Beam–beam interactions	H⁻ injection into accelerators
Space-charge effects	$\Lambda/\bar{\Lambda}$ decays
Plasma oscillations	Electron cooling

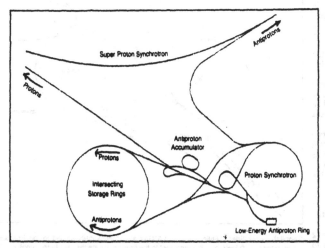

CERN p̄p collider. Antiprotons are cooled in the Antiproton Accumulator, collected, transferred to the Proton Synchrotron, accelerated to 26 GeV/c, bunched and sent to the Super Proton Synchrotron. Meanwhile 26-GeV protons are injected into the SPS at 26 GeV/c. Both beams are then accelerated to 270 GeV/c and allowed to collide.
Figure 6

bunched and transferred to the Freezer. Here the antiprotons are cooled further while more antiprotons are added. When some 10^{11} antiprotons have been obtained, they will be transferred back to the Precooler, reaccelerated to 8 GeV and then injected into the Main Ring. After further acceleration, the antiprotons are transferred into the Superconducting Ring. Protons are accelerated in the conventional fashion in the Main Ring and transferred to the Superconducting Ring. Then both beams are simultaneously accelerated to collision energy.

The target station and Freezer are already under construction as part of the Fermilab R&D program. The $18-million (including contingencies and escalation) Precooler is being designed and is scheduled for completion in 1983. The Precooler is part of the Tevatron I package, which is in the DOE FY 1981 budget request. This package also includes the refrigeration and experimental areas needed for p̄p collisions at 1000 GeV on 1000 GeV.

The Fermilab research and development project is a collaboration of Argonne National Laboratory, Lawrence Berkeley Laboratory, Fermilab, Institute for Nuclear Physics at Novosibirsk, USSR and the University of Wisconsin.

Table 3 compares the parameters of the CERN and Fermilab schemes. Note that both machines are expected to reach a luminosity of 10^{30} cm^{-2}sec^{-1}. This appears to be adequate to carry out an exciting research program at CERN and Fermilab. However, considerable experience is likely to be

required with these systems before the higher luminosity is reached.

Intermediate Vector Bosons

When the first p̄p machine operates it will be the beginning of the study of ultra high-energy interactions and very likely of the weak interaction at very high energy. Hideki Yukawa first predicted in the mid-1930's that the exchange of a massive object could be responsible for the weak force.[14] These particles have been named the Intermediate Vector Boson. The gauge theory of Sheldon Glashow, Steven Weinberg and Abdus Salam,[15] which received remarkable support with the discovery of weak neutral currents[16] at CERN and Fermilab in 1973, incorporated three intermediate bosons—W^+, W^- and Z^0— and provides predictions for the static and dynamic properties of these particles. It is crucial that these properties be measured by some technique. Table 4 gives a list of these properties, using the current experimental estimates of the Weinberg angle.

We now turn to the mechanism of producing such massive bosons in high-energy collisions. The intermediate vector boson particles decay by the channels

$$Z \to \text{lepton} + \text{antilepton}$$
$$Z \to \text{quark} + \text{antiquark}$$

Thus the inverse process, the fusion of $\ell\bar{\ell}$ or $q\bar{q}$, provides a reaction to produce the intermediate vector boson, and very high energy e^+e^- or quark–antiquark collisions can result in IVB production.[13,17,18] Antiproton–proton collisions involve copious quark–antiquark collisions and thus provide a mechanism for production of the IVB.

However, to study the complete properties of the IVB it is necessary to produce the IVB in reactions other than quark–antiquark fusion as well. Figure 8 shows the processes that we expect will be crucial to discover and measure the properties of these particles. The cross sections for these reactions have been calculated by several groups and are shown[17] in figure 9. These calculations illustrate that antiproton–proton colliders have several characteristic advantages in this regard. A p̄p collider with luminosity 10^{30}–10^{31} cm^{-2}sec^{-1} results in copious numbers of IVB per day. (See figure 7.) Two such properties, among others, are observable:

▶ p̄p→$W^\pm$ + all. Measurement of the decay lepton with respect to the beam direction will show a forward-backward asymmetry.[18]

▶ p̄p→$W^+ + \gamma$ + all. Measurement of the photon angular distribution with respect to the p̄ direction is sensitive to the anomalous magnetic moment of the W.[19]

Table 3: Properties of the CERN/Fermilab Schemes

	Fermilab	CERN
Energy (GeV)	1000	270
Number of protons (N_p)	1.2×10^{13}	6×10^{11}
Number of antiprotons ($N_{\bar{p}}$)	10^{11} (5 h)	6×10^{11} (24 h)
Number of bunches	12	6
Low beta at interaction $\beta^* = \sqrt{(\beta_x \beta_y m)}$	1.5	2.2
Proton emittance, horizontal (mrad)	$2.6\pi \times 10^{-6}$	$3.5\pi \times 10^{-6}$
Proton emittance, vertical (mrad)	$2.6\pi \times 10^{-6}$	$3.5\pi \times 10^{-6}$
Antiproton emittance, horizontal (mrad)	$1.0\pi \times 10^{-6}$	$3.8\pi \times 10^{-6}$
Antiproton emittance, vertical (mrad)	$1.0\pi \times 10^{-6}$	$1.9\pi \times 10^{-6}$
Luminosity (cm^{-2} sec^{-1})	$> 10^{30}$	10^{30}

Table 4 lists the other properties of the IVB that are to be measured. In nearly all cases it appears that these measurements are accessible to experimental techniques using the antiproton colliders being constructed at CERN and Fermilab. Note that the IVB pair production is predicted to rise. (See figure 7.)[20]

The Higgs Boson plays a crucial role in the gauge theory because it is necessary to eliminate several infinities. Unfortunately the mass and decay modes are not yet well defined.

The next generation of antiproton-proton collider should provide an even greater "source" of IVB's and thus allow more refined measurements of the properties of these particles. We believe that these machines are almost unique to the study of the IVB in much the same way as special rings have been built to study the $(g-2)$ of the electron and muon.

Special particle detectors covering nearly 4π solid angle are necessary to detect and study the properties of the IVB. Unfortunately we have no space to describe these detectors except to note that such detectors are in an advanced stage of construction at CERN. One such detector, the UA1 detector, is expected to be quite "universal" and well matched to "W/Z physics." It employs a large dipole magnetic field and complete particle tracking as well as electron and muon identification and momentum measurements.

Future antiproton–proton colliders

The antiproton–proton colliders now under construction at CERN and Fermilab were first suggested to allow the discovery of the intermediate vector bosons and to study their properties.[11] In that case it was shown that a luminosity of roughly 10^{30} cm^{-2} sec^{-1} provides an adequate event rate, if suitable detectors are employed. This goal has been incorporated as a design criterion for the two machines being constructed. If higher energy $\bar{p}p$ colliders are to be constructed the design luminosity must be set by either physics goals or by the inherent limitations in the luminosity that can be achieved in such machines. This limitation may arise from beam–beam interaction or from the total intensity of the antiproton source.

Consider first the expected cross sections for high-energy pointlike reactions

$$\sigma^{"}_{pt} \simeq \frac{4\pi\alpha^2}{3Q^2} \simeq \frac{87}{s} \times 10^{-33} \ \text{cm}^2$$

for the electro-weak cross section where s is the center-of-mass energy squared for the collision and α the fine-structure constant. For strong interaction cross sections

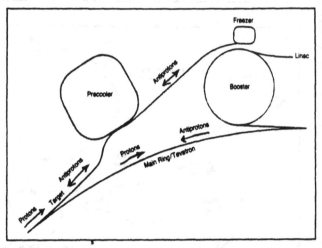

Fermilab $\bar{p}p$ collider. Antiprotons at 4.5 GeV are stochastically cooled in the Precooler, decelerated to 200 MeV, bunched and sent to the Freezer, where they are further cooled. The antiproton beam is then sent back to the Precooler, reaccelerated to 8 GeV and sent to the Main Ring. After further acceleration, the beam goes to the Superconducting Ring. Meanwhile a proton beam is accelerated and transferred to the Superconducting Ring, where it collides with the antiproton beam.

Figure 7

$$\sigma^{'}_{pt} \simeq \left(\frac{\alpha_s}{\alpha}\right)^2 \left(\frac{87}{s}\right) \times 10^{-33} \ \text{cm}^2$$

where α_s is the strong-interaction coupling constant and $(\alpha_s/\alpha)^2 >> 1$. The cross section falls rapidly with center-of-mass energy, and the luminosity of a $\bar{p}p$ machine required to produce one event per day for $\sqrt{s} = 10$ TeV is 10^{32} cm^{-2} sec^{-1}.

There is an obvious moral to all

this—high-energy $\bar{p}p$ machines should provide for the production of W and Z with modest luminosity but may need high luminosity to reach into new areas of very high momentum transfer pointlike collisions, where new physics may be hiding.

Table 1 gives a comparison of the luminosity of a machine at 5–10 TeV with that of the present generation of machines. Such machines could be

Table 4: Properties of Intermediate Bosons

Boson	Property	Prediction from SU(2)×U(1) theory	Technique to measure	Expected resolution
W'	Mass	79.5 GeV	$W' \rightarrow e' + \nu_e$	~10%
	Width	2.6 GeV	$W \rightarrow l + \nu +$ missing $p_\perp$ measurement	~100%
	Anomalous magnetic moment, K	$K = +1$	$\bar{p}p \rightarrow W' + \gamma +$ all	$\delta K/K \sim 20\%$
	WW coupling strength	Yang–Mills	$\bar{p}p \rightarrow W' + W +$ all	
	Principal decay modes	quark–antiquark and lepton pairs	$\bar{p}p \rightarrow W' \rightarrow l + \nu$ $\searrow q + \bar{q}$	
	Higgs Boson coupling	Semi-strong coupling between W and H	$\bar{p}p \rightarrow W +$ Higgs Boson (H) + all	
Z^0	Mass	90 GeV	$Z^0 \rightarrow e' \cdot e'$ or $\mu' \cdot \mu$	~5%
	Width	2.6 GeV	$Z^0 \rightarrow e' \cdot e$	~50%
	WZ/ZZ coupling	Yang–Mills	$\bar{p}p \rightarrow W + Z +$ all $Z + Z +$ all	
	Principal decay modes	lepton, antilepton quark, antiquark pairs	$Z \rightarrow l/l$, $\bar{b}b$, $\bar{t}t$	
	Rare decay modes	gluon pairs, neutrino pairs	$e \cdot e \rightarrow Z^0$ Factory	
	Higgs Boson coupling	Semi-strong coupling between Z and H	$\bar{p}p \rightarrow Z^0 + H^0 +$ all $e \cdot e \rightarrow Z^0 + H^0$	

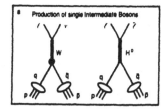

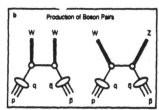

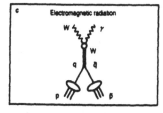

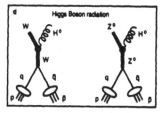

Production and decay of Intermediate Vector Bosons. Properties measured are (a) mass, width and decay modes; (b) coupling constant, g_{WW}; (c) anomalous magnetic moment of Intermediate Bosons; (d) coupling constants, g_{WWH} and g_{ZZH}. Figure 8

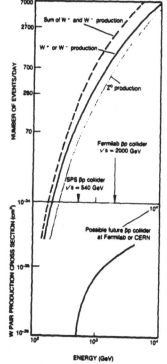

Production of Intermediate Vector Bosons as a function of energy in $\bar{p}p$ colliders. The ordinate for the top curves is events/day; the ordinate for the bottom curve is cross section. Note that the cross section for W pair production is 3×10^{-37} cm^2 for the energy of the Isabelle pp collider. Figure 9

constructed in the latter half of the 1980's in the large tunnel being constructed at CERN for the LEP machine or at Fermilab as a new "site filling" ring. There are also plans in the USSR for a 3-TeV $\bar{p}p$ machine. However it is likely that a much more intense antiproton source will be required than for the present generation of $\bar{p}p$ colliders. Thus the present generation of antiproton sources should lead to more intense sources for the higher energy machines, in the same manner that the ion sources in the 1930's are now standard equipment at Fermilab and CERN but are correspondingly more intense. Low-energy antiproton facilities will likely also become of increasing interest and importance.[21]

• • •

We would like to thank J. Adams, D. Berley, R. Billinge, G. Budker, N. Dikansky, F. Kreinan, P. McIntyre, F. Mills, D. Mohl, G. Petrucci, S. Van Der Meer and F. Sacherer, N. Skrinsky, L. Thorndahl and D. Young for many enlightening discussions about beam cooling and antiproton–proton colliders.

References

1. G. I. Budker, Atomnaya Energiya **22**, 346 (1967).

2. The feasibility of cooling high-energy beams using electrons is described in D. Cline, A. Garren, H. Herr, F. E. Mills, C. Rubbia, A. Ruggiero, D. Young, "High Energy Electron Cooling to Improve the Luminosity and Lifetime in Colliding Beam Machines." SLAC-PUB-2278, March 1979.

3. S. Van Der Meer, "Stochastic Cooling of Betatron Oscillations in the ISR," CERN-ISR, PO/72/31 (1972).

4. G. Carron, L. Thorndahl, "Stochastic Cooling of Momentum Spread by Filter Techniques," CERN Report 2RF/78-12 (1979); D Möhl, G. Petrucci, L. Thorndahl, S. Van Der Meer, CERN/PS/AA 79-23 (1979).

5. G. I. Budker *et al.*, "New Experimental Results of Electron Cooling," Preprint 76-32, Nucl. Phys. Inst., Novosibirsk. Presented to the All Union High-Energy Accelerator Conference, Moscow, October 1976. G. I. Budker *et al.*, Particle Accelerators 7, 2044 (1976), G. I. Budker *et al.*, "Experimental Study of Electron Cooling," IYaF Preprint 76-33, Nucl. Phys. Inst., Novosibirsk (1976). (Translated by Brookhaven National Laboratory, BNL-TR-634.)

6. D. Möhl, "Stochastic Cooling," CERN Report PS/D7/78-75 (1978).

7. M. Bell *et al.*, "Electron Cooling Experiment at CERN," CERN EP/79-96, 3 Sept. 1979.

8. F. Sacherer, "Stochastic Cooling Theory," CERN Report IST/TH78-11.

9. "The Fermilab Electron Cooling Experiment," Fermilab report, August, 1978.

10. "The CERN $\bar{p}p$ Design Report," CERN/PS/AA 78-3 Report (1978).

11. C. Rubbia, P. McIntyre, D. Cline, "Producing Massive Neutral Intermediate Vector Bosons with Existing Accelerators," in Proceedings of the International Neutrino Conference Aachen, H. Faissner, H. Reithler, P. Zerwas (Eds.), 1976. D. Cline, P. McIntyre, F. Mills, C. Rubbia, Fermilab Internal Report TM-689 (1976).

12. "Tevatron Phase I Report," Fermilab Report (1979).

13. D. Cline *et al.*, Harvard/Wisconsin proposal to Fermilab 492 (1976).

14. H. Yukawa, Proc. Phys.-Math. Soc. Japan 17, 48 (1935).

15. S. L. Glashow, Nucl. Phys. **22**, 579 (1961). S. Weinberg, Phys. Rev. Lett. **19**, 1264 (1967). A. Salam, in Elementary Particle Theory: Relativistic Groups and Analyticity (Nobel Symposium No. 8), N. Svartholm, ed. (Almqvist and Wiksell, Stockholm, 1968), page 367.

16. F. J. Hasert *et al.*, Phys. Lett. **46B**, 121 (1973); 46B, 138 (1973); Nucl. Phys B73, 1 (1974). A. Benvenuti *et al.*, Phys. Rev. Lett. 32, 800 (1974); B. Aubert *et al.*, ibid., 32, 1454 (1974); 32, 1457 (1974).

17. C. Quigg, Rev. Mod. Phys. 49, 297 (1977); L. B. Okun, M. B. Voloshin, Nucl. Phys. B120, 459 (1977); R. F. Peierls, T. L. Trueman, L.-L. Wang, Phys. Rev. D16, 1397 (1977).

18. F. Paige, "Updated Estimates of W Production in pp and $\bar{p}p$ Interactions," in Proceedings of the Topical Workshop on New Particles in Super High-Energy Collisions, Madison, Wisc., 1979.

19. K. O. Mikaelian, M. A. Samuel, Phys. Rev. Lett. 43, 746 (1979).

20. R. W. Brown, K. O. Mikaelian, Phys. Rev. B19, 922 (1979).

21. U. Gastaldi, K. Kilian, G. Plass, "A Low Energy Antiproton Facility at CERN," CERN/PSCC/79-17. □

FIRST PROTON–ANTIPROTON COLLISIONS IN THE CERN SPS COLLIDER

The Staff of the CERN Proton–Antiproton Project
CERN, Geneva, Switzerland

Received 6 November 1981

Protons and antiprotons have been stored simultaneously in the CERN SPS for several hours. Their interactions at 54 GeV in the centre-of-mass system have been observed by three different experiments.

The CERN antiproton project (fig. 1) has been described in recent papers [1,2]. Antiprotons, produced in a target by 26 GeV protons from the CERN Proton Synchrotron (PS) are collected at 3.5 GeV/c, stored and stochastically cooled in the antiproton accumulator ring (AA). After about one day of collecting and cooling, the antiprotons are injected into the PS, accelerated up to 26 GeV and then transported to the

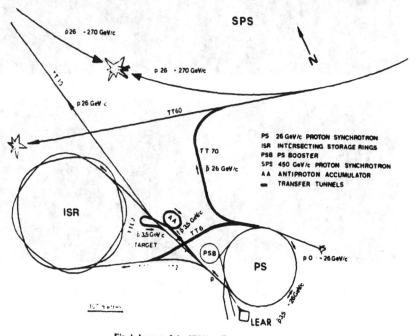

Fig. 1. Layout of the CERN p–p̄ complex.

CERN Super Proton Synchrotron (SPS), where they are further accelerated up to 270 GeV together with counterrotating protons. Commissioning of the SPS as p–p̄ collider started in June 1981 after a shutdown of one year needed to modify the SPS ring for this additional mode of operation and to construct two underground experimental areas.

We can now report successful storage of protons and antiprotons at 270 GeV with lifetimes of several hours. Typically two bunches of 5×10^{10} protons each were colliding against one bunch of about 10^9 antiprotons, giving an initial luminosity of 2×10^{25} cm^{-2}s^{-1} per interaction point in these first runs.

Proton–antiproton collisions have been observed in both experimental areas. Fig. 2 shows the tracks of outgoing particles produced by a 540 GeV p–p̄ collision as seen by the UA5 streamer chambers. The UA5 detector [3] consists of two 6 m long streamer chambers, situated immediately above and below the beam pipe and triggered by external planes of scintillation hodoscopes. From an analysis of previous background runs and from the fact that larger numbers of tracks emerge in both directions from a vertex, which lies inside the region where the bunches are known to cross, it is concluded that the event shown results from a beam–beam interaction.

Fig. 2. Photograph of a p–p̄ event as seen by the upper and lower UA5 streamer chambers. The chambers are mounted immediately above and below the beam pipe, so as to observe tracks down to less than 1° production angle. The sensitive volume per chamber is $6 \times 1.25 \times 0.5$ m^3.

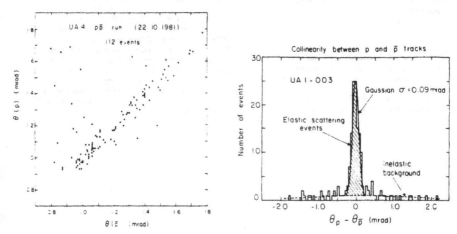

Fig. 3. p–p̄ elastic scattering events. (a) Distribution of proton versus antiproton scattering angles for 112 events observed by experiment UA4. (b) Distribution of the measured difference of the p and p̄ scattering angles for 168 events observed by experiment UA1.

Further evidence for beam–beam interactions is given by a clear signal of p–p̄ elastic scattering observed by two collaborations UA1 [4] and UA4 [5] in both interaction areas. The experimental set-up is similar for the two experiments. Drift chamber telescopes are placed symmetrically some 20 m (UA1) or 40 m (UA4) away from the crossing point to observe trajectories of particles scattered in the angular range 1–2 mrad. Tracks measured in the telescopes are extrapolated through the machine quadrupoles back to the interaction region to obtain the scattering angles θ_p and $\theta_{\bar{p}}$. Figs. 3a and 3b show plots of elastic scattering angles measured by experiments UA4 and UA1 respectively. The clear grouping of events around the line $\theta_p = \theta_{\bar{p}}$ in fig. 3a and the sharp peak at $(\theta_p - \theta_{\bar{p}}) = 0$ in fig. 3b correspond to collinear, i.e. elastic events. The background signal from inelastic events is relatively low because of the momentum filtering effect of the machine quadrupoles between the interaction point and each telescope.

The SPS is the first proton machine which uses a small number of short bunches (~0.5 m) to maximize the rate of p–p̄ collisions around the centre of the detectors. In such a machine it is necessary to reduce the noise in the radiofrequency system to an extremely low level to limit the diffusion of particles out of the buckets with a corresponding reduction in lifetime of the luminosity. The recent results indicate that this problem can be overcome in the SPS.

Another problem related to the use of tightly bunched beams is the maximum beam–beam tune shift tolerable in the head-on collisions of bunched proton and antiproton beams. In the p–p̄ collisions reported here, the beam–beam space-charge tune-shift of the antiproton bunch induced by each of the two counter-rotating proton bunches has been calculated to be 2.5×10^{-3}. No evidence of a catastrophic loss or emittance growth has been recorded at this level, although more detailed measurements are needed before final conclusions can be drawn.

In these preliminary runs, no attempt was made to achieve a high luminosity. Within the limits of the beam–beam tune shift already obtained, it should be possible to increase the p̄-intensity by a factor 50 in the near future. Furthermore, the application of the low beta scheme which has already been successfully tested with protons, should increase the luminosity by nearly another factor 50.

The further development of the p–p̄ colliding beam facility towards its design luminosity of 10^{30} cm^{-2}s^{-1}

will require the storage of up to six bunches, each with an intensity of about 10^{11} particles, in both beams stored in the SPS.

References

[1] Producing massive neutral intermediate vector bosons with existing accelerators, C. Rubbia, P. McIntyre and D. Cline. Proc. Intern. Neutrino Conf. (Aachen, 1976) p. 683.
[2] The CERN p–p̄ complex, J. Gareyte, Proc. Intern. Conf. on High-energy accelerators at Geneva (1980) p. 79.
[3] The UA5 Streamer Chamber Experiment at the SPS p–p̄ collider, Bonn–Brussels–Cambridge–CERN–Stockholm Collab., Phys. Scr. 33 (1981) 642.
[4] A 4π solid angle detector for the SPS used as a proton–antiproton collider at a centre of mass energy of 540 GeV, Proposal of the Aachen–Annecy–Birmingham–CERN–College de France–Queen Mary College–Riverside–Rutherford–Roma–Saclay–Wien Collab. CERN/SPSC/78-06.
[5] The measurement of elastic scattering and of the total cross section at the CERN p–p̄ collider, Proposal of the Amsterdam–CERN–Genova–Napoli–Pisa Collab., CERN/SPSC/78-105.

EXPERIMENTAL OBSERVATION OF ISOLATED LARGE TRANSVERSE ENERGY ELECTRONS WITH ASSOCIATED MISSING ENERGY AT $\sqrt{s}$ = 540 GeV

UA1 Collaboration, CERN, Geneva, Switzerland

G. ARNISON [j], A. ASTBURY [j], B. AUBERT [b], C. BACCI [i], G. BAUER [1], A. BÉZAGUET [d], R. BÖCK [d], T.J.V. BOWCOCK [f], M. CALVETTI [d], T. CARROLL [d], P. CATZ [b], P. CENNINI [d], S. CENTRO [d], F. CERADINI [d], S. CITTOLIN [d], D. CLINE [1], C. COCHET [k], J. COLAS [b], M. CORDEN [c], D. DALLMAN [d], M. DeBEER [k], M. DELLA NEGRA [b], M. DEMOULIN [d], D. DENEGRI [k], A. Di CIACCIO [i], D. DiBITONTO [d], L. DOBRZYNSKI [g], J.D. DOWELL [c], M. EDWARDS [c], K. EGGERT [a], E. EISENHANDLER [f], N. ELLIS [d], P. ERHARD [a], H. FAISSNER [a], G. FONTAINE [g], R. FREY [h], R. FRÜHWIRTH [l], J. GARVEY [c], S. GEER [g], C. GHESQUIÈRE [g], P. GHEZ [b], K.L. GIBONI [a], W.R. GIBSON [f], Y. GIRAUD-HÉRAUD [g], A. GIVERNAUD [k], A. GONIDEC [b], G. GRAYER [j], P. GUTIERREZ [h], T. HANSL-KOZANECKA [a], W.J. HAYNES [j], L.O. HERTZBERGER [2], C. HODGES [h], D. HOFFMANN [a], H. HOFFMANN [d], D.J. HOLTHUIZEN [2], R.J. HOMER [c], A. HONMA [f], W. JANK [d], G. JORAT [d], P.I.P. KALMUS [f], V. KARIMÄKI [e], R. KEELER [f], I. KENYON [c], A. KERNAN [h], R. KINNUNEN [e], H. KOWALSKI [d], W. KOZANECKI [h], D. KRYN [d], F. LACAVA [d], J.-P. LAUGIER [k], J.-P. LEES [b], H. LEHMANN [a], K. LEUCHS [a], A. LÉVÊQUE [k], D. LINGLIN [b], E. LOCCI [k], M. LORET [k], J.-J. MALOSSE [k], T. MARKIEWICZ [d], G. MAURIN [d], T. McMAHON [c], J.-P. MENDIBURU [g], M.-N. MINARD [b], M. MORICCA [i], H. MUIRHEAD [d], F. MULLER [d], A.K. NANDI [j], L. NAUMANN [d], A. NORTON [d], A. ORKIN-LECOURTOIS [g], L. PAOLUZI [i], G. PETRUCCI [d], G. PIANO MORTARI [i], M. PIMIÄ [e], A. PLACCI [d], E. RADERMACHER [a], J. RANSDELL [h], H. REITHLER [a], J.-P. REVOL [d], J. RICH [k], M. RIJSSENBEEK [d], C. ROBERTS [j], J. ROHLF [d], P. ROSSI [d], C. RUBBIA [d], B. SADOULET [d], G. SAJOT [g], G. SALVI [f], G. SALVINI [i], J. SASS [k], J. SAUDRAIX [k], A. SAVOY-NAVARRO [k], D. SCHINZEL [f], W. SCOTT [j], T.P. SHAH [j], M. SPIRO [k], J. STRAUSS [l], K. SUMOROK [c], F. SZONCSO [l], D. SMITH [h], C. TAO [d], G. THOMPSON [f], J. TIMMER [d], E. TSCHESLOG [a], J. TUOMINIEMI [e], S. Van der MEER [d], J.-P. VIALLE [d], J. VRANA [g], V. VUILLEMIN [d], H.D. WAHL [l], P. WATKINS [c], J. WILSON [c], Y.G. XIE [d], M. YVERT [b] and E. ZURFLUH [d]

Aachen [a]–Annecy (LAPP) [b]–Birmingham [c]–CERN [d]–Helsinki [e]–Queen Mary College, London [f]–Paris (Coll. de France) [g] –Riverside [h]–Rome [i]–Rutherford Appleton Lab. [j]–Saclay (CEN) [k]–Vienna [l] Collaboration

Received 23 January 1983

We report the results of two searches made on data recorded at the CERN SPS Proton–Antiproton Collider: one for isolated large-E_T electrons, the other for large-E_T neutrinos using the technique of missing transverse energy. Both searches converge to the same events, which have the signature of a two-body decay of a particle of mass ~80 GeV/c^2. The topology as well as the number of events fits well the hypothesis that they are produced by the process $\bar{p} + p \rightarrow W^\pm + X$, with $W^\pm \rightarrow e^\pm + \nu$; where $W^\pm$ is the Intermediate Vector Boson postulated by the unified theory of weak and electromagnetic interactions.

[1] University of Wisconsin, Madison, WI, USA.
[2] NIKHEF, Amsterdam, The Netherlands.

1. Introduction. It is generally postulated that the beta decay, namely (quark) → (quark) + e± + ν is mediated by one of two charged Intermediate Vector Bosons (IVBs), W+ and W− of very large masses. If these particles exist, an enhancement of the cross section for the process (quark) + (antiquark) → e± + ν should occur at centre-of-mass energies in the vicinity of the IVB mass (pole), where direct experimental observation and a study of the properties of such particles become possible. The CERN Super Proton Synchrotron (SPS) Collider, in which proton and antiproton collisions at $\sqrt{s}$ = 540 GeV provide a rich sample of quark −antiquark events, has been designed with this search as the primary goal [1].

Properties of IVBs become better specified within the theoretical frame of the unified weak and electromagnetic theory and of the Weinberg−Salam model [2]. The mass of the IVB is precisely predicted [3]:

$$M_{W^\pm} = (82 \pm 2.4) \text{ GeV}/c^2$$

for the presently preferred [4] experimental value of the Weinberg angle $\sin^2\theta_W = 0.23 \pm 0.01$. The cross section for production is also reasonably well anticipated [5]

$$\sigma(p\bar{p} \rightarrow W^\pm \rightarrow e^\pm + \nu) \simeq 0.4 \times 10^{-33} \, k \text{ cm}^2 \, ,$$

where k is an enhancement factor of ~1.5, which can be related to a similar well-known effect in the Drell−Yan production of lepton pairs. It arises from additional QCD diagrams in the production reaction with emission of gluons. In our search we have reduced the value of k by accepting only those events which show no evidence for associated jet structure in the detector.

2. The detector. The UA1 apparatus has already been extensively described elsewhere [6]. Here we concentrate on those aspects of the detector which are relevant to the present investigation.

The detector is a transverse dipole magnet which produces a uniform field of 0.7 T over a volume of 7 × 3.5 × 3.5 m³. The interaction point is surrounded by the central detector (CD): a cylindrical drift chamber volume, 5.8 m long and 2.3 m in diameter, which yields a bubble-chamber quality picture of each pp̄ interaction in addition to measuring momentum and specific ionization of all charged tracks.

Momentum precision for high-momentum particles is dominated by a localization error inherent to the

system ($\leqslant$100 μm) and the diffusion of electrons drifting in the gas (proportional to $\sqrt{l}$ and about 350 μm after l = 22 cm maximum drift length). This results in a typical relative accuracy of ±20% for a 1 m long track at p = 40 GeV/c, and in the plane normal to the magnetic field. The precision, of course, improves considerably for longer tracks. The ionization of tracks can be measured by the classical method of the truncated mean of the 60% lowest readings to an accuracy of 10%. This allows an unambiguous identification of narrow, high-energy particle bundles (e+e− pairs or pencil jets) which cannot be resolved by the drift chamber digitizings.

The central section of electromagnetic and hadronic calorimetry has been used in the present investigation to identify electrons over a pseudorapidity interval $|\eta| < 3$ with full azimuthal coverage. Additional calorimetry, both electromagnetic and hadronic, extends to the forward regions of the experiment, down to 0.2° (for details, see table 1).

The central electromagnetic calorimeters consist of two different parts:

(i) 48 semicylindrical modules of alternate layers of scintillator and lead (gondolas), arranged in two cylindrical half-shells, one on either side of the beam axis with an inner radius of 1.36 m. Each module extends over approximately 180° in azimuth and measures 22.5 cm in the beam direction. The light produced in each of the four separate segmentations in depth is seen by wavelength shifter plates on each side of the counter, in turn connected to four photomultipliers (PMs), two at the top and two at the bottom. Light attenuation is exploited in order to further improve the calorimetric information: the comparison of the pulse heights of the top and bottom PM of each segment gives a measurement of the azimuthal angle ϕ for localized energy depositions, $\Delta\phi$ (rad) = 0.3/$[E(\text{GeV})]^{1/2}$. A similar localization along the beam direction is possible using the complementary pairing of PMs. The energy resolution for electrons using all four PMs is $\Delta E/E = 0.15/[E(\text{GeV})]^{1/2}$.

(ii) 64 petals of end-cap electromagnetic shower counters (bouchons), segmented four times in depth, on both sides of the central detector at 3 m distance from the beam crossing point. The position of each shower is measured with a position detector located inside the calorimeter at a depth of 11 radiation lengths, i.e. after the first two segments. It consists of

Table I
Calorimetry.

Calorimeter	Angular coverage θ (deg)	Thickness		Cell size		Sampling step	· Segmentation in depth	Resolution
		No. rad. lengths	No. abs. lengths	Δθ (deg)	Δφ (deg)			
barrel EM: gondolas	25 −155	26.4/sin θ	1.1/sin θ	5	180	1.2 mm Pb 1.5 mm scint.	3.3/6.5/10.1/6.5 X_0	$0.15/\sqrt{E}$
hadr.: c's	25 −155	−	5.0/sin θ	15	18	50 mm Fe 10 mm scint.	2.5/2.5 λ	$0.8/\sqrt{E}$
end-caps EM: bouchons	5 − 25	27/cos θ	1.1/cos θ	20	11	4 mm Pb 6 mm scint.	4/7/9/7 X_0	$0.12/\sqrt{E_T}$
hadr.: I's	155 −175	−	7.1/cos θ	5	10	50 mm Fe 10 mm scint.	3.5/3.5 λ	$0.8/\sqrt{E}$
calcom EM	0.7− 5	30	1.2	4	· 45	3 mm Pb 3 mm scint.	4 × 7.5 X_0	$0.15/\sqrt{E}$
hadr.	175 −179.3	−	10.2	−	−	40 mm Fe 8 mm scint.	6 × 1.7 λ	$0.8/\sqrt{E}$
very forward EM	0.2− 0.7	24.5	1.0	0.5	90	3 mm Pb 6 mm scint.	5.7/5.3/5.8/7.7 X_0	$0.15/\sqrt{E}$
hadr.	179.3−179.8	−	5.7	0.5	90	40 mm Fe 10 mm scint.	5 × 1.25 λ	$0.8/\sqrt{E}$

two planes of orthogonal proportional tubes of 2 × 2 cm² cross section and it locates the centre of gravity of energetic electromagnetic showers to ±2 mm in space. The attenuation length of the scintillator has been chosen to match the variation of sin θ over the radius of the calorimeters, so as to directly measure in first approximation $E_T = E \sin \theta$ rather than the true energy deposition E, which can, however, be determined later, using the information from the position detector. This technique permits us to read out directly from the end-cap detectors the amount of transverse energy deposited, without reconstruction of the event topology.

3. Electron identification. Electromagnetic showers are identified by their characteristic transition curve, and in particular by the lack of penetration in the hadron calorimeter behind them. The performance of the detectors with respect to hadrons and electrons has been studied extensively in a test beam as a function of the energy, the angle of incidence, and the location of impact. The fraction of hadrons (pions) delivering an energy deposition E_c below a given threshold in the hadron calorimeter is a rapidly falling function of energy, amounting to about 0.3% for $p \simeq 40$ GeV/c

and $E_c < 200$ MeV. Under these conditions, 98% of the electrons are detected.

4. Neutrino identification. The emission of one (or more) neutrinos can be signalled only by an apparent visible energy imbalance of the event (missing energy). In order to permit such a measurement, calorimeters have been made completely hermetic down to angles of 0.2° with respect to the direction of the beams. (In practice, 97% of the mass of the magnet is calorimetrized.) It is possible to define an energy flow vector ΔE, adding vectorially the observed energy depositions over the whole solid angle. Neglecting particle masses and with an ideal calorimeter response and solid-angle coverage, momentum conservation requires $\Delta E = 0$. We have tested this technique on minimum bias and jet-enriched events for which neutrino emission ordinarily does not occur. The transverse components ΔE_y and ΔE_z exhibit small residuals centred on zero with an rms deviation well described by the law $\Delta E_{y,z} = 0.4(\Sigma_i |E_T^i|)^{1/2}$, where all units are in GeV and the quantity under the square root is the scalar sum of all transverse energy contributions recorded in the event (fig. 1). The distributions have gaussian shape and no prominent tails. The longitudinal component

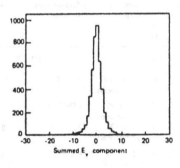

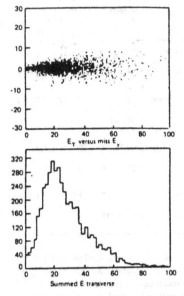

Fig. 1. The missing transverse energy in the y direction $[\Delta E_y \; (GeV)]$ plotted versus the scalar sum of missing transverse energy $[E_T \; (GeV)]$ for minimum bias triggers. The y-axis is pointing up vertically.

of energy ΔE_x is affected by the energy flow escaping through the 0° singularity of the collider's beam pipe and it cannot be of much practical use. We remark that, like neutrinos, high-energy muons easily penetrate the calorimeter and leak out substantial amounts of energy. A muon detector, consisting of stacks of eight planes of drift chambers, surrounds the whole apparatus and has been used to identify such processes, which are occurring at the level of 1 event per nanobarn for $\Delta E_{y,z} \geqslant 10$ GeV.

5. Data-taking and initial event selections. The present work is based on data recorded in a 30-day period during November and December 1982. The integrated luminosity after subtraction of dead-time and other instrumental inefficiencies was 18 nb^{-1}, corresponding to about 10^9 collisions between protons and antiprotons at $\sqrt{s} = 540$ GeV.

For each beam–beam collision detected by scintillator hodoscopes, the energy depositions in all calorimeter cells after fast digitization were processed, in the time prior to the occurrence of the next beam–beam crossing, by a fast arithmetic processor in order to rec-

ognize the presence of a localized electromagnetic energy deposition, namely of at least 10 GeV of transverse energy either in two gondola elements or in two bouchon petals. In addition, we have simultaneously operated three other trigger conditions: (i) a jet trigger, with $\geqslant 15$ GeV of transverse energy in a localized cluster [1] of electromagnetic and hadron calorimeters; (ii) a global E_T trigger, with >40 GeV of total transverse energy from all calorimeters with $|\eta| < 1.4$; and (iii) a muon trigger, namely at least one penetrating track with $|\eta| < 1.3$ pointing to the diamond.

The electron trigger rate was about 0.2 event per second at the (peak) luminosity $L = 5 \times 10^{28}$ cm^{-2} s^{-1}. Collisions with residual gas or with vacuum chamber walls were completely negligible, and the apparatus in normal machine conditions yielded an almost pure sample of beam–beam collisions. In total, 9.75×10^5 triggers were collected, of which 1.4×10^5 were char-

---

[1] We define a cluster as: (i) a group of eight gondolas and the two hadron calorimeter elements immediately behind; or (ii) a quadrant of bouchon elements (8) with the corresponding hadron calorimeters.

acterized by an electron trigger flag.

Event filtering by calorimetric information was further perfected by off-line selection of 28 000 events with $E_T > 15$ GeV in two gondolas, or $E_T > 15$ GeV in two bouchon petals with valid position-detector information. These events were finally processed with the central detector reconstruction. Of these events there are 2125 with a good quality, vertex-associated charged track of $p_T > 7$ GeV/c. This sample will be used for the subsequent analysis of events in the gondolas.

6. Search for electron candidates. We now require three conditions in succession in order to ensure that the track is isolated, namely to reject the debris of jets:

(i) The fast track ($p_T > 7$ GeV/c) as recorded by the central detector must hit a pair of adjacent gondolas with transverse energy $E_T > 15$ GeV (1106 events).

(ii) Other charged tracks, entering the same pair of gondolas, must not add up to more than 2 GeV/c of transverse momenta (276 events).

(iii) The ϕ information from pulse division from gondola phototubes must agree within 3σ with the impact of the track (167 events).

Next we introduce two simple conditions to enhance its electromagnetic nature:

(iv) The energy deposition E_c in the hadronic calorimeters aimed at by the track must not exceed 600 MeV (72 events).

(v) The energy deposited in the gondolas E_{gon} must match the measurement of the momentum of the track p_{CD}, namely $|1/p_{CD} - 1/E_{gon}| < 3\sigma$.

At this point only 39 events are left, which were individually examined by physicists on the visual scanning and interactive facility Megatek. The surviving events break up cleanly into three classes, namely 5 events with no jet activity [2], 11 with a jet opposite

---

[2] The definition of a jet is based on the UA1 standard algorithm, applied separately on the calorimetry and on the central detector data. Positive results on either set are taken as evidence for a jet. In the calorimetry a four-vector (k_i, E_i) pointing to the interaction vertex is associated with each struck cell. Working in the transverse plane, all vectors with $k_T > 2.5$ GeV are ordered and are used as potential jet initiators. They are combined if their separation in phase space satisfies the cut $\Delta R = [(\Delta\eta)^2 + (\Delta\phi)^2]^{1/2} < 1$ (with ϕ in radians). The remaining soft particles are added to the nearest jet in $\Delta\eta$ and $\Delta\phi$, provided the relative p_T is < 1 GeV and $\Delta\theta < 45°$. A jet is considered valid if $E_T^{jet} > 10$ GeV. This same procedure is used for central detector tracks with appropriately adjusted parameters.

to the track within a 30° angle in ϕ, and 23 with two jets (one of which contains the electron candidate) or clear e^+e^- conversion pairs. A similar analysis performed on the bouchon has led to another event with no jets. The classes of events have striking differences. We find that whilst events with jet activity have essen-

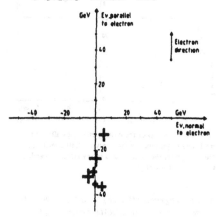

a EVENTS WITHOUT JETS

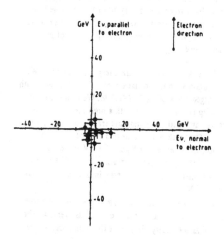

b EVENTS WITH JETS

Fig. 2. The missing transverse energy (E_ν) is plotted vectorially against the electron direction for the events yielded by the electron search: (a) without jets, (b) with jets.

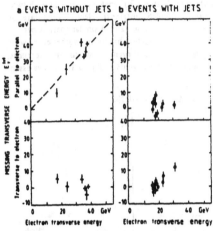

Fig. 3. The components of the missing energy parallel and perpendicular to the electron momentum plotted versus the electron energy for the events found in the electron search: (a) without jets, (b) with jets.

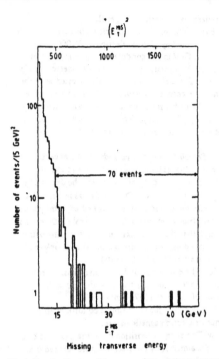

Fig. 4. The distribution of the square of the missing transverse energy for those events which survive the cuts requiring association of the central detector isolated track and a struck gondola in the missing-energy search. The five jetless events from the electron search are indicated.

tially no missing energy (fig. 2b) [*3], the ones with no jets show evidence of a missing transverse energy of the same magnitude as the transverse electron energy (fig. 3a), with the vector momenta almost exactly balanced back-to-back (fig. 2a). In order to assess how significant the effect is, we proceed to an alternative analysis based exclusively on the presence of missing transverse energy.

7. Search for events with energetic neutrinos. We start again with the initial sample of 2125 events with a charged track of $p_T > 7$ GeV/c. We now move to pick up validated events with a high missing transverse energy and with the candidate track not part of a jet:

(i) The track must point to a pair of gondolas with deposition in excess of $E_T > 15$ GeV and no other track with $p_T > 2$ GeV/c in a 20° cone (911 events).

(ii) Missing transverse energy imbalance in excess of 15 GeV.

Only 70 events survive these simple cuts, as shown in fig. 4. The previously found 5 jetless events of the gondolas are clearly visible. At this point, as for the

*3 The 11 events with an electron and a jet exhibit a p_T^{-4} spectrum with the highest event at $p_T = 32$ GeV/c.

electron analysis, we process the events at the interactive facility Megatek:

(iii) The missing transverse energy is validated, removing those events in which jets are pointing to where the detector response is limited, i.e. corners, light-pipe ducts going up and down. Some very evident, big secondary interactions in the beam pipe are also removed. We are left with 31 events, of which 21 have $E_c > 0.01 E_{gon}$ and 10 events in which $E_c < 0.01 E_{gon}$.

(iv) We require that the candidate track be well isolated, that there is no track with $p_T > 1.5$ GeV in a cone of 30°, and that $E_T < 4$ GeV for neutrals in neighbouring gondolas at similar ϕ angle. Eighteen events survive: ten with $E_c \neq 0$ and eight with $E_c = 0$.

The events once again divide naturally into the two classes: 11 events with jet activity in the azimuth op-

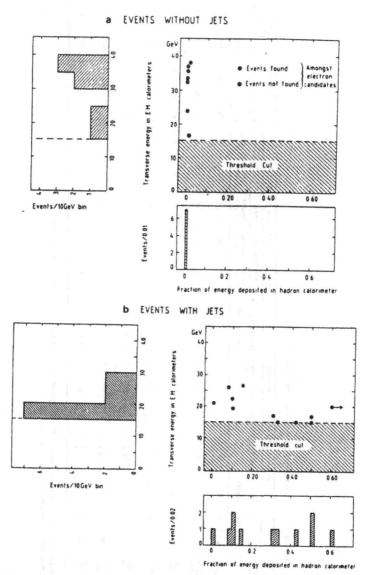

Fig. 5. A plot of the transverse energy in the EM calorimeters versus the fraction of energy deposited in the hadron calorimeters for events which survive the missing-energy search: (a) without jets, (b) with jets.

Table 2
Main parameters of electron events with a large missing transverse energy.

Run, event	\multicolumn{10}{c}{Properties of the electron track}	\multicolumn{6}{c}{Calorimeter information}	\multicolumn{4}{c}{General event topology}
	E_T (GeV)	E (GeV)	p (GeV/c)	Δp a)	q	dE/dx 1/I_0 b)	y b)	Track No.	Length (m)	Sagitta (mm)	Sample 1 (GeV)	Sample 2 (GeV)	Sample 3 (GeV)	Sample 4 (GeV)	E_{had} (GeV)	E_{tot} (GeV)	Missing E_T (GeV)	Δφ c) (deg.)	Charged tracks	[$|E_T|$] (GeV)
A 2958 1279	26	42	33.8	+6.3/−4.6	−	1.22 ±0.2	+1.1	36	1.36	1.7	4	55	3	0.2	.0	278	24.4 ± 4.6	179	65	81
B 3522 214	17	46	47.5	+8.2/−6.1	−	1.37 ±0.16	+1.7	18	1.64	1.5	2	32	10	0.5	0	296	10.8 ± 4.0	219	49	60
C 3524 197	34	48	21.0	+21.8/−7.3	−	1.37 ±0.3	−0.8	26	1.25	2.11	1	30	14	0.2	0	367	41.3 ± 3.6	187	21	68
D 3610 760	38	40	33.4	+11.0/−11.1	−	1.44 −0.34	+0.3	9	0.98	0.75	3	9	26	2.2	0.4	111	40.0 ± 2.0	181	10	47
E 3701 305	37	37	50.2	+121.3/−22.8	+	1.54 ±0.28	−0.1	12	0.95	0.4	1	18	17	0.9	0	363	35.5 ± 4.3	173	39	87
F 4017 838	37	70	55.1	+6.6/−5.3	−	1.30 ±0.26	+1.4	3	2.01	2.0	19	48	5	0.3	0	177	32.3 ± 2.4	179	14	49
G 5262 1108	40	40	6.7	+1.9/−1.2	−	1.23 ±0.28	0.0	21	0.85	3.0	2	22	15	0.9	0	218	33.4 ±2.9	172	21	63

a) Including 200 μm systematic error. b) y is defined as positive in the direction of the outgoing p̄.
c) Angle between electron and missing energy (neutrino).

posite to the track, and 7 events without detectable jet structure. If we now examine E_c, we see that these two classes are strikingly different, with large E_c for the events with jets (fig. 5b) and negligible E_c for the jetless ones (fig. 5a). We conclude that whilst the first ones are most likely to be hadrons, the latter constitute an electron sample.

We now compare the present result with the candidates of the previous analysis based on electron signature. We remark that five out of the seven events constitute the previous final sample (fig. 5a). Two new events have been added, eliminated previously by the test of energy matching between the central detector and the gondolas. Clearly the same physical process that provided us with the large-p_T electron delivers also high-energy neutrinos. The selectivity of our apparatus is sufficient to isolate such a process from either its electron or its neutrino features individually. If (ν_e, e) pairs and (ν_τ, τ) pairs are both produced at comparable rates, the two additional new events can readily be explained since missing energy can arise equally well from ν_e and ν_τ. Indeed, closer inspection of these events shows them to be compatible with the τ hypothesis, for instance, $\tau^- \to \pi^- \pi^0 \nu_\tau$ with leading π^0. However, our isolation requirements on the charged track strongly biases against most of the τ decay modes.

8. Detailed description of the electron–neutrino events. The main properties of the final sample of six events (five gondolas, one bouchon) are given in table 2 and marked A through F. The event G is a τ candidate. One can remark that both charges of the electrons are represented. The successive energy depositions in the gondola samples are consistent with test beam findings. All but event D have no energy deposition in the hadron calorimeter; event D has a 400 MeV visible, 1% leakage beyond 26.4 radiation lengths. Test beam measurements show that this is a possible fluctuation. Multiplicity of the events is widely different: event F (fig. 6b, fig. 7b) has a small charged multiplicity (14), whilst event A (fig. 6a, fig. 7a) is very rich in particles (65). Event B is the bouchon event, and it has a number of features which must be mentioned. A 100 MeV/c track emerges from the vacuum chamber near the exit point of the electron track, which might form a part of an asymmetric electron pair with the candidate. The initial angle between the two tracks would then be $11°$, not incompatible with this hypothesis once Coulomb

scattering and measurement errors of the two tracks are taken into account. There is also some activity in the muon detector opposite to the electron candidate; the muon track is unmeasurable in the central detector. For these reasons we prefer to limit our final analysis to the events in the gondolas, although we believe that everything is still consistent with event B being a good event.

9. Background evaluations. We first consider possible backgrounds to the electron signature for events with no jets. Missing energy (neutrino signature) is not yet advocated. We have taken the following into consideration:

(1) A high-p_T charged pion (hadron) misidentified as an electron, or a high-p_T charged pion (hadron) overlapping with one or more π^0.

The central detector measurement obviously gives only the momentum p of the charged pion. In addition, the electromagnetic detectors can accumulate an arbitrary amount of electromagnetic energy from π^0's, which would simulate the electron behaviour. Since gondolas are thick enough to absorb the electromagnetic cascade, the energy deposition in the hadron calorimeter is dominated by the punch-through of the charged pion of momentum p measured in the central detector, for which rejection tables exist from test beam results. In our 18 nb^{-1} sample we have searched for single-track events with $p_T > 20$ GeV/c, no associated jet, $E_c > 600$ MeV to ensure hadronic signature, and a reasonable energy balance (within 3 SD) between the charged track momentum measurement and the sum of hadronic and electromagnetic energy depositions. We have found no such event. Once the measured pion rejection table is folded in, this background is entirely negligible. A further test against pile-up is given by the matching in the x-direction between the charged track of the central detector and the centroid of the energy depositions in the gondolas, and which is very good for all events.

(2) High-p_T π^0, η^0, or γ internally (Dalitz) or externally converted to an e^+e^- pair with one leg missed. The number of isolated EM conversions (π^0, η, γ, etc.) per unit of rapidity has been directly measured as a function of E_T in the bouchons, using the position detectors over the interval 10–40 GeV. From this spectrum, the Bethe–Heitler formula for pair creation, and the Kroll–Wada formula for Dalitz pairs [7], the ex-

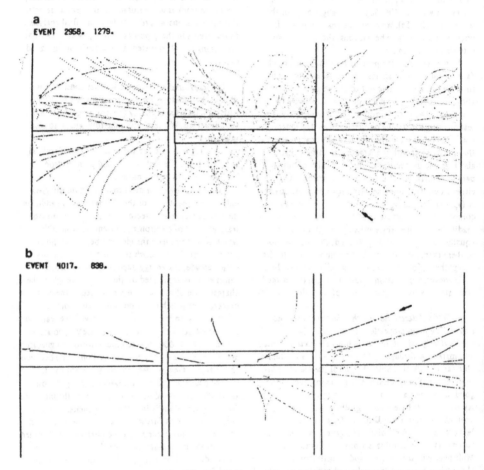

a
EVENT 2958. 1279.

b
EVENT 4017. 838.

Fig. 6. The digitization from the central detector for the tracks in two of the events which have an identified, isolated, well-measured high-p_T electron: (a) high-multiplicity, 65 associated tracks; (b) low-multiplicity, 14 associated tracks.

pected number of events with a "single" $e^\pm$ with p_T > 20 GeV/c is 0.2 p_0 (GeV), largely independent of the composition of the EM component; p_0 is the effective momentum below which the low-energy leg of the pair becomes undetectable. Very conservatively, we can take p_0 = 200 MeV/c (curvature radius 1.2 m) and conclude that this background is negligible.

(3) Heavy quark associated production, followed by pathological fragmentation and decay configuration, such that $Q_1 \rightarrow e(\nu X)$ with the electron leading and the rest undetected, and $Q_2 \rightarrow \nu(\ell X)$, with the neutrino leading and the rest undetected. In 5 nb^{-1} we have observed one event in which there is a muon and an electron in separate jets, with $p_T^{(\mu)}$ = 4.4 GeV/c and

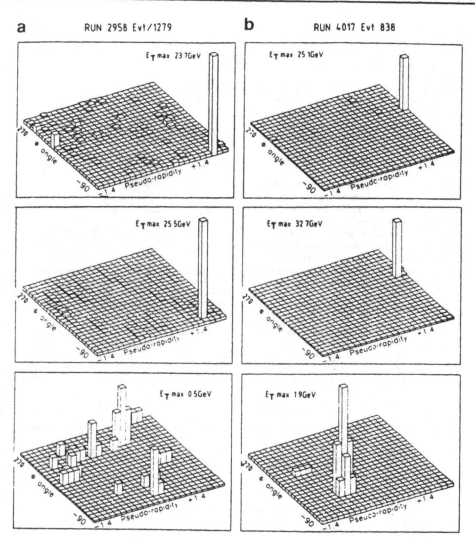

a RUN 2958 Evt/1279

b RUN 4017 Evt 838

Fig. 7. The energy deposited in the cells of the central calorimetry and the equivalent plot for track momenta in the central detector for the two events of fig. 6. The top diagram shows the electromagnetic cells, the middle shows the central detector tracks, and the bottom plot, with a very much increased sensitivity, shows the energy in the hadron calorimeter. The plots reveal no hadronic energy behind the electron and no jet structure; (a) high-multiplicity; (b) low-multiplicity.

$p_T^{(e)} = 13.3$ GeV/c. Requiring (i) extrapolation to the energy of the events, (ii) fragmentation functions for leading lepton, and (iii) a detection hole for all remaining particles, makes the rate of these background events negligible.

In conclusion, we have been unable to find a background process capable of simulating the observed high-energy electrons. Thus we are led to the conclusion that they are electrons. Likewise we have searched for backgrounds capable of simulating large-E_T neutrino events. Again, none of the processes considered appear to be even near to becoming competitive.

10. Comparison between events and expectations from W decays.

The simultaneous presence of an electron and (one) neutrino of approximately equal and opposite momenta in the transverse direction (fig. 8) suggests the presence of a two-body decay, $W \to e + \nu_e$. The main kinematical quantities of the events are given in table 3. A lower, model-independent bound to the W mass m_W can be obtained from the transverse mass, $m_T^2 = 2 p_T^{(e)} p_T^{(\nu)} (1 - \cos \phi_{\nu e})$, remarking that $m_W \geqslant m_T$ (fig. 9). We conclude that:

$$m_W > 73 \text{ GeV}/c^2 \quad (90\% \text{ confidence level}).$$

A better accuracy can be obtained from the data if one assumes W decay kinematics and standard V − A couplings. The transverse momentum distribution of the W at production also plays a role. We can either (i) extract it from the events (table 3); or, (ii) use theoretical predictions [8].

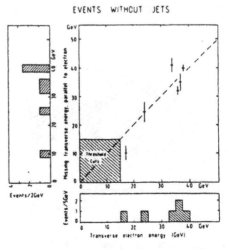

EVENTS WITHOUT JETS

Fig. 8. The missing transverse energy component parallel to the electron, plotted versus the transverse electron energy for the final six electron events without jets (5 gondolas, 1 bouchon) All the events in the gondolas appear well above the threshold cuts used in the searches.

As one can see from fig. 10, there is good agreement between two extreme assumptions of a theoretical model [8] and our observations. By requiring no associated jet, we may have actually biased our sample towards the narrower first-order curve. Fitting of the in-

Table 3
Transverse mass and transverse momentum of a W decaying into an electron and a neutrino computed from the events of table 2.

Run, event	$p_T^{(e)}$ of electron (GeV/c)	$p_T^{(\nu)}$ = missing E_T (GeV)	Transverse mass (GeV/c)^2	$p_T^{(W)} = |p_T^{(e)} + p_T^{(\nu)}|$ (GeV)
A 2958 1279	24 ± 0.6	24.4 ± 4.6	48.4 ± 4.6	0.6 ± 4.6
B 3522 214	17 ± 0.4	10.9 ± 4.0	26.5 ± 4.6	10.8 ± 4.0
C 3524 197	34 ± 0.8	41.3 ± 3.6	74.8 ± 3.4	8.6 ± 3.7
D 3610 760	38 ± 1.0	40.0 ± 2.0	78.0 ± 2.2	2.1 ± 2.2
E 3701 305	37 ± 1.0	35.5 ± 4.3	72.4 ± 4.5	4.7 ± 4.4
F 4017 838	36 ± 0.7	32.3 ± 2.4	68.2 ± 2.6	3.8 ± 2.5

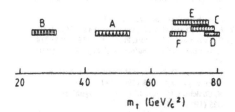

Fig. 9. The distribution of the transverse mass derived from the measured electron and neutrino vectors of the six electron events

clusive electron spectrum and using full QCD smearing gives $m_W = (74^{+4}_{-4})$ GeV/c^2. The method finally used is the one of correcting, on an event-to-event basis, for the transverse W motion from the $(E_\nu - E_e)$ imbalance, and using the Drell–Yan predictions with no smearing. The result of a fit on electron angle and energy and neutrino transverse energy with allowance for systematic errors, is

$$m_W = (81^{+5}_{-5}) \text{ GeV}/c^2 \, ,$$

in excellent agreement with the expectation of the Weinberg–Salam model [2].

We find that the number of observed events, once detection efficiencies are taken into account, is in

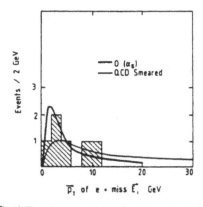

Fig. 10. The transverse momentum distribution of the W derived from our events, using the electron and missing-energy vectors. This is compared with the theoretical predictions of Halzen et al. [8] for W production without $|O(\alpha_s)|$ and with QCD smearing.

agreement with the cross-section estimates based on structure functions, scaling violations, and the Weinberg–Salam parameters for the W particle [5].

We gratefully acknowledge J.B. Adams and L. Van Hove, CERN Directors-General during the initial phase of the project and without whose enthusiasm and support our work would have been impossible. The success of the collider run depended critically upon the superlative performance of the whole of the CERN accelerator complex, which was magnificently operated by its staff.

We are thankful to the management and staff of CERN and of all participating Institutes who have vigorously supported the experiment.

The following funding Agencies have contributed to this programme:

Fonds zur Förderung der Wissenschaftlichen Forschung, Austria.

Valtion luonnontieteellinen toimikunta, Finland.

Institut National de Physique Nucléaire et de Physique des Particules and Institut de Recherche Fondamentale (CEA), France.

Bundesministerium für Forschung und Technologie, Germany.

Istituto Nazionale di Fisica Nucleare, Italy.

Science and Engineering Research Council, United Kingdom.

Department of Energy, USA.

Thanks are also due to the following people who have worked with the collaboration in the preparation and data collection on the runs described here,

F. Bernasconi, F. Cataneo, A.-M. Cnops, L. Dumps, J.-P. Fournier, A. Micolon, S. Palanque, P. Quéru, P. Skimming, G. Stefanini, M. Steuer, J.C. Thevenin, H. Verweij and R. Wilson.

References

[1] C. Rubbia, P. McIntyre and D. Cline, Proc. Intern. Neutrino Conf. (Aachen, 1976) (Vieweg, Braunschweig, 1977) p. 683;
Study Group, Design study of a proton–antiproton colliding beam facility, CERN/PS/AA 78-3 (1978), reprinted in Proc. Workshop on Producing high-luminosity, high-energy proton–antiproton collisions (Berkeley, 1978) report LBL-7574, UC34, p. 189;
The staff of the CERN proton–antiproton project, Phys. Lett. 107B (1981) 306.

[2] S. Weinberg, Phys. Rev. Lett. 19 (1967) 1264;
A. Salam, Proc. 8th Nobel Symp. (Aspenäsgården, 1968)
(Almqvist and Wiksell, Stockholm, 1968) p. 367.

[3] A. Sirlin, Phys. Rev. D22 (1980) 971;
W.J. Marciano and A. Sirlin, Phys. Rev. D22 (1980) 2695;
C.H. Llewellyn Smith and J.A. Wheater, Phys. Lett. 105B
(1981) 486.

[4] For a review, see: M. Davier, Proc. 21st Intern. Conf. on
High-energy physics (Paris, 1982), J. Phys. (Paris) 43
(1982) C3-471.

[5] F.E. Paige, Proc. Topical Conf. on the Production of new
particles at super-high energies (University of Wisconsin,
Madison, 1979);
L.B. Okun and M.B. Voloshin, Nucl. Phys. B120 (1977)
459;
C. Quigg, Rev. Mod. Phys. 94 (1977) 297;
J. Kogut and J. Shigemitsu, Nucl. Phys. B129 (1977) 461;
R. Horgan and M. Jacob, Proc. CERN School of Physics
(Malente, Fed. Rep. Germany, 1980), CERN 81-04, p. 65;
R.F. Peierls, T. Trueman and L.L. Wang, Phys. Rev. D16
(1977) 1397.

[6] UA1 proposal, A 4π solid-angle detector for the SPS used
as a proton–antiproton collider at a centre-of-mass energy
of 540 GeV, CERN/SPSC 78-06 (1978);

M. Barranco Luque et al., Nucl. Instrum. Methods 176
(1980) 175;
M. Calvetti et al., Nucl. Instrum. Methods 176 (1980)
255;
K. Eggert et al., Nucl. Instrum. Methods 176 (1980) 217,
233;
A. Astbury, Phys. Scr. 23 (1981) 397.

[7] H.M. Kroll and W. Wada, Phys. Rev. 98 (1955) 1355.

[8] F. Halzen and D.M. Scott, Phys. Lett. 78B (1978) 318;
P. Aurenche and F. Lindfors, Nucl. Phys. B185 (1981)
301;
F. Halzen, A.D. Martin, D.M. Scott and M. Dechants-
reiter, Phys. Lett. 106B (1981) 147;
M. Chaichian, M. Hayashi and K. Yamagishi, Phys. Rev.
D25 (1982) 130;
A. Martin, Proc. Conf. on Antiproton–proton collider
physics (Madison, 1981) AIP Proc. No. 85 (American
Institute of Physics, New York, 1982) p. 216;
F. Halzen, A.D. Martin and D.M. Scott, Phys. Rev. D25
(1982) 754;
V. Barger and R.J.N. Phillips, University of Wisconsin
preprint MAD/PH/78 (1982).

OBSERVATION OF SINGLE ISOLATED ELECTRONS OF HIGH TRANSVERSE MOMENTUM
IN EVENTS WITH MISSING TRANSVERSE ENERGY AT THE CERN p̄p COLLIDER

The UA2 Collaboration

M. BANNER [f], R. BATTISTON [1,2], Ph. BLOCH [f], F. BONAUDI [b], K. BORER [a], M. BORGHINI [b],
J.-C. CHOLLET [d], A.G. CLARK [b], C. CONTA [e], P. DARRIULAT [b], L. Di LELLA [b], J. DINES-HANSEN [c],
P.-A. DORSAZ [b], L. FAYARD [d], M. FRATERNALI [e], D. FROIDEVAUX [b], J.-M. GAILLARD [d],
O. GILDEMEISTER [b], V.G. GOGGI [e], H. GROTE [b], B. HAHN [a], H. HÄNNI [a], J.R. HANSEN [b],
P. HANSEN [c], T. HIMEL [b], V. HUNGERBÜHLER [b], P. JENNI [b], O. KOFOED-HANSEN [c],
E. LANÇON [f], M. LIVAN [b,e], S. LOUCATOS [f], B. MADSEN [c], P. MANI [a], B. MANSOULIÉ [f],
G.C. MANTOVANI [1], L. MAPELLI [b], B. MERKEL [d], M. MERMIKIDES [b], R. MØLLERUD [c],
B. NILSSON [c], C. ONIONS [b], G. PARROUR [b,d], F. PASTORE [b,e], H. PLOTHOW-BESCH [b,d],
M. POLVEREL [f], J.-P. REPELLIN [d], A. ROTHENBERG [b], A. ROUSSARIE [f], G. SAUVAGE [d],
J. SCHACHER [a], J.L. SIEGRIST [b], H.M. STEINER [b,3], G. STIMPFL [b], F. STOCKER [a], J. TEIGER [f],
V. VERCESI [e], A. WEIDBERG [b], H. ZACCONE [f] and W. ZELLER [a]

[a] Laboratorium für Hochenergie physik, Universität Bern, Sidlerstrasse 5, Bern, Switzerland
[b] CERN, 1211 Geneva 23, Switzerland
[c] Niels Bohr Institute, Blegdamsvej 17, Copenhagen, Denmark
[d] Laboratoire de l'Accélérateur Linéaire, Université de Paris-Sud, Orsay, France
[e] Dipartimento di Fisica Nucleare e Teorica, Università di Pavia and INFN, Sezione di Pavia,
    Via Bassi 6, Pavia, Italy
[f] Centre d'Etudes nucléaires de Saclay, France

Received 15 February 1983

We report the results of a search for single isolated electrons of high transverse momentum at the CERN p̄p collider. Above 15 GeV/c, four events are found having large missing transverse energy along a direction opposite in azimuth to that of the high-p_T electron. Both the configuration of the events and their number are consistent with the expectations from the process $\bar{p} + p \rightarrow W^\pm + $ anything, with $W \rightarrow e + \nu$, where $W^\pm$ is the charged Intermediate Vector Boson postulated by the unified electroweak theory.

1. Introduction. The very successful operation of the CERN p̄p Collider at the end of 1982, with peak luminosities of $\sim 5 \times 10^{28}$ cm^{-2} s^{-1}, has allowed the UA2 experiment to collect data corresponding to a total integrated luminosity of ~ 20 nb^{-1}. According to current expectations [1], these data should contain

[1] Gruppo INFN del Dipartimento di Fisica dell'Università di Perugia, Italy.
[2] Also at Scuola Normale Superiore, Pisa, Italy.
[3] On leave from Department of Physics, University of California, Berkeley, CA, USA.

approximately four events of the type

$$\bar{p} + p \rightarrow W^\pm + \text{anything}$$
$$\quad\quad\quad \hookrightarrow e^\pm + \nu(\bar{\nu}), \qquad (1)$$

where $W^\pm$ is the charged Intermediate Vector Boson (IVB) which mediates the weak interaction between charged currents [2]. In fact it was the search for such particles, and for the neutral IVB, the Z^0, that motivated the transformation of the CERN Super Proton Synchrotron (SPS) into a p̄p collider operating at a

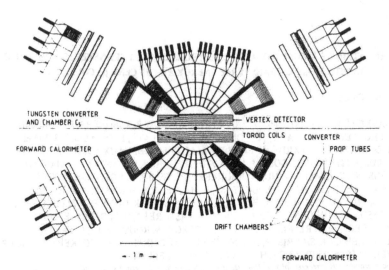

Fig. 1. The UA2 detector: Schematic cross section in the vertical plane containing the beam.

total centre-of-mass energy $\sqrt{s}$ = 540 GeV [3].

We report here the results of a search for single electrons of high transverse momentum (p_T), which are expected to originate from reaction (1). Because the neutrino from W decay is not detected, the events from reaction (1) are expected to show a large missing transverse energy (of the order of the electron p_T) along a direction opposite in azimuth to that of the electron. Results on this subject have been reported earlier [4,5].

The experimental apparatus is shown in fig. 1. At the centre of the apparatus a system of cylindrical chambers (the vertex detector) measures charged particle trajectories in a region without magnetic field. The vertex detector consists of: (a) two drift chambers with measurement of the charge division on a total of 12 wires per track; (b) four multi-wire proportional chambers having cathode strips with pulse height read-out, at ±45° with respect to the wires. These chambers are used to determine the position of the event vertex along the beam line with a precision of ±1 mm.

The vertex detector is surrounded by a highly segmented electromagnetic and hadronic calorimeter (the central calorimeter), which covers the polar angle in-

terval 40° < θ < 140° and the azimuthal range 30° < ϕ ≤ 330°. In the present experiment the remaining azimuthal interval (±30° around the horizontal plane) is covered by a large angle magnetic spectrometer (LAMS), which includes a lead-glass wall, to measure charged and neutral particle production [6] over $\Delta\phi$ = 30°.

The forward regions (20° < θ < 37.5° and 142.5° < θ < 160°, respectively), are each equipped with twelve toroidal magnet sectors, with an average field integral of 0.38 T m, followed successively by: (a) drift chambers; (b) multitube proportional chambers immediately behind a converter to detect showers; (c) electromagnetic calorimeters.

More details on the apparatus will be discussed whenever they are relevant to the present investigation.

2. Electron search in the central calorimeter. The central calorimeter is segmented into 200 cells, each covering 15° in ϕ and 10° in θ and built in a tower structure pointing to the centre of the interaction region. The cells are segmented longitudinally into a 17 radiation length thick electromagnetic compartment (lead-scintillator) followed by two hadronic compartments (iron-scintillator) of two absorption lengths each. The light from each compartment is collected by two BBQ-doped light guides on opposite sides of the cell.

All calorimeters have been calibrated in a 10 GeV beam from the CERN PS, using incident electrons and muons. The stability of the calibration has since been monitored using a light flasher system and a ^{60}Co source. The systematic uncertainty in the energy calibration for the data discussed here is less than ±2% for the electromagnetic calorimeter.

The response of the calorimeter to electrons, and single and multi-hadrons, has been measured at the CERN PS and SPS machines using beams from 1 to 70 GeV/c. In particular, both longitudinal and transverse shower development have been studied as well as the effect of particles impinging near the cell boundaries. The energy resolution for electrons is measured to be $\sigma_E/E = 0.14/\sqrt{E}$ (*E* in GeV). Details of the construction and performance of the calorimeter are reported elsewhere [7].

In the angular region covered by the central calorimeter a cylindrical tungsten converter, 1.5 radiation lengths thick, followed by a cylindrical proportional chamber, is located just after the vertex detector. This chamber, named C_5 (see fig. 1) has cathode strips at ±45° to the wires. We measure the pulse heights on the cathode strips and the charge division on the wires.

In order to implement a trigger sensitive to high-p_T electrons, the gains of the photomultipliers were adjusted so that their signals were proportional to the transverse energy. The eight signals from any 2 × 2 matrix of adjacent electromagnetic cells were then linearly added and their sum was required to exceed a given threshold (typically set at 8 GeV). In order to suppress background from sources other than $\bar{p}p$ collisions, we required a coincidence with two additional signals obtained from scintillator arrays surrounding the vacuum chamber downstream of the interaction point. These arrays were used in an experiment to measure the $\bar{p}p$ total cross section [8]: they gave a coincidence signal in more than 98% of the non-diffractive $\bar{p}p$ collisions.

The full data sample recorded using the W trigger corresponds to a total integrated luminosity of 19.0 nb^{-1}. A first event selection is made by searching for clusters of energy with the configuration expected from isolated electrons. A cluster is obtained by joining all

cells of the electromagnetic calorimeter which share a common side and contain at least 0.5 GeV, and the cluster energy is defined as the sum of all energies contained in the three compartments of both the cluster cells and the surrounding ones. The following requirements are then applied: (i) the cluster must be contained within a 2 × 2 cell matrix; (ii) no more than 10% of the cluster energy must be contained in the electromagnetic compartment of the surrounding cells; (iii) no more than 10% of the cluster energy must be contained in the hadronic compartments; (iv) the cluster centroid must not fall in an edge cell. This last requirement reduces the solid angle acceptance by 25%. Tests with electron beams from the SPS have shown that more than 95% of all isolated electrons in the energy range 10–80 GeV pass the cuts (i) to (iii).

For each cluster satisfying all four requirements a transverse energy E_T' is calculated using the position of the cluster centroid and assuming that the event vertex was in the centre of the apparatus. The events with $E_T' > 15$ GeV (363 events) are fully reconstructed and the exact location of the event vertex (as measured in the vertex detector) is used to obtain the correct value of the transverse energy E_T. The E_T distribution for these events is shown in fig. 2a. The fall-off for $E_T < 17$ GeV results from the selection requirement $E_T' > 15$ GeV applied without knowledge of the exact event vertex. For $E_T > 17$ GeV the distribution of fig. 2a has no threshold bias. There are 7 events with $E_T > 30$ GeV, the highest E_T value being 40.3 GeV.

This sample is further reduced by requiring that one, and only one charged particle track reconstructed in the vertex detector points to the energy cluster. To take the calorimeter cell size into account we define the quantity

$$\Delta^2 = (\delta\theta/10°)^2 + (\delta\phi/15°)^2, \qquad (2)$$

where $\delta\theta(\delta\phi)$ is the polar (azimuthal) separation between a track and the line joining the event vertex to the cluster centroid. Studies with electron beams from the SPS have shown that $\langle\Delta^2\rangle$ is ~0.13 for high-energy electrons, with 95% having $\Delta^2 < 0.4$. We require that the track with the smallest Δ^2 value has $\Delta^2 < 1$, together with the further condition that no other charged particle track be present in a cone of 10° half-aperture around it. A total of 96 events survive these cuts, with the E_T distribution shown in fig. 2b.

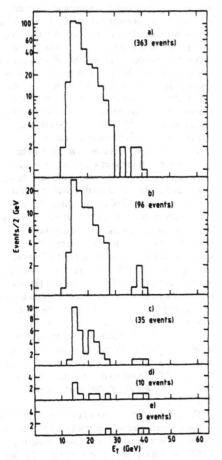

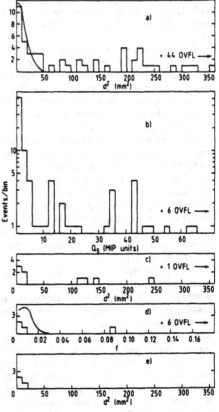

Fig. 2. Transverse energy distribution of the event samples in the central calorimeter: (a) After requirements on energy cluster. (b) After association with a track. (c) After association of the track with a shower in the tungsten converter. (d) After requiring only one such shower. (e) After further cuts on the quality of the track-energy cluster matching. This is the final electron sample.

Fig. 3. (a) Distribution of d^2, the distance between the track and the shower in the tungsten converter. The smooth curve represents the distribution measured using high-energy electrons from the SPS (arbitrary normalisation). (b) Distribution of the charge Q_5 (MIP units), observed in chamber C_5 after the tungsten converter. (c) Distribution of d^2 for the sample of 10 events satisfying the isolation criteria. (d) Distribution of f, the quality parameter for the energy cluster shape, as defined in eq. (3). The curve represents the distribution measured using high-energy electrons from the SPS (arbitrary normalisation). (e) Distribution of d^2 for the three electron candidates.

We then require that the track produces a shower in the tungsten converter, with an associated charge cluster in chamber C_5, as expected in the case of electrons. Fig. 3a shows the distribution of d^2, the square of the distance between the track and the closest charge cluster centroid in C_5. Clusters associated with a track

appear as a clear peak superimposed on a flat background.

This background has two origins: (a) chamber C_5 is only ~60% efficient for minimum ionising particles

(MIPs) and whenever the charge cluster generated by a MIP is lost, a cluster not associated with the track is used to calculate d^2; (b) the track-associated cluster may be merged with a nearby cluster (e.g. from photon conversion) and the resulting cluster may have its centroid displaced with respect to the track.

Fig. 3b shows the distribution of Q_5, the charge of the closest cluster with $d^2 < 500$ mm^2, measured in units of the most probable charge generated by a MIP. We have studied the response of chamber C_5 to electrons and pions by exposing a similar device, including a converter and a calorimeter, to a beam from the CERN SPS. In the energy range from ~20 to ~60 GeV, electron clusters in such a chamber were found to have a broad charge distribution, peaked around $Q_5 \simeq 20$, with more than 90% of the electrons satisfying the condition $Q_5 > 4$. Furthermore, the d^2 distribution of these electrons has $\sigma = 14$ mm^2, with 97% having $d^2 < 50$ mm^2 (see also fig. 3a).

The condition $Q_5 > 4$, when applied to our sample, reduces the number of events from 96 to 35. The E_T-distribution for these events is shown in fig. 2c.

We note that, although the requirement $d^2 < 50$ mm^2 is adequate to select charge clusters associated with electrons, we used the looser cut $d^2 < 500$ mm^2 in order to be able to estimate the background contamination in the final event sample.

In order to further ensure the condition that the electron candidate is isolated, we require that no other charge clusters are present in C_5 in a cone of 10° half-aperture around the track. Such clusters could result from the conversion of high-energy photons accompanying the charged particle in the tungsten. We estimate that the loss of events due to radiative corrections [9] is less than 6%.

Ten events satisfy this requirement. The E_T and d^2 distributions for these events are shown in fig. 2d and 3c, respectively.

As a final selection criterion we use the high segmentation of the central calorimeter to check if the shape of the energy cluster is consistent with that expected from an isolated electron impinging along the direction of the observed track. To this purpose we define a quality factor f as follows

$$f = \frac{1}{E} \left(\sum_i (E_i - \widetilde{E}_i)^2 \right)^{1/2}, \tag{3}$$

where E is the measured cluster energy, E_i is the energy measured in cell i, $\widetilde{E}_i$ is the energy predicted for cell i under the assumption that the observed charged particle track is an electron of energy E, and the sum is extended to all cells of the cluster and those for which $\widetilde{E}_i$ is different from zero. The energies $\widetilde{E}_i$ are obtained from the energy distributions measured in the calorimeter using electron beams from the CERN SPS.

Fig. 3d shows the f distribution for the sample of 10 events. Three events appear to satisfy the condition $f < 0.05$. Such a condition is satisfied by more than 95% of single, isolated electrons, as verified with electron beams from the SPS (see also fig. 3d).

As a further check that the events with $f < 0.05$ are indeed electrons, we estimate the lateral position of the electromagnetic shower in the calorimeter cell from the pulse height ratio of the two photomultipliers and we compare it to the track impact point. We find good agreement (within ±2.5 mm, as expected) for the three events, whereas for the remaining seven events there is a difference of more than 5 mm between these measurements.

The d^2 distribution for the three surviving events is shown in fig. 3e. These events all have $d^2 < 50$ mm^2, as expected for charge clusters in C_5 which are associated with a track. The E_T distribution for these events, which represent our final sample of electron candidates, is shown in fig. 2e.

As a check of this analysis all of the initial sample of 363 events was carefully scanned by physicists using a high resolution graphics terminal. The same three events were found.

There are three main sources of background contamination:

(a) Single, isolated high-p_T π^0- or η-mesons undergoing Dalitz decay, or high-p_T photons (both single and from $\pi^0(\eta) \rightarrow \gamma\gamma$ decay) converting in the vacuum chamber wall. Conversions in the detector material were excluded by requiring that the track coordinate were found in at least one of the two innermost chambers of the vertex detector. An upper limit to the number of single π^0- or η-mesons or single photons is obtained from the original sample of 363 events (fig. 2a) by selecting those events with no track pointing to the energy cluster and with at most one charge cluster in chamber C_5 in front of the energy cluster. There are 5 such events with $E_T > 25$ GeV, of which 1 satisfies the requirement $f < 0.05$ [see eq. (3)], as expected for high-energy isolated π^0- or η-mesons or single photons. Taking into account the probability for photon conversion in the vacuum chamber wall (1.3%) and the Dalitz decay branching ratio, the expected number of background events from this source is <0.04 for $E_T > 25$ GeV.

(b) Single high-p_T charged hadrons interacting in the tungsten converter and depositing a large fraction of their energy (>90%) in the electromagnetic calorimeter. Using high-energy pion beams between 30 and 60 GeV from the CERN SPS we have found that only <1/400 of the pions satisfy these requirements. Assuming that the ratio of charged particles to π^0-mesons at high p_T is 3, we estimate that the expected number of background events from this source is $3/400 \simeq 0.01$ events for $E_T > 25$ GeV.

(c) "Overlap" events, consisting of either at least one high-p_T photon accompanied by a charged particle or of several high-p_T photons of which one converts. In these events the charge cluster observed in chamber C_5 is not correlated with the track, giving rise to a flat background in both d^2 and f variables (see figs. 3c and 3d). Since these variables are correlated, we estimate the overlap background by extrapolating the distribution of the 10 events of figs. 3c and 3d in the d^2, f plane to the region where electrons are expected ($d^2 < 50$ mm^2, $f < 0.05$). This method provides an estimate of 0.6 background events with $E_T > 15$ GeV. Taking into account the E_T distribution of the original event sample (fig. 2a), we estimate this background to be approximately 0.1 event for $E_T > 25$ GeV.

In conclusion, the total background contribution to the three electron candidates amounts to less than 0.2 events for $E_T > 25$ GeV. Furthermore, background sources would have an E_T-distribution similar to that of fig. 2a, whereas the distribution of the three electron candidates is inconsistent with it. This is an independent indication that the background contribution is indeed small.

3. Electron search in the forward detectors. Each of the two forward detectors (see fig. 1) is divided into twelve identical sectors covering 30° in ϕ and 17.5° in θ. A sector consists of:

– Three drift chambers located after the magnetic field region. Each chamber consists of three planes, with wires at $-7°$, $0°$, $+7°$, respectively, with respect to the magnetic field direction.

– A 1.4 radiation-lengths thick lead–iron converter, followed by two chambers of proportional tubes (MTPC). Each chamber consists of two layers of 20 mm diameter proportional tubes, staggered by a tube radius and equipped with pulse height measurement. There is a 77° angle between the tubes of the two MTPCs, with the tubes of the first one being parallel to the magnetic field direction. This device localises electromagnetic showers with a precision of $\lesssim 10$ mm, as verified with the data themselves.

– An electromagnetic calorimeter consisting of lead-scintillator counters assembled in ten independent cells, each covering 15° in ϕ and 3.5° in θ. Each cell is subdivided into two independent longitudinal sections, 24 and 6 radiation lengths thick, respectively. The light from each section is collected by two BBQ-doped light guides om opposite sides of the cell.

More than 98% of an electron shower is contained in the first section, as verified experimentally up to 80 GeV using electron beams from the CERN SPS. High-energy hadrons, on the contrary, deposit relatively large amounts of energy in the second section which is used, therefore, as a hadron veto. The energy resolution for electrons has been measured to be $\sigma_E/E = 0.15/\sqrt{E}$ (E in GeV). The calibration technique used for these calorimeter modules is similar to that for the central calorimeter.

A trigger sensitive to high-p_T electrons was implemented by requiring that the transverse energy deposited in any 2×2 matrix of adjacent cells in the same sector exceed a threshold (typically set at 8 GeV), in coincidence with the small angle scintillator arrays described in section 2. The full data sample recorded with this trigger corresponds to a total integrated luminosity of 16 nb^{-1}.

The forward calorimeter cell size is much larger than the lateral extension of an electromagnetic shower; consequently an electron is expected to deposit its energy in a cluster consisting of at most two adjacent cells. Furthermore, because of the geometry of the calorimeter, these cells must have the same azimuth. A total of 761 events with a cluster transverse energy E_T above 15 GeV has been selected from the full data sample. The E_T distribution of these events is shown in fig. 4a.

We next require that a charged particle track reconstructed in the drift chamber points to the cluster

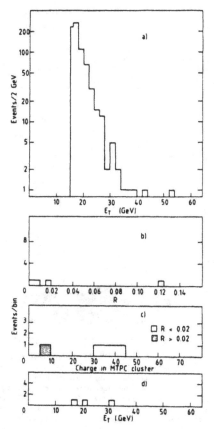

Fig. 4. (a) E_T-distribution of the sample of 761 events selected from energy requirements in the forward calorimeters. (b) Distribution of R, the ratio between the energy deposited in the hadron veto section of the calorimeter and the cluster energy. (c) Distribution of the charge of the track-associated cluster measured in the MTPCs after the lead–iron converter (MIP units). (d) Transverse energy distribution for the final sample of electron candidates.

centroid. The centroid position along an axis parallel to the magnetic field direction in that sector (the x-axis) is obtained with an accuracy $\sigma = 25$ mm from the pulse height ratio of the two photomultipliers in the cell. We require, therefore, that the x-coordinate

of the track impact point on the calorimeter be within ±70 mm of the cluster centroid.

Because the magnetic field deflects charged particles in the plane defined by the particle direction and the beam line, the track projection in a plane perpendicular to the beam line is required to extrapolate close to the beam axis and to match a track found in the vertex detector within the reconstruction errors. The momentum resolution achieved so far is $\Delta(1/p) = 0.02 \; (\text{GeV}/c)^{-1}$.

We then require that the track matches the position of a shower in the lead–iron converter, as measured by the MTPCs. The distance between the impact point of the track on the MTPCs and the location of the charge cluster in the proportional tubes must not exceed 60 mm in x and 30 mm in the orthogonal coordinate.

The next condition is that the momentum and the shower energy measured in the calorimeter agree within errors. This is done by imposing the condition $|p^{-1} - E^{-1}| < 3 \; \sigma$, where σ is obtained by adding in quadrature the contributions of the momentum and energy resolutions quoted above.

In order to further reduce overlap background and to eliminate electrons originating from photon conversions in the vacuum chamber, we require that only one track points to the energy cluster in the calorimeter, and that no additional MTPC cluster is present in front of it. Background due to photon conversions in the detector material is suppressed by requiring the presence of a hit in at least one of the first two chambers in the vertex detector.

In order to ensure that the electron is isolated, we require that the total energy in the calorimeter cells surrounding the energy cluster does not exceed 3 GeV.

Four events survive all these conditions.

The ratio, R, of the energy deposited in the second section of the calorimeter to that deposited in the first section, is shown in fig. 4b. Three events satisfy the electron criterion, $R < 0.02$, whereas one has $R > 0.1$, as expected for hadrons. As a further check, we plot the charge distributions of the clusters measured in the MTPCs for the two classes of events separately (fig. 4c). The large charges associated with the events having $R < 0.02$ are consistent with the behaviour of electrons.

The transverse energy distribution of the electron candidates is shown in fig. 4d.

Backgrounds in this sample are expected to originate from the same sources as described in section 2. However, in this case the background from conversions of high-E_T photons is smaller because the magnetic field separates unresolved e^+e^- pairs.

Concerning the overlap background, we note that soft charged hadrons accompanying high-E_T photons are rejected by the requirement that the measured particle momentum and the energy deposition agree within errors.

The background contribution from high-p_T charged hadrons can be estimated from fig. 4b by extrapolating the distribution of events with $R > 0.02$ to the electron region, $R < 0.02$. There is one event having $R > 0.02$ which we use to estimate that less than 0.2 background events have $E_T > 15$ GeV.

4. Missing transverse momentum. We now verify another feature of reaction (1); namely, the presence of an undetected neutrino with transverse momentum similar in magnitude to that of the electron but opposite in azimuth.

In order to estimate the missing transverse momentum carried away by the neutrino we reconstruct the total momentum vector from the available calorimetric and spectrometric measurements. To each calorimeter cell, including those in the lead glass wall, we assign a vector p_i with magnitude equal to the energy deposited in the cell and direction along the line joining the event vertex to the cell centre. In the forward detectors, when both momentum and the corresponding energy deposition are measured, we use the larger value in order to avoid double counting. The total momentum vector $P = \Sigma_i p_i$ is then projected onto the direction of the electron transverse momentum vector p_{eT} to define the missing transverse momentum P_T^{miss}

$$P_T^{miss} = (P \cdot p_{eT})/E_T , \qquad (4)$$

where $E_T = |p_{eT}|$, and where P includes the electron momentum.

The ratio P_T^{miss}/E_T is shown in fig. 5a for the electron candidates found both in the central calorimeter and in the forward detectors. The events with $P_T^{miss}/E_T \approx 1$ are consistent with reaction (1).

The E_T distribution of the four events with $P_T^{miss}/E_T > 0.8$ is shown in fig. 5b. These electron candidates are those having the highest E_T values.

The two electron candidates with $P_T^{miss}/E_T < 0.5$

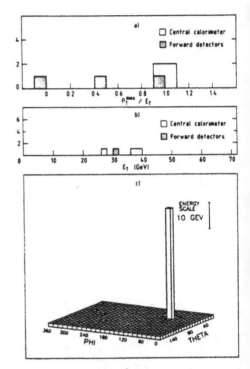

Fig. 5. (a) Distribution of P_T^{miss}/E_T, the ratio between the total missing transverse momentum and the transverse energy of the electron candidate. (b) Transverse energy distribution of the electron candidates observed in events with large missing transverse momentum ($P_T^{miss}/E_T > 0.8$). (c) The cell energy distribution as a function of polar angle θ and azimuth ϕ for the electron candidate with the highest E_T value (event C of table 1).

for which $E_T = 16.3$ and 22.6 GeV, respectively, have both been found in the forward detectors. We note that the isolation criteria used here are looser than those used for the central calorimeter. In particular, we have not applied the condition that no other particle is observed within a $10°$ half-aperture cone around the electron track, because the higher particle density expected in the angular region of the two forward detectors would reduce the detection efficiency for reaction (1).

Table 1
Parameters of the central calorimeter electrons.

Event	E_T (GeV)	$\dfrac{E_{had}}{E}$ [a)]	E_{had2} [a)] (GeV)	$\dfrac{P_T^{miss}}{E_T}$	ΣE_T [b)] (GeV)	θ [c)] (deg.)	ϕ (deg.)	d^2 (mm²)	q_s (MIPs)	f	Δz [d)] (mm)
A	27.7	0.04	0	1.03	13.5	99.7	173.1	12	31	0.003	1.3
B	38.5	0.03	0	0.93	8.5	131.0	248.7	4	95	0.006	-0.7
C	40.3	0.04	0	1.01	4.9	50.1	84.7	6	9	0.002	-2.2

a) Energy deposited in the second hadronic compartment of the cluster cells.
b) The sum extends over all particles excepting the electron.
c) $\theta = 0$ corresponds to the direction of the incident protons.
d) Distance between the track and the predicted impact in the calorimeter, calculated from the pulse height ratio of the two photomultipliers.

Table 2
Parameters of the forward detector electron with missing transverse momentum.

Event	E_T (GeV)	R	E (GeV)	Electron charge sign	$\dfrac{P_T^{miss}}{E_T}$	ΣE_T [a)] (GeV)	θ [b)] (deg.)	ϕ (deg.)	Calorimeter-MTPC match (mm)	Track-MTPC match (mm)	Charge in MTPC (MIPs)
D	30.9	0.004	62.3	+	1.00	15.9	150.3	268.3	68	12	> 38

a) The sum extends over all particles excepting the electron.
b) $\theta = 0$ corresponds to the direction of the incident protons.

In order to verify that the large missing transverse momentum of the four electron candidates is not dominated by the limited coverage of the detectors, we have studied a sample of background events obtained from the raw W-trigger sample. This control sample is dominated by events containing a high-E_T jet, which are not expected to show a large missing transverse momentum [10]. We find that the fraction of such events which satisfies the condition $P_T^{miss}/E_T > 0.8$ is only ~20% for both the central calorimeter and the forward detectors.

A list of relevant parameters for the four events with $P_T^{miss}/E_T > 0.8$ is given in tables 1,2. For each event the total transverse energy, ΣE_T, of all particles excluding the electron is also shown in tables 1,2. It should be compared with the corresponding value of ~8 GeV for minimum bias events.

Fig. 5c shows the cell energy distribution in θ and ϕ for the highest E_T event (event C in table 1). The only

significant energy flow within the detector acceptance is carried by the electron candidate. The other three events have the same spectacular configuration.

5. Conclusions. The application of the various selection criteria used in the analysis to identify single isolated electrons has reduced the original event sample to four events with $E_T > 15$ GeV and large missing transverse momentum. The number of events, accounting approximately for acceptance and detection efficiency, and their configuration are consistent with expectations from reaction (1).

The assumption that these four electrons indeed originate from the decay W → eν, allows us to estimate the W mass for a given W momentum distribution, assuming W decay kinematics and standard V−A coupling. This is done by fitting the E_T distributions predicted for the four measured θ angles to the measured E_T values.

We find that the best fit value of M_W is rather insensitive to the particular choice of the W momentum distribution, provided that the transverse momentum p_{WT} is much less than $M_W c$. We use gaussian distributions for the three components of the W momentum p_W and vary $(\langle p_{WL}^2 \rangle)^{1/2}$ between 50 and 100 GeV/c, and $\langle p_{WT} \rangle$ between 3 and 8 GeV, where p_{WL} is the longitudinal momentum. The ranges of values used for $(\langle p_{WL}^2 \rangle)^{1/2}$ and $\langle p_{WT} \rangle$ reflect the limits of reasonable production models [1].

The results of the fit gives

$$M_W = (80^{+10}_{-6})\,\text{GeV}/c^2 \,,$$

where the quoted error takes into account the effect of varying the W momentum distribution.

This value agrees with the expectation of the electro-weak theory [11].

This experiment would have been impossible without the collective effort of the staffs of the relevant CERN accelerators, whom we gratefully acknowledge.

We deeply thank the technical staffs of the Institutes collaborating in UA2 for their invaluable contributions. We are particularly indebted to C. Bruneton, D. Burckhart, W. Carena, J.-P. Dufey, F. Gagliardi, G. Mornacchi, A. Rimoldi, M Sciré, D. Sendall, A. Silverman, A. Vascotto and V. White for their contributions to the on-line data acquisition system and to the data analysis, to G. Bertalmio, G. Bosc, F. Bourgeois, M. Dialinas, G. Dubail, A. Hrisoho, C. Lamprecht, F. Impellizzeri, G. Reiss, B. Rossini, P. Wicht and their teams for major contributions to the construction of the detector and to L. Bonnefoy, J.-M. Chapuis, Y. Cholley, G. Dubois-Dauphin, G. Fumagalli, G. Gurrieri, M. Hess, G. Iuvino, M. Lemoine, A. Sigrist, G. Souchère and A.Vicini for their invaluable technical contribution.

We are grateful to the UA4 collaboration for providing the signals from the small-angle scintillator arrays.

Financial supports from the Schweizerischer National-fonds zur Förderung der Wissenschaftlichen Forschung to the Bern group, from the Danish Natural Science Research Council to the Niels Bohr Institute group, from the Institut National de Physique Nucléaire et de Physique des Particules to the Orsay group, from the Istituto Nazionale di Fisica Nucleare to the Pavia group, and from the Institut de Recherche Fondamentale (CEA) to the Saclay group are acknowledged.

References

[1] L.B. Okun' and M.B. Voloshin, Nucl. Phys. B120 (1977) 459;
C. Quigg, Rev. Mod. Phys. 94 (1977) 297;
J. Kogut and J. Shigemitsu, Nucl. Phys. B129 (1977) 461;
R.F. Peierls, T. Trueman and L.L. Wang, Phys. Rev. D16 (1977) 1397;
F.E. Paige, BNL – 27066 (1979);
F. Rapuano, Lett. Nuovo Cimento 26 (1979) 219;
R. Horgan and M. Jacob, CERN 81 –04 (1981) p. 65.

[2] S.L. Glashow, Nucl. Phys. 22 (1961) 579;
S. Weinberg, Phys. Rev. Lett. 19 (1967) 1264;
A. Salam, Proc. 8th Nobel Symp. (Aspenäsgården, 1968) (Almqvist and Wiksell, Stockholm) p. 367.

[3] C. Rubbia, P. Mc Intyre and D. Cline, Proc. Intern. Neutrino Conf. (Aachen, 1976) (Vieweg, Braunschweig, 1977) p. 683;
S. Van der Meer, Stochastic damping of betatron oscillations, CERN/ISR-PO/72–31 (1972).
The Staff of the CERN proton-antiproton project, Phys. Lett. 107B (1981) 306.

[4] M. Banner et al., Preliminary searches for hadron jets and for large transverse momentum electrons at the SPS p̄p Collider, presented Third Topical Workshop on Proton–antiproton collider physics (Rome, January 1983) CERN-EP/83–23 (1983).

[5] G. Arnison et al., Phys. Lett. 122B (1983) 103;
C. Rubbia, Report Third Topical Workshop on Proton-antiproton collider physics (Rome, January 1983).

[6] M. Banner et al., Phys. Lett. 115B (1982) 59;
M. Banner et al., Phys. Lett. 121B (1983) 187;
M. Banner et al., Phys Lett. 122B (1983) 322.

[7] A.G. Clark, Proc. Intern. Conf. on Instrumentation for colliding beam physics, SLAC-250 (1982);
UA2 Collab., Status and first results from the UA2 experiment, presented 2nd Intern. Conf. on Physics in collisions (Stockholm, June 1982), to be published in Phys. Scr.

[8] R. Battiston et al., Phys. Lett. 117B (1982) 126.

[9] E. Calva-Tellez et al., Lett. Nuovo Cimento 4 (1972) 619.

[10] M. Banner et al., Phys. Lett. 118B (1982) 203.

[11] For a review see: J. Ellis et al., Ann. Rev. Nucl. Part. Sci. 32(1982) 443.

EXPERIMENTAL OBSERVATION OF LEPTON PAIRS OF INVARIANT MASS AROUND 95 GeV/c^2 AT THE CERN SPS COLLIDER

UA1 Collaboration, CERN, Geneva, Switzerland

G. ARNISON [j], A. ASTBURY [j], B. AUBERT [b], C. BACCI [i], G. BAUER [l], A. BÉZAGUET [d],
R. BÖCK [d], T.J.V. BOWCOCK [f], M. CALVETTI [d], P. CATZ [b], P. CENNINI [d], S. CENTRO [d],
F. CERADINI [d,i], S. CITTOLIN [d], D. CLINE [l], C. COCHET [k], J. COLAS [b], M. CORDEN [c],
D. DALLMAN [d,l], D. DAU [2], M. DeBEER [k], M. DELLA NEGRA [b,d], M. DEMOULIN [d],
D. DENEGRI [k], A. Di CIACCIO [i], D. DiBITONTO [d], L. DOBRZYNSKI [g], J.D. DOWELL [c],
K. EGGERT [a], E. EISENHANDLER [f], N. ELLIS [d], P. ERHARD [a], H. FAISSNER [a], M. FINCKE [2],
G. FONTAINE [g], R. FREY [h], R. FRÜHWIRTH [l], J. GARVEY [c], S. GEER [g], C. GHESQUIÈRE [g],
P. GHEZ [b], K. GIBONI [a], W.R. GIBSON [f], Y. GIRAUD-HÉRAUD [g], A. GIVERNAUD [k], A. GONIDEC [b],
G. GRAYER [j], T. HANSL-KOZANECKA [a], W.J. HAYNES [j], L.O. HERTZBERGER [3], C. HODGES [h],
D. HOFFMANN [a], H. HOFFMANN [d], D.J. HOLTHUIZEN [3], R.J. HOMER [c], A. HONMA [f], W. JANK [d],
G. JORAT [d], P.I.P. KALMUS [f], V. KARIMÄKI [e], R. KEELER [f], I. KENYON [c], A. KERNAN [h],
R. KINNUNEN [e], W. KOZANECKI [h], D. KRYN [d,g], F. LACAVA [i], J.-P. LAUGIER [k], J.-P. LEES [b],
H. LEHMANN [a], R. LEUCHS [a], A. LÉVÊQUE [k,d], D. LINGLIN [b], E. LOCCI [k], J.-J. MALOSSE [k],
T. MARKIEWICZ [d], G. MAURIN [d], T. McMAHON [c], J.-P. MENDIBURU [g], M.-N. MINARD [b],
M. MOHAMMADI [l], M. MORICCA [i], K. MORGAN [h], H. MUIRHEAD [4], F. MULLER [d], A.K. NANDI [j],
L. NAUMANN [d], A. NORTON [d], A. ORKIN-LECOURTOIS [g], L. PAOLUZI [i], F. PAUSS [d],
G. PIANO MORTARI [i], E. PIETARINEN [e], M. PIMIÄ [e], A. PLACCI [d], J.P. PORTE [d],
E. RADERMACHER [a], J. RANSDELL [h], H. REITHLER [a], J.-P. REVOL [d], J. RICH [k],
M. RIJSSENBEEK [d], C. ROBERTS [j], J. ROHLF [d], P. ROSSI [d], C. RUBBIA [d], B. SADOULET [d],
G. SAJOT [g], G. SALVI [f], G. SALVINI [i], J. SASS [k], J. SAUDRAIX [k], A. SAVOY-NAVARRO [k],
D. SCHINZEL [d], W. SCOTT [j], T.P. SHAH [j], M. SPIRO [k], J. STRAUSS [l], J. STREETS [c],
K. SUMOROK [d], F. SZONCSO [l], D. SMITH [h], C. TAO [3], G. THOMPSON [f], J. TIMMER [d],
E. TSCHESLOG [a], J. TUOMINIEMI [e], B. Van EIJK [3], J.-P. VIALLE [d], J. VRANA [g],
V. VUILLEMIN [d], H.D. WAHL [l], P. WATKINS [c], J. WILSON [c], C. WULZ [l], G.Y. XIE [d],
M. YVERT [b] and E. ZURFLUH [d]

Aachen [a] *–Annecy (LAPP)* [b] *–Birmingham* [c] *–CERN* [d] *–Helsinki* [e] *–Queen Mary College, London* [f] *–*
Paris (Coll. de France) [g] *–Riverside* [h] *–Rome* [i] *–Rutherford Appleton Lab.* [j] *–Saclay (CEN)* [k] *–Vienna* [l] *Collaboration*

Received 6 June 1983

We report the observation of four electron–positron pairs and one muon pair which have the signature of a two-body decay of a particle of mass ~95 GeV/c^2. These events fit well the hypothesis that they are produced by the process $\bar{p} + p \rightarrow Z^0 + X$ (with $Z^0 \rightarrow \ell^+ + \ell^-$), where Z^0 is the Intermediate Vector Boson postulated by the electroweak theories as the mediator of weak neutral currents.

[1] University of Wisconsin, Madison, WI, USA.
[2] University of Kiel, Fed. Rep. Germany.
[3] NIKHEF, Amsterdam, The Netherlands.
[4] Visitor from the University of Liverpool, England.

1. Introduction. We have recently reported the observation of large invariant mass electron–neutrino pairs [1] [*1] produced in high-energy collisions at the CERN Super Proton Synchrotron (SPS) [3]. The most likely interpretation of these events is that they are the leptonic decays of charged intermediate vector bosons W^+ and W^- mediating ordinary weak interactions.

We have now extended our search to their neutral partner Z^0, responsible for neutral currents. As in our previous work, production of intermediate vector bosons is achieved with proton–antiproton collisions at $\sqrt{s}$ = 540 GeV in the UA1 detector [4], except that now we search for electron and muon pairs rather than for electron–neutrino coincidences. The process is then:

$$\bar{p} + p \to Z^0 + X$$
$$\hookrightarrow e^+ + e^- \quad \text{or} \quad \mu^+ + \mu^- . \qquad (1)$$

The paper is based on an early analysis of a sample of collisions with an integrated luminosity of 55 nb^{-1}. In this event sample, 27 $W^\pm \to e^\pm \nu$ events have been recorded [5] [*2]. According to minimal SU(2) × U(1), the Z^0 mass is predicted to be [6] [*3] m_{Z^0} = 94 ± 2.5 GeV/c^2. The reaction (1) is then approximately a factor of 10 less frequent than the corresponding $W^\pm$ leptonic decay channels [9] [*4]. A few events of type (1) are therefore expected in our muon or electron samples. Evidence for the existence of the Z^0 in the range of masses accessible to the UA1 experiment can also be drawn from weak-electromagnetic interference experiments at the highest PETRA energies, where deviations from point-like expectations have been reported [*5].

[*1] See also the corresponding result by the UA2 Collaboration [2].

[*2] Since the run is at present continuing at the CERN SPS Collider, the paper is likely to contain an event sample significantly larger than what is reported here. The result for the mass of the W is m_W = (81 ± 2) GeV/c^2.

[*3] For latest parameters, see ref. [7]. The values used in this paper come from ref. [8]. The two-parameter fit to ν_μ and $\bar{\nu}_\mu$ data yields ρ = 1.02 ± 0.026 and $\sin^2\theta_W$ (m_W) = 0.236 ± 0.030 at the W mass. The values for the W and Z masses are then m_W = $(79.3^{+5.5}_{-4.7})$ GeV/c^2 and m_Z = (89.9 ± 4.4) GeV/c^2. Imposing ρ = 0.99 (theoretical) gives m_W = $(83.0^{+3.0}_{-2.8})$ GeV/c^2 and m_Z = $(93.8^{+2.5}_{-2.4})$ GeV/c.

[*4] All cross sections are calculated in the leading log approximation, assuming SU(2) × U(1).

[*5] For a summary of the results of the CELLO, JADE, MARK J and TASSO Collaborations, see ref. [10]. The result quoted is m_Z = (76^{+21}_{-11}) GeV/c^2, or a 2 SD effect from m_Z = ∞.

This paper deals with four $e^+ e^-$ pairs and one $\mu^+\mu^-$ pair, consistent with a common value of invariant mass and with the general expectations for lepton pairs from Z^0 decay.

2. Detector. The UA1 apparatus has already been described [1]. We limit our discussion to those components which are relevant to the identification and measurement of muons and electrons.

The momenta of charged tracks are determined by deflection in the central dipole magnet generating a field of 0.7 T over a volume of 7 × 3.5 × 3.5 m^3. Tracks are recorded by the central detector (CD) [11], a cylindrical volume of drift chambers 5.8 m in length and 2.3 m in diameter, surrounding the beam crossing region. Accuracy for high-momentum tracks is dominated by the localization error of the electrons drifting in the gas, it is about 100 μm close to the anode wires and 350 μm after 22 cm, the longest drift path. At this stage we find no evidence of significant additional systematic errors, even for the highest momentum tracks [*6]. Ionization can be measured to an accuracy of about ±10% for a 1 m long track. This allows identification of narrow, high-energy particle bundles (e^+e^- pairs), even if they cannot be resolved by the digitizings.

The large-angle section of electromagnetic and hadronic calorimetry [1] extends to angles of about 5° with respect to the beam pipe, and it consists of lead/scintillator stacks followed by the instrumented iron of the magnet yoke used as a hadron calorimeter. Additional calorimetry [1], both electromagnetic and hadronic, extends to forward regions, down to 0.2°. Electrons of the present sample have been recorded by the central section of the e.m. calorimetry, consisting of 48 semicylindrical lead/scintillator modules, with an inner radius of 1.36 m, arranged in two cylindrical

---

[*6] A large number of cosmic-ray muons traversing both the upper and the lower elements of the central detector were studied, comparing momenta determinations. Events were recorded continuously during the run and within the beam crossing gate (50 ns). Additional data were collected with beams off. This was done in order to ensure the absence of positive ion distortions in the tracks. Apart from the overall timing, which is fitted from the track as long as there is at least one drift volume crossing, all other calibration constants are identical to those used for actual events. No evidence for systematic effects, beyond statistical contributions from individual drift distance measurements, are needed to account for momentum measurements up to p > 50 GeV/c.

half-shells, one on either side of the beam axis. Each module (gondola) extends over approximately 180° in azimuth and measures 22.5 cm in the beam direction. Light produced in each of the four separate segmentations in depth (3.3/6.6/9.9/6.6 X_0) is seen by wavelength shifter plates on each side of the stack, which are in turn connected to four photomultipliers (PMs), two at the top and two at the bottom. Light attenuation is exploited in order to further improve the calorimeter information for localized energy depositions: top and bottom PMs give the azimuthal angle ϕ with error $\Delta\phi$ (rad) = $0.3/[E\,(\text{GeV})]^{1/2}$. Likewise, localization of the coordinate x (along the beam direction) is determined, using the appropriate PMs pairings, to an accuracy Δx (cm) = $6.3/[E\,(\text{GeV})]^{1/2}$ for normal incident tracks. Very inclined tracks have a substantially worse localization. Energy resolution, using all four segments and PMs, is $\Delta E/E = 0.15/[E\,(\text{GeV})]^{1/2}$. The properties of these detectors were extensively investigated in test beams. Energy calibrations are performed periodically with a strong, collimated ^{60}Co source and detailed scans over the whole detector surface.

High-energy particles are identified by their behaviour as they traverse the calorimeters. Isolated electrons are identified by their characteristic transition curve, and in particular by the lack of penetration in the hadron calorimeter behind them. The performance of the detectors with respect to hadrons and electrons has been studied extensively in a test beam as a function of the energy, the angle of incidence, and the position of impact. The fraction of hadrons (pions) delivering an energy deposition E_c below a given threshold in the hadron calorimeter is a rapidly falling function of energy, amounting to about 0.3% for $p \simeq 40$ GeV/c and $E_c < 200$ MeV. Under these conditions, 98% of electrons are detected.

Isolated muons traverse the calorimeters and the added absorbers without deviations beyond those of multiple scattering and without significant energy deposition in excess of ionization losses in the four e.m. calorimeter and two hadron calorimeter segments. In order to detect muon tracks, 50 large drift chambers [12], nearly 4 m $\times$ 6 m in size, surround the whole detector, covering a very large area of ~500 m^2. Each chamber consists of two orthogonal layers of drift tubes with two planes per projection. We have chosen a staggered arrangement for adjacent planes, thus resolving the left–right ambiguity and at the same time compensating for the inefficiency from the dead spaces between tubes. The extruded aluminium drift tubes have a cross section of 45 mm $\times$ 150 mm, leading to a maximum drift length of 70 mm. An average spatial resolution of 300 μm has been achieved throughout the sensitive volume of the tubes. In order to reach a good angular resolution of the muon tracks, a second set of four planes is placed 60 cm away from the first one. This long lever arm was chosen in order to reach an angular resolution of few milliradians, comparable to the multiple scattering angle of high-energy muons, typically 3 mrad at 40 GeV/c.

An independent momentum determination of muons traversing the magnetized iron yoke is performed using the known position of the interaction vertex and the track coordinates after the iron. With an appropriate algorithm which reduces effects of multiple scattering [*7], excellent agreement is found between momentum measurements in the central detector and the magnetized iron for vertical cosmic-ray muons in the momentum range 10–50 GeV/c. The relative precision of the momentum measurement within the iron is found to be $\Delta p/p \simeq 0.20$, in agreement with expectations, mainly due to multiple scattering. Cosmic-ray muons were also used to verify the relative alignments of the central and muon detectors.

The calorimeters have been made completely hermetic down to angles of 0.2° with respect to the direction of the beams. About 97% of the mass of the magnet is calorimetrized. Adding the energy depositions vectorially over the whole solid angle [1], and adding muons, under ideal conditions and with no neutrino emission, one should observe $\Delta E = 0$. In practice [1] transverse energy components exhibit small gaussian residuals centred on zero with rms deviations well described by the formula $\Delta E_{y,z} = 0.4\,(\Sigma_i |E_T^i|)^{1/2}$, where all units are in GeV. The longitudinal component ΔE_x is of little use, since it is strongly affected by energy flow escaping undetected through the beam pipes.

[*7] Correlations between multiple scatterings have been removed using the known relationship between angle and displacement: $\langle x_{\text{ms}}^2\rangle = (k')^2 \langle \phi_{\text{ms}}^2\rangle$ and $\langle x\phi_{\text{ms}}\rangle = k\langle\phi_{\text{ms}}^2\rangle$. From known properties of materials, $k' = 2.40$, $k = 1.60$ (units: cm, rad). Cosmic-ray data give $k' = 3.0$ for the azimuth and 3.9 for the dip angle.

3. Event selection and data analysis. The present work is based on a four-week period of data-taking during the months of April and May 1983. The integrated luminosity after subtraction of dead-time and other instrumental inefficiencies was 55 nb^{-1}. As in our previous work [1], four types of trigger were operated simultaneously:

(i) An "electron trigger", namely at least 10 GeV of transverse energy deposited in two adjacent elements of the electromagnetic calorimeters covering angles larger than 5° with respect to the beam pipes.

(ii) A "muon trigger", namely at least one penetrating track detected in the muon chambers with pseudorapidity $|\eta| \lesssim 1.3$ and pointing in both projections to the interaction vertex within a specified cone of aperture ±150 mrad. This is accomplished by a dedicated set of hardware processors filtering the patterns of the muon tube hits.

(iii) A "jet trigger", namely at least 20 GeV of transverse energy in a localized calorimeter cluster [*8].

(iv) A global "E_T trigger", with >50 GeV of total transverse energy from all calorimeters with $|\eta| < 1.4$.

Events for the present paper were further selected by the so-called "express line", consisting of a set of four 168E computers [13] operated independently in real time during the data-taking. A subsample of events with $E_T \geq 12$ GeV in the electromagnetic calorimeters and dimuons are selected and written on a dedicated magnetic tape. These events have been fully processed off-line and further subdivided into four main classes: (i) single, isolated electromagnetic clusters with $E_T > 15$ GeV and missing energy events with $E_{miss} > 15$ GeV, in order to extract $W^{\pm} \rightarrow e^{\pm}\nu$ events [1,5]; (ii) two or more isolated electromagnetic clusters with $E_T > 25$ GeV for $Z^0 \rightarrow e^+e^-$ candidates; (iii) muon pair selection to find $Z^0 \rightarrow \mu^+\mu^-$ events; and (iv) events with a track reconstructed in the central detector, of transverse momentum within one standard deviation, $p_T \gtrsim 25$ GeV/c, in order to evaluate some of the background contributions. We will discuss these different categories in more detail.

4. Events with two isolated electron signatures. An electron signature is defined as a localized energy

_____
[*8] The jet cluster is defined as in ref. [1], namely six electromagnetic cells and two hadronic cells immediately behind. Energy responses of calorimeters for hadrons and electrons are somewhat different.

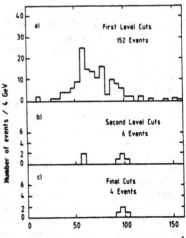

Fig. 1. Invariant mass distribution (uncorrected) of two electromagnetic clusters: (a) with $E_T > 25$ GeV; (b) as above and a track with $p_T > 7$ GeV/c and projected length >40 cm pointing to the cluster. In addition, a small energy deposition in the hadron calorimeters immediately behind (<0.8 GeV) ensures the electron signature. Isolation is required with $\Sigma\,p_T$ < 3 GeV/c for all other tracks pointing to the cluster. (c) The second cluster also has an isolated track.

deposition in two contiguous cells of the electromagnetic detectors with $E_T > 25$ GeV, and a small (or no) energy deposition (<800 MeV) in the hadron calorimeters immediately behind them. The isolation requirement is defined as the absence of charged tracks with momenta adding up to more than 3 GeV/c of transverse momentum and pointing towards the electron cluster cells. The effects of the successive cuts on the invariant electron–electron mass are shown in fig. 1. Four e^+e^- events survive cuts, consistent with a common value of (e^+e^-) invariant mass. They have been carefully studied using the interactive event display facility MEGATEK. One of these events is shown in figs. 2a and 2b. The main parameters of the four events are listed in tables 1 and 3. As one can see from the energy deposition plots (fig. 3), their dominant feature is of two very prominent electromagnetic energy depositions. All events appear to balance the visible total transverse energy components; namely, there is no evidence for the emission of energetic neutrinos. Ex-

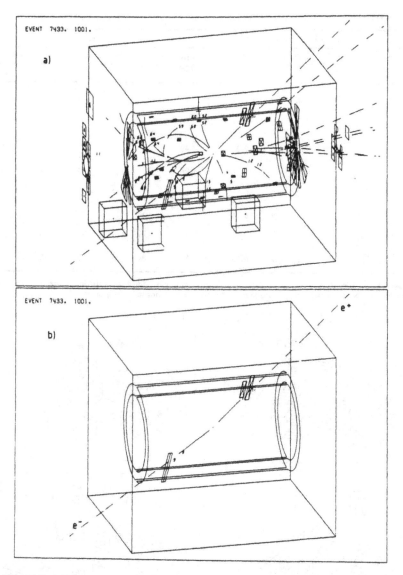

Fig. 2. (a) Event display. All reconstructed vertex associated tracks and all calorimeter hits are displayed. (b) The same, but thresholds are raised to $p_T > 2$ GeV/c for charged tracks and $E_T > 2$ GeV for calorimeter hits. We remark that only the electron pair survives these mild cuts.

Table 1
Properties of the individual electrons of the pair events.

Run, event	Drift chamber measurement						Shower counter measurement					
	p (GeV)	Δp a) (GeV)	Q	dE/dx b)	y c)	ϕ (deg)	E_{tot} (GeV)	Electromagnetic samples (GeV)				Had. energy
								S_1	S_2	S_3	S_4	H
A 7433 1001	33	$+9$ -6	+	1.8 ± 0.3	1.01	144	44	14	27	3	0.0	0.0
	63	$+23$ -13	-	1.7 ± 0.2	-1.19	-31	48	6	37	4	0.2	0.0
B 7434 746	27	$+19$ -8	+	1.6 ± 0.3	-0.36	131	42	2	18	20	1.3	0.1
	93	$+66$ -28	-	1.8 ± 0.2	-1.45	-60	102	42	56	4	0.2	0.0
C 6059 1010	32	$+11$ -6	+	1.3 ± 0.2	0.64	67	61	1	37	22	0.6	0.0
	9	$+1$ -1	-	1.4 ± 0.1	0.24	-121	48	1	23	23	1.3	0.0
D 7739 1279	d)	d)	d)	d)	-0.19	169	51	1	13	34	2.4	0.0
	50	$+50$ -17	-	1.5 ± 0.2	-0.79	-9	55	8	38	9	0.0	0.1

a) $\pm 1\sigma$ including systematic errors.
b) Ionization loss normalized to minimum ionizing pion.
c) The rapidity y is defined as positive in the direction of outgoing $\bar{p}$.
d) Unmeasured owing to large dip angle.

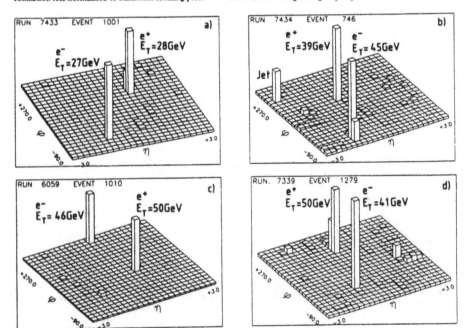

Fig. 3. Electromagnetic energy depositions at angles $>5°$ with respect to the beam direction for the four electron pairs.

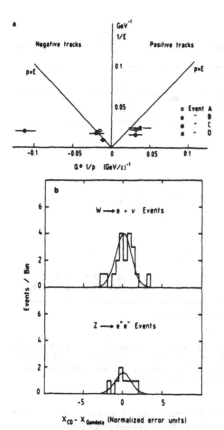

Fig. 4. (a) Magnetic deflection in $1/p$ units compared to the inverse of the energy deposited in the electromagnetic calorimeters. Ideally, all electrons should lie on the $1/E = 1/p$ line. (b) Normalized deviation between the track hit as computed from the central detector and calorimetry centroids. The deviations have been measured in test beam runs, for (i) $W \to e\nu$ events and (ii) $Z^0 \to e^+e^-$ candidates. The continuous line is a unit variance gaussian.

cept for one track of event D which travels at less than 15° parallel to the magnetic field, all tracks are shown in fig. 4a, where the momenta measured in the central detector are compared with the energy deposition in the electromagnetic calorimeters. All tracks but one have consistent energy and momentum measurements. The low-momentum track of event C is interpreted as being due to a hard bremsstrahlung process, either

internal or in the corrugated vacuum chamber and detector walls. We have estimated the probability that one of the electrons in the sample radiates at least 70% of its energy and found it to be $\sim\frac{1}{4}$ [*9], assuming that the "average" thickness is traversed by all the tracks. Furthermore, sensitive checks of the correctness of the electron assignment for the tracks can be obtained by comparing the impact of the electron tracks on the e.m. calorimeters as measured by the central detector with the centroid of the energy deposition measured from the ratio of PM signals. This test is particularly sensitive along the *x*-direction (see fig. 4b) and it appears to be entirely consistent with expectations based on the electron charge assignments of the two tracks. The average invariant mass of the pairs, combining the four consistent values, is (95.2 ± 2.5) GeV/c^2 (table 3).

5. Events with two muon tracks. Events from the dimuon trigger flag have been submitted to the additional requirement that there is at least one muon track reconstructed off line in the muon chambers, and with one track in the central detector of reasonable projected length ($\geqslant 40$ cm) and $p_T \geqslant 7$ GeV/c. Only 42 events survive these selection criteria. Careful scanning of these events has led to only one clean dimuon event, with two "isolated" tracks (fig. 5). Most of the events are due to cosmics. Parameters are given in tables 2 and 3. Energy losses in the calorimeters traversed by the two muon tracks are well within expectations of ionization losses of high-energy muons (fig. 6a). The position in the coordinate and the angles at the exit of the iron absorber (fig. 6b) are in agreement with the extrapolated track from the central detector, once multiple scattering and other instrumental effects have been calibrated with $p > 50$ GeV cosmic-ray muons traversing the same area of the appara-

[*9] The inner bremsstrahlung has been calculated according to ref. [14]. On the same subject, see ref. [15]. We are grateful to F.A. Berends for his assistance with this difficult subject. We find that the probability that either electron of the $Z \to e^+e^-$ decay emits more than 70% of the initial particle energy into photons is about 2.3%. External radiation has also been calculated. It amounts to about 3.5% for the average thickness traversed (0.1 radiation length traversed by both electrons). However, large fluctuations can occur as a result of critical angles due to corrugations in the vacuum pipe, especially around 70°. The faulty track of event C has an angle of emission close to this value.

Table 2
Properties of the muons of the dimuon event 6600-222.

	Track parameters				Muon identification				
					Normalized ionization I/I_0		λ_{abs}	μ/CD matching d)	
Q	p (GeV/c)	ℓ (cm)	y	ϕ (°)	e.m. calorimeter	Hadron calorimeter		Position (cm)	Angle (mrad)
	$58.8 {+8 \atop -6}$ a)								
+	60.3 ± 10.8 b)								
	$59.2 {+6.4 \atop -5.2}$ c)	170	1.19	-27.6	0.8 ± 0.5	1.2 ± 0.5	10.2	$\Delta X_1 = -1.3 \pm 1.9$ $\Delta X_2 = 11.6 \pm 10.7$	$\Delta\phi = -2 \pm 6$ $\Delta\Lambda = 11 \pm 14$
	$63.6 {+30 \atop -15}$ a)								
_	43.1 ± 6.2 b)								
	$46.1 {+6.1 \atop -5.7}$ c)	80	-0.28	119.1	1.2 ± 0.9	1.6 ± 0.8	11.1	$\Delta X_1 = -0.1 \pm 8.0$ $\Delta X_2 = -8.0 \pm 8.5$	$\Delta\phi = -6 \pm 3$ $\Delta\Lambda = -9 \pm 14$

Momentum determination: a) Central detector and μ chamber (statistical errors only);
b) Transverse momentum balance;
c) Weighted average of (a) and (b).
μ-CD matching: d) Difference between the extrapolated CD track and the track measured in the μ chambers (see fig. 6).
Remarks: The acceptance of the single muon trigger starts at a transverse momentum of about 2.5 GeV/c and reaches its full efficiency of 97% at 5.5 GeV/c. The geometrical acceptance of the dimuon trigger, used in this analysis, reduced the acceptance for Z^0 events to about 30%.

tus. There are two ways of measuring momenta, either in the central detector or using the muon detector. Both measurements give consistent results. Furthermore, if no neutrino is emitted (as suggested by the electron events which exhibit no missing energy), the recoil of the hadronic debris, which is significant for this event, must be equal to the transverse momentum of the ($\mu^+\mu^-$) pair by momentum conservation. The directions of the two muons then suffice to calculate the momenta of the two tracks. Uncertainties of muon parameters are then dominated by the errors of calorimetry. As shown in table 2, this determination is in agreement with magnetic deflection measurements. The invariant mass of the ($\mu^+\mu^-$) pair is found to be $m_{\mu\mu} = (95.5 \pm 7.3)$ GeV/c^2, in excellent agreement with that of the four electron pairs (see table 3).

6. Background estimates. The most striking feature

of the events is their common value of the invariant mass (fig. 8); values agree within a few percent and with expectations from experimental resolution. Detection efficiency is determined by the energy thresholds in the track selection, 15 GeV/c for $e^\pm$ and 7 GeV/c for $\mu^\pm$. Most "trivial" sources of background are not expected to exhibit such a clustering at high masses. Also, most backgrounds would have an equal probability for (eμ) pairs, which are not observed. Nevertheless, we have considered several possible spurious sources of events:

(i) Ordinary large transverse momentum jets which fragment into two apparently isolated, high-momentum tracks, both simulating either muons or electrons. To evaluate this effect, events with (hadronic) tracks of momenta compatible with $p_T > 25$ GeV/c were also selected in the express line. After requiring that the track is isolated, one finds one surviving event

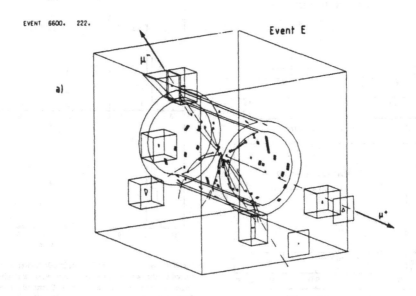

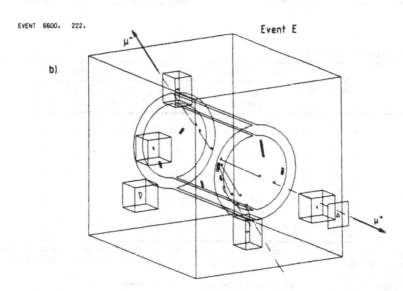

Fig. 5. Display for the high-invariant-mass muon pair event: (a) without cuts and (b) with $p_T > 1$ GeV thresholds for tracks and $E_T > 0.5$ GeV for calorimeter hits.

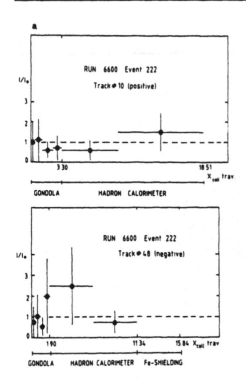

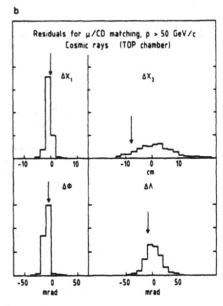

Fig. 6. (a) Normalized energy losses in calorimeter cells traversed by the two muon tracks. (b) Arrows show residuals in angle and position for the muon track. Distributions come from cosmic-ray calibration with $p > 50$ GeV/c.

Table 3
Mass and energy properties of lepton pair events.

Run, event	Lepton pair properties			General event properties			
	Mass [a] (GeV/c^2)	P_T (GeV/c)	x_F [b]	E_{tot} (GeV)	$\Sigma\|E_T\|$ (GeV)	Missing E_T (GeV)	Charged tracks
A 7433 1001	91 ± 5	2.9 ± 0.9	0.02 ± 0.01	274	82	2.1 ± 3.6	27
B 7434 746	97 ± 5	7.9 ± 1.2	0.39 ± 0.01	494	149	9.3 ± 5.0	67
C 6059 1010	98 ± 5	8.0 ± 1.5	0.17 ± 0.01	412	143	3.3 ± 4.8	38
D 7339 1279	95 ± 5	8.4 ± 1.4	0.17 ± 0.01	493	157	0.8 ± 5.0	54
E 6600 ($\mu\mu$) 222	95 ± 8	24 ± 5	0.14 ± 0.02	278[c]	128[c]	3.4 ± 5.9[c]	28

[a] These errors have been scaled up arbitrarily to 5 GeV to represent the present level of uncertainty in the overall calibration of the e.m. calorimeter which will be recalibrated completely at the end of the present run. This scale factor is not included in the error bars plotted in fig. 8.

[b] x_F is defined as the longitudinal momentum of the dilepton divided by beam energy.

[c] Includes the muon energies.

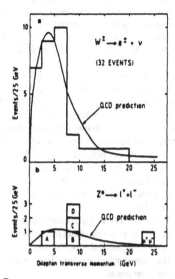

Fig. 7. Transverse momentum spectra: (a) for W → eν events, and (b) $Z^0 \to \ell^+\ell^-$ candidates. The lines represent QCD predictions (ref. [20]).

with transverse energy $\simeq$ 25 GeV in a sample corresponding to 30 nb^{-1}. Including the probability that this track simulates either a muon ($\sim$2 × 10^{-3}) or an electron ($\sim$6 × 10^{-3}), we conclude that this effect is negligible [*10]. Note that two tracks (rather than one) are needed to simulate our events (probabilities must be squared!) and that the invariant mass of the events is much higher than the background. The background is expected to fall approximately like m^{-5} according to the observed jet—jet mass distributions [16].

(ii) Heavy-flavoured jets with subsequent decay into leading muons or electrons. In the 1982 event sample (11 nb^{-1}), two events have been observed with a single isolated muon of $p_T > 15$ GeV and one electron event with $p_T > 25$ GeV/c. Some jet activity in the opposite hemisphere is required. One event exhibits also a significant missing energy. Once this is taken

---

[*10] Electron—pion discrimination has been measured in a test beam in the full energy range and angles of interest. The muon tracks have the following probabilities: (i) no interaction: 2 × 10^{-5} (4 × 10^{-5}); (ii) interaction but undetected by the calorimeter and geometrical cuts: 10^{-4} (4 × 10^{-4}); (iii) decay: 10^{-3} (0.7 × 10^{-3}). Numbers within parenthesis refer to negative tracks.

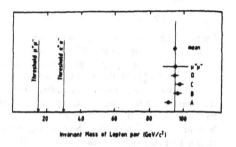

Fig. 8. Invariant masses of lepton pairs.

into account they all have a total (jet+jet+lepton+ neutrino) transverse mass of around 80 GeV/c^2, which indicates that they are most likely due to heavy-flavour decay of W particles. This background will be kinematically suppressed at the mass of our five events. Nevertheless, if the fragmentation of the other jet is also required to give a leading lepton and no other visible debris, this background contributes at most to 10^{-4} events. Monte Carlo calculations using ISAJET lead to essentially the same conclusion [9].

(iii) Drell—Yan continuum. The estimated number and the invariant mass distribution make it negligible [17,9].

(iv) W^+W^- pair production is expected to be entirely negligible at our energy [18,9].

(v) Onium decay from a new quark, of mass compatible with the observation ($\sim$95 GeV/c^2). Cross sections for this process have been estimated by different authors [19], and they appear much too small to account for the desired effect.

In conclusion, none of the effects listed above can produce either the number or the features of the observed events.

7. Dilepton events as Z^0 leptonic decays. All the observations are in agreement with the hypothesis that events are due to the production and decay of the neutral intermediate vector boson Z^0 according to reaction (1). The transverse momentum distribution is shown in fig. 7, compared with the observed distributions for the $W^{\pm} \to e\nu$ events [5] and with QCD calculations [20]. The muon event and one of the electron events (event B) have visible jet structure. Other events are instead apparently structureless.

From our observation, we deduce a mass value for

the Z^0 particle,

$$m_{Z^0} = (95.2 \pm 2.5)\ GeV/c^2 \ .$$

The half width based on the four electron events is 3.1 GeV/c^2 ($<5.1\ GeV/c^2$ at 90% CL), consistent with expectation from the experimental resolution and the natural Z^0 width [6–8], $\Gamma_{Z^0} = 3.0$ GeV. At this point it is important to stress that the final calibration of the electromagnetic calorimeters is still in progress and that small scale shifts are still possible, most likely affecting both the $W^\pm$ and Z^0 mass values. No e.m. radiative corrections have been applied to the masses [14,15].

We now compare our result with the prediction of standard SU(2) X U(1). Employing the renormalized weak mixing angle $\sin^2\theta_w\ (m_W)$ defined by modified minimal subtraction, we find to O(α):

$$\sin^2\theta_w\ (m_W) = (38.5\ GeV/m_W)^2\ .$$

From our preliminary result [5] we find

$$\sin^2\theta_w\ (m_W) = 0.226 \pm 0.011\ ,$$

in excellent agreement with the extrapolation from the world low-energy data [+3] $\sin^2\theta_w\ (m_W) = (0.236 \pm 0.030)$. If we then parametrize the Z^0 mass with the well-known formula $m_{Z^0}^2 = m_W^2/\rho\cos^2\theta_w\ (m_W)$, we find $\rho = 0.94 \pm 0.06$ (see fig. 9), in excellent agreement with the prediction of the minimal model, where one

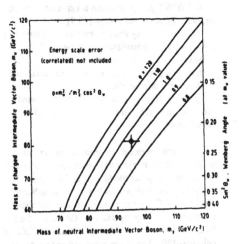

Fig. 9. Mass of charged versus neutral intermediate vector bosons. Values of parameters ρ and $\sin^2\theta_w\ (m_W)$ according to SU(2) X U(1) can be associated with our measured point.

usually assumes that $\rho = 1$. Potential deviations from this value could come from higher Higgs representations, additional fermion generations, dynamical symmetry effects, etc. Within the accuracy of our result, none of these effects needs to be invoked.

The continued success of the collider and the steady increase in luminosity which have made this result possible, depend critically upon the superlative performance of the whole CERN accelerator complex, which was magnificently operated by its staff. We thank W. Kienzle who, as coordinator, balanced very effectively the sometimes conflicting interests of the physicists and accelerator staff. We have received enthusiastic support from the Director General, H. Schopper, and his Directorate, for the results emerging from the SPS Collider programme.

We are thankful to the management and staff of CERN and of all participating Institutes who have vigorously supported the experiment.

The following funding Agencies have contributed to this programme:
Fonds zur Förderung der Wissenschaftlichen Forschung, Austria.
Valtion luonnontieteellinen toimikunta, Finland.
Institut National de Physique Nucléaire et de Physique des Particules and Institut de Recherche Fondamentale (CEA), France.
Bundesministerium für Forschung und Technologie, Germany.
Istituto Nazionale di Fisica Nucleare, Italy.
Science and Engineering Research Council, United Kingdom.
Department of Energy, USA.

Thanks are also due to the following people who have worked with the collaboration in the preparation and data collection on the runs described here: O.C. Allkofer, F. Bernasconi, F. Cataneo, R. Del Fabbro, L. Dumps, D. Gregel, G. Stefanini and R. Wilson.

The whole UA1 Collaboration is very grateful for the unfailing help and friendly advice of Mme Mirella Keller.

References

[1] UA1 Collab., G. Arnison et al., Phys. Lett. 122B (1983) 103.
[2] UA2 Collab., G. Banner et al., Phys. Lett. 122B (1983) 476.

[3] The staff of the CERN proton–antiproton project, Phys. Lett. 107B (1981) 306;
C. Rubbia, P. McIntyre and D. Cline, Proc. Intern. Neutrino Conf. (Aachen, 1976) (Vieweg, Braunschweig, 1977) p. 683;
Study Group, Design study of a proton–antiproton colliding beam facility, CERN/PS/AA 78-3 (1978), reprinted in: Proc. Workshop on Producing high-luminosity, high-energy proton–antiproton collisions (Berkeley, 1978), report LBL-7574, UC34c, p. 189.

[4] UA1 proposal: A 4π solid-angle detector for the SPS used as a proton–antiproton collider at a centre-of-mass energy of 540 GeV, CERN/SPSC 78-06 (1978);
M. Barranco Luque et al., Nucl. Instrum. Methods 176 (1980) 175;
M. Calvetti et al., Nucl. Instrum. Methods 176 (1980) 255;
K. Eggert et al., Nucl. Instrum. Methods 176 (1980) 217, 223;
A. Astbury et al., Phys. Scr. 23 (1981) 397.

[5] UA1 Collab., G. Arnison et al., in preparation.

[6] S. Weinberg, Phys. Rev. Lett. 19 (1967) 1264;
A. Salam, Proc. 8th Nobel Symposium (Aspenäsgården, 1968) (Almqvist and Wiksell, Stockholm, 1968) p. 367;
S.L. Glashow, Nucl. Phys. 22 (1961) 579.

[7] W.J. Marciano and Z. Parsa, Proc. Particles and fields summer study on elementary particle physics and future facilities (Snowmass, CO, 1982) (AIP, New York, 1983) p. 155.

[8] J. Kim et al., Rev. Mod. Phys. 53 (1981) 211;
I. Liede and M. Roos, Nucl. Phys. B167 (1980) 397.

[9] F.E. Paige and S.D. Protopopescu, ISAJET program, BNL 29777 (1981);
R. Horgan and M. Jacob, Proc. 1980 CERN School of Physics (Malente), CERN report 81-04 (1981).

[10] M. Davier, Proc. 21st Intern. Conf. on High-energy physics (Paris, 1982) [J. Phys. (Paris) 43 (1982), No. 12], p. C3-471.

[11] M. Calvetti et al., IEEE Trans. Nucl. Sci. NS-30 (1983) 71;
M. Barranco-Luque et al., Nucl. Instrum. Methods 176 (1980) 175;
M. Calvetti et al., The UA1 central detector, Proc. Intern. Conf. on Instrumentation for colliding beam physics (SLAC, Stanford, 1982) (SLAC-250, Stanford, 1982) p. 16;
The UA1 Collaboration, The UA1 data acquisition system, Proc. Intern. Conf. on Instrumentation for colliding beam physics (SLAC, Stanford, 1982) (SLAC-250, Stanford, 1982) p. 151.

[12] K. Eggert et al., Nucl. Instrum. Methods 176 (1980) 217.

[13] J.T. Carroll, S. Cittolin, M. Demoulin, A. Fucci, B. Martin, A. Norton, J.-P. Porte, P. Ross and K.M. Storr, Data Acquisition using the 168E, Paper presented at: Three-day in-depth review on the impact of specialized processors in elementary particle physics (Padua, 1983), to be published.

[14] F.A. Berends and R. Kleiss, On the process $\bar{q}q \rightarrow Z \rightarrow \ell^+\ell^-(\gamma)$, preprint Instituut-Lorentz, Leiden University, to be published;
F.A. Berends et al., Nucl. Phys. B202 (1982) 63.

[15] G. Passarino and M. Veltman, Nucl. Phys. B160 (1979) 151;
M. Greco et al., Nucl. Phys. B171 (1980) 118; B197 (1982) 543 (E);
V.N. Baier et al., Phys. Rep. 78 (1981) 293.

[16] UA1 Collab., J. Sass, Study of jets in pp̄ collisions with UA1 calorimetry, presented 18th Rencontre de Moriond, Proc. 18th Rencontre de Moriond (La Plagne, 1983) (Éditions Frontières, Dreux, 1983), in preparation;
UA1 Collab., V. Vuillemin, Jet fragmentation in the SPS pp̄ Collider, presented 18th Rencontre de Moriond, Proc. 18th Rencontre de Moriond (La Plagne, 1983) (Éditions Frontières, Dreux, 1983), in preparation.

[17] S.D. Drell and T.M. Yan, Phys. Rev. Lett. 25 (1970) 316;
F. Halzen and D.H. Scott, Phys. Rev. D18 (1978) 3378;
S. Pakvasa, M. Dechantsreiter, F. Halzen and D.M. Scott, Phys. Rev. D20 (1979) 2862.

[18] R. Kinnunen, Proc. Proton–Antiproton collider physics Workshop (Madison, WI, 1981) (University of Wisconsin, Madison, WI, 1982);
R.W. Brown and K.O. Mikaelian, Phys. Rev. D19 (1979) 922;
R.W. Brown, D. Sahdev and K.O. Mikaelian, Phys. Rev. D20 (1979) 1164.

[19] T.K. Gaisser, F. Halzen and E.A. Paschos, Phys. Rev. D15 (1977) 2572;
R. Baier and R. Rückl, Phys. Lett. 102B (1981) 364;
F. Halzen, Proc. 21st Intern. Conf. on High-energy physics (Paris, 1982) [J. Phys. (Paris) 43 (1982) No. 12], p. C3-381;
F.D. Jackson, S. Olsen and S.H.H. Tye, Proc. Particles and fields summer study on elementary particle physics and future facilities (Snowmass, CO, 1982) (AIP, New York, 1983) p. 175.

[20] P. Aurenche and R. Kinnunen, QCD predictions for weak vector boson production in pp̄ collisions, in preparation;
P. Aurenche and J. Lindfors, Nucl. Phys. B185 (1981) 274;
P. Chiappetta and M. Greco, Nucl. Phys. B199 (1982) 77, and references therein.

EVIDENCE FOR $Z^0 \rightarrow e^+e^-$ AT THE CERN p̄p COLLIDER

The UA2 Collaboration

P. BAGNAIA [b], M. BANNER [f], R. BATTISTON [1,2], Ph. BLOCH [f], F. BONAUDI [b], K. BORER [a],
M. BORGHINI [b], J.-C. CHOLLET [d], A.G. CLARK [b], C. CONTA [e], P. DARRIULAT [b], L. Di LELLA [b],
J. DINES-HANSEN [c], P.-A. DORSAZ [b], L. FAYARD [d], M. FRATERNALI [e], D. FROIDEVAUX [b],
G. FUMAGALLI [e], J.-M. GAILLARD [d], O. GILDEMEISTER [b], V.G. GOGGI [e], H. GROTE [b], B. HAHN [a],
H. HÄNNI [a], J.R. HANSEN [b], P. HANSEN [c], T. HIMEL [b], V. HUNGERBÜHLER [b], P. JENNI [b],
O. KOFOED-HANSEN [c], E. LANÇON [f], M. LIVAN [b,e], S. LOUCATOS [f], B. MADSEN [c], P. MANI [a],
B. MANSOULIÉ [f], G.C. MANTOVANI [1], L. MAPELLI [b,3], B. MERKEL [d], M. MERMIKIDES [b],
R. MØLLERUD [c], B. NILSSON [c], C. ONIONS [b], G. PARROUR [b,d], F. PASTORE [e], H. PLOTHOW-BESCH [b],
M. POLVEREL [f], J.-P. REPELLIN [d], A. RIMOLDI [e], A. ROTHENBERG [b], A. ROUSSARIE [f],
G. SAUVAGE [d], J. SCHACHER [a], J.L. SIEGRIST [b], H.M. STEINER [b,4], G. STIMPFL [b], F. STOCKER [a],
J. TEIGER [f], V. VERCESI [e], A.R. WEIDBERG [b], H. ZACCONE [f], J.A. ZAKRZEWSKI [b,5] and
W. ZELLER [a]

[a] Laboratorium für Hochenergiephysik, Universität Bern, Sidlerstrasse 5, Bern, Switzerland
[b] CERN, 1211 Geneva 23, Switzerland
[c] Niels Bohr Institute, Blegdamsvej 17, Copenhagen, Denmark
[d] Laboratoire de l'Accélérateur Linéaire, Université de Paris-Sud, Orsay, France
[e] Dipartimento di Fisica Nucleare e Teorica, Università di Pavia and INFN, Sezione di Pavia, Via Bassi 6, Pavia, Italy
[f] Centre d'Etudes Nucléaires de Saclay, France

Received 11 August 1983

From a search for electron pairs produced in p̄p collisions at $\sqrt{s} = 550$ GeV we report the observation of eight events which we interpret as resulting from the process $\bar{p} + p \rightarrow Z^0 +$ anything, followed by the decay $Z^0 \rightarrow e^+ + e^-$ or $Z^0 \rightarrow e^+ + e^- + \gamma$, where Z^0 is the neutral Intermediate Vector Boson postulated by the unified electroweak theory. We use four of these events to measure the Z^0 mass

$$M_Z = 91.9 \pm 1.3 \pm 1.4 \text{ (systematic) GeV}/c^2.$$

1. Introduction. The primary goal of the experimental program at the CERN p̄p Collider has been to search for the massive Intermediate Vector Bosons (IVB), which are postulated to mediate the electroweak interaction [1].

The recent observation of single isolated electrons with high transverse momentum in events with missing transverse energy [2,3] is consistent with the process $\bar{p} + p \rightarrow W^\pm +$ anything, followed by the decay $W^\pm \rightarrow e^\pm + \nu(\bar{\nu})$, where W is the charged IVB.

We report here the observation in the UA2 detector of eight events which we interpret in terms of the reaction

$$\bar{p} + p \rightarrow Z^0 + \text{anything}$$
$$\hookrightarrow e^+ + e^- \text{ or } e^+ + e^- + \gamma. \quad (1)$$

[1] Gruppo INFN del Dipartimento di Fisica dell'Università di Perugia, Italy.
[2] Also at Scuola Normale Superiore, Pisa, Italy.
[3] On leave from INFN, Pavia, Italy.
[4] On leave from Department of Physics, University of California, Berkeley, CA, USA.
[5] On leave from Institute of Physics, University of Warsaw, Poland.

where Z^0 is the neutral IVB. The observation of these events, which have been found in a data sample corresponding to a total integrated luminosity of 131 nb^{-1}, agrees with the SU(2) X U(1) model and with the recent results of the UA1 experiment [4].

2. The detector. The experimental apparatus [5] is shown in fig. 1. At the centre of the apparatus a system of cylindrical chambers (the vertex detector) measures charged particle trajectories in a region without magnetic field. The vertex detector consists of: (a) four multi-wire proportional chambers having cathode strips with pulse height read-out at ±45° to the wires; (b) two drift chambers with measurement of the charge division on a total of 12 wires per track. These chambers are used both to obtain tracking information and to evaluate the most likely ionisation *I* (in units of equivalent minimum ionising particles, mip) associated with each track. From the reconstructed charged particle tracks the position of the event vertex is determined with a precision of ±1 mm in all directions.

The vertex detector is surrounded by an electromagnetic and hadronic calorimeter [6] (the central

calorimeter), which covers the polar angle interval $40° < \theta < 140°$ and the full azimuth. This calorimeter is segmented into 240 cells, each covering 15° in ϕ and 10° in θ and built in a tower structure pointing to the centre of the interaction region. The cells are segmented longitudinally into a 17 radiation lengths thick electromagnetic compartment (lead-scintillator) followed by two hadronic compartments (iron-scintillator) of ~2 absorption lengths each. The light from each compartment is channelled to two photomultipliers (PMs) by means of BBQ-doped light guides on opposite sides of the cell.

In the angular region covered by the central calorimeter a cylindrical tungsten converter, 1.5 radiation lengths thick, followed by a cylindrical proportional chamber, is located just after the vertex detector. This chamber, named C5 (see fig. 1), has cathode strips at ±45° to the wires. We measure the pulse height on the cathode strips and the charge division on the wires. This device localises electromagnetic showers initiated in the tungsten with a precision of 3 mm.

For the first 15 nb^{-1} of integrated luminosity, collected during the Autumn of 1982, the azimuthal coverage of the central calorimeter was only 300°. The

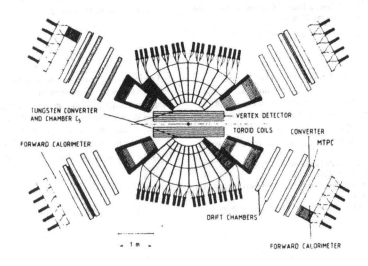

Fig. 1. Schematic detector assembly (cut in a plane containing the beam line).

remaining interval (±30° around the horizontal plane) was covered by a magnetic spectrometer which included a lead-glass wall, to measure charged and neutral particle production [7,8].

The two forward regions ($20° < \theta < 37.5°$ and $142.5° < \theta < 160°$), are each equipped with twelve toroidal magnet sectors with an average field integral of 0.38 Tm. Each sector is instrumented with:

(a) Three drift chambers [9] located after the magnetic field region. Each chamber consists of three planes, with wires at $-7°$, $0°$, $+7°$, with respect to the magnetic field direction.

(b) A 1.4 radiation lengths thick lead–iron converter, followed by a chamber [10] consisting of two pairs of layers of 20 diameter proportional tubes (MTPC), staggered by a tube radius and equipped with pulse height measurement. There is a 77° angle between the tubes of the two pairs of layers, with the tubes of the first one being parallel to the magnetic field direction. This device localises electromagnetic showers with a precision of ≤ 8 mm.

(c) An electromagnetic calorimeter consisting of lead-scintillator counters assembled in ten independent cells, each covering 15° in ϕ and 3.5° in θ. Each cell is subdivided into two independent longitudinal sections, 24 and 6 radiation lengths thick, the latter providing rejection against hadrons. The light from each section is collected by two BBQ-doped light guides on opposite sides of the cell.

In order to implement a trigger sensitive to electrons of high transverse momentum, the PM gains in all calorimeters were adjusted so that their signals were proportional to the transverse energy. Because of the cell dimensions, electromagnetic showers initiated by electrons may be shared among adjacent cells. Trigger thresholds were applied, therefore, to linear sums of signals from matrices of 2×2 cells, rather than to individual cells. In the central calorimeter, all possible 2×2 matrices were considered; in the two forward ones, we included only those made up of cells belonging to the same sector. A trigger signal was generated whenever the transverse energy deposition in at least two such matrices, separated in azimuth by more than 60°, exceeded a threshold corresponding to a transverse energy deposition of 3.5 GeV.

All calorimeters have been calibrated in a 10 GeV beam from the CERN PS, using incident electrons and muons. The stability of the calibration has since been monitored using a light flasher system, a Co^{60} source and a measurement of the average energy flow into each module for unbiased $\bar{p}p$ collisions [6]. The systematic uncertainty in the energy calibration of the electromagnetic calorimeters for the data discussed here amounts to an average value of ±1.5%. The cell-to-cell calibration uncertainty has a distribution with an rms of 3%.

The response of the calorimeters to electrons, and to single and multi-hadrons, has been measured at the CERN PS and SPS using beams from 1 to 70 GeV/c. In particular, both longitudinal and transverse shower developments have been studied, as well as the effect of particles impinging near the cell boundaries. The energy resolution for electrons is measured to be $\sigma_E/E = 0.14/\sqrt{E}$ (E in GeV).

3. Data analysis. The full data sample consists of approximately 7×10^5 triggers, which correspond to an integrated luminosity $L = 131$ nb^{-1}.

An initial selection is made by rejecting all events which are identified as due to sources other than $\bar{p}p$ collisions (<10% of the entire sample, mainly beam–gas background and cosmic rays). In the surviving events, a search is made for configurations consistent with the presence of electrons among the collision products. An electron is identified from the observation of:

(a) A track measured in the wire chambers.

(b) A large signal detected in the preshower counters (C5 in the central detector or the MTPCs in the two forward regions).

(c) An energy deposition in the calorimeters with small lateral sizes and limited penetration into the hadronic compartments.

And from the quality of the matching in space among these three properties.

Since the primary goal of this analysis is the detection of process (1), we first reduce the data sample by requiring the total electromagnetic transverse energy to exceed 30 GeV and the presence of a pair of energy clusters having an invariant mass in excess of 50 GeV/c^2 as calculated in the following way.

In the central calorimeter clusters are obtained by joining all electromagnetic cells which share a common side and contain at least 0.5 GeV. A contribution from the cells having at least one side in common with a cluster cell is also added.

The forward calorimeter clusters consist of at most two adjacent cells having the same azimuth (here the cell is far from the interaction point and much larger than the lateral extension of an electromagnetic shower. and the dead region between cells at different azimuths is too large to allow clustering across it).

In both cases the cluster energy E_{cl} is defined as $E_{cl} = E_{em} + E_{had}$ where E_{em} is the sum of the energies deposited in the electromagnetic compartments of the cluster cells and E_{had} is the corresponding sum for the hadronic compartments.

The invariant mass is calculated under the assumption that the event vertex is at the centre of the apparatus. We use the cluster centroids to define the momenta.

The remaining data sample contains 7427 events.

These events are then fully reconstructed and their invariant mass M is calculated again. this time taking into account the exact position of the event vertex. The difference between this new value and the previous one does not exceed 2 GeV/c^2.

At this stage the event sample is dominated by two-jet events [11]. However. while E_{cl} measures correctly the energy of jets produced in the central region. it is in general a gross underestimate of that of forward jets, for which the calorimeter thickness is only 88% of an absorption length. As a consequence. the sample contains many more events having both clusters in the central calorimeter than events with at least one cluster in the forward regions, because the jet momentum distribution falls off steeply with increasing jet transverse momentum [11].

In order to select events with similar characteristics in the central and forward regions and to enhance the electron signal. we further reduce the sample by requiring that both clusters have a small lateral size in the electromagnetic compartment of the calorimeter and a limited energy leakage in the hadronic compartment.

For clusters in the central calorimeter. cluster sizes R_θ, R_ϕ are calculated from the cluster centroid and the values of the angles θ and ϕ at the cell centres. weighted by their energy depositions. The conditions R_θ. $R_\phi < 0.5$ cell sizes are required.

In the two forward calorimeters we require that the sum of the energies deposited in the cells adjacent to the cluster cells does not exceed 3 GeV.

The condition that the showers have only a small

energy leakage in the hadronic compartments of the calorimeters is applied by requiring that the ratio $H = E_{had}/E_{cl}$ does not exceed a value H_0. equal to 0.02 for the forward calorimeters, and $0.023 + 0.034 \times \ln E_{cl}$, where E_{cl} is in GeV, for the central one.

The cuts applied at this stage are very loose and are satisfied by more than 95% of isolated electrons between 10 and 80 GeV, as verified experimentally using test beam data. They reduce the event sample to 24 events. whose invariant mass distribution is shown in fig. 2a. There are 12 events with both clusters in the central region. 8 events with one cluster in the central and the other in the forward regions. and 4 events with both clusters in the forward regions.

The sample with both clusters in the central region has been reduced by a factor ~430 by the cuts on cluster size and hadronic leakage.

In the following step we define a series of additional criteria for electron identification. We use measurements of the response of various parts of the de-

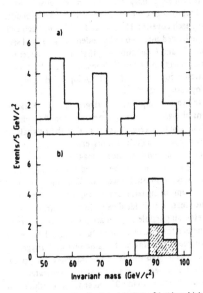

Fig. 2. Invariant mass distributions (a) of the 24 pairs which pass cut 1 of table 1, (b) of the eight of these 24 pairs for which all cuts of table 1 are satisfied by at least one electron. The three events in which both electrons pass all cuts of table 1 are cross-hatched.

Table 1
Electron identification criteria.

REQUIREMENTS	CENTRAL REGION		FORWARD REGIONS	
	Description	$\eta(\%)$	Description	$\eta(\%)$
1. Presence of a calorimeter cluster (preselection)	Electromagnetic cluster size $R_{\theta,\phi} < 0.5$ Hadronic leakage : $H < .023 + .034 \ln E_{cl}$	98	Cluster size $\leqslant 2$ cells Energy in adjacent cells $E_{adj} < 3$ GeV Hadronic leakage $H < 0.02$	98
2. Presence of a track	Reconstructed in the vertex detector in both transverse and longitudinal projections	85	Reconstructed a) in the vertex detector in transverse projection with $N \geqslant 1$ signal in the two inner chambers b) in the forward drift chambers. Azimuth difference between a) and b) $\Delta\phi(a,b)$ less than 40 mrad	92
3. Track/cluster match	Track impact to cluster centroid distance $\Delta < 1$ cell size. Compare the energy distribution observed in the 3×3 cells centered on the impact cell to that expected for an electron incident along the track. Light sharing between phototubes of impact cell and hadronic leakage are taken into account. Require that the probability P that the energy distribution for an electron is farther from the mean than that observed is larger than 0.01	95	Shower position calculated from light sharing between impact cell phototubes must be consistent with the track impact point to within $\Delta x \leqslant 70$ mm	98
4. Presence of a preshower counter signal	Measured in C5 as a coincidence in space of charge clusters measured on the anode wires and the inner and outer cathode strips. Its charge q_s must exceed 4 mip.	>90	Measured in MTPC in both coordinate a) planes. Its charge Q(MTPC) must exceed 6 mip	95
5. Track/preshower position match	Within $d \leqslant 7$ mm measured on the C5 surface	94	Within $\Delta x, \Delta y < 50$ mm in each direction	100
6. Momentum measurement			Momentum p measured in spectrometer and energy E measured in calorimeter must satisfy $|p^{-1}-E^{-1}| / \sigma(p^{-1}-E^{-1}) < 3$	99

a) This does not include a 5% inefficiency due to a $\Delta\phi = 18°$ azimuthal region in which C5 was not operational because of electrical breakdown.

tector to isolated electrons [6,9,10] from which we evaluate approximate cut efficiencies η. The cuts are described in table 1 for both the central and forward regions. The efficiencies resulting from the simultaneous application of all selection criteria are at least as large as the products of the individual efficiencies η, namely 67% for the central region and 83% for the forward regions.

We have studied the effect of applying these criteria to our original event sample. We find that the rejection power of any one of these cuts enters as the square when applied to both clusters simultaneously. Furthermore the shape of the mass distribution shows no dependence, within statistics, on the combination of cuts used. These two observations provide a simple method to estimate the background contribution from two-jet events to any mass region in a sample of events surviving a given combination of cuts.

Fig. 2b shows the mass distribution for the events of fig. 2a with at least one cluster satisfying all of the electron identification criteria. There are eight events in this plot which cluster around a mass value of ~ 90 GeV/c^2. A list of relevant parameters for these events (named A–H) is given in tables 2–4. For clusters passing cut 3 of table 1 the cluster energy has been corrected to account for the calorimeter response as a function of the electron incidence angle and impact point. The corrected energy value, together with the measured track direction, has been used to calculate the invariant mass plotted in figs. 2a and 2b.

An upper limit on the background contribution to the eight events in fig. 2b can be inferred from fig. 2a under the assumption that background events have the same mass distribution in both samples. By comparing the event populations above and below 80 GeV/c^2 we find an upper limit (90% CL) of 0.32 background

Table 2
Event parameters.

Event	A	B	C	D	E	F	G	H
Pair Configuration C = central F = forward	CF	CC	CF	CF	CF	CC	CC	CC
Pair Mass (GeV/c^2)	90.7 ±2.1	95.2 ±3.4	89.7 ±2.8	89.1 ±3.2	94.0 ±2.9	89.3 ±4.9	83.2 a) ±2.6	88.3 a) ±2.6
Pair Transverse Momentum (GeV/c)	5.0	11.9	1.4	2.4	4.6	5.0	7.9	6.2

a) Ignoring additional energy measured in neighbour cells.
   Its inclusion results in a mass increase of ∿ 3 GeV/c^2.

events to the signal in fig. 2b. However, this allows for up to 2.3 low mass events ($M < 80$ GeV/c^2) in fig. 2b – while in fact we observe none – and may result in a substantial overestimate of the background.

A better estimate implies a more realistic evaluation of the expected number of low mass background events in fig. 2b. This can be done starting from a much richer event sample by simply releasing cut 1 (hadronic leakage and cluster size) on one of the two clusters and evaluating the rejection power of cuts 2–6 (applied to reduce the sample of fig. 2a to that of fig. 2b) on the other cluster. We have checked that the absence of significant correlation between the fragmentations of the two jets in background events makes this procedure legitimate. Also we have taken advantage of the fact that all events in fig. 2b have at least one central cluster to restrict the operation of releasing cut 1 to central clusters exclusively. In this way we estimate a background contribution of 0.03 events to the signal of fig. 2b.

If we apply the electron identification criteria to both clusters, only three events (A–C, shown as cross-hatched areas in fig. 2b) survive, with an estimated background of 2×10^{-4} events above a mass of 80 GeV/c^2. For two of them (A and C) one of the electrons is in the forward regions, the other electrons are in the central region. Events A and B are interpreted as resulting from reaction (1). Event C consists of two electrons and a well separated high energy photon (or unresolved photons such as from the γγ decay of a π^0 or η meson). The invariant mass value in table 2 and in fig. 2b is calculated for the three particles. We have estimated [11] that $Z^0 \to e^+e^-\gamma$ decays with a

[11] We thank R. Petronzio for assistance in making this calculation (to first order in α).

photon at least as hard as the observed one, and with e^+e^- opening angles equal to, or smaller than the measured one occur approximately once every 200 $Z^0 \to e^+e^-$ decays.

Fig. 3a shows the longitudinal view of event A in the plane containing the central electron. Figs. 3b and 4a show the cell energy distribution in θ and ϕ for events A and C. The transverse view of event C, indicating the presence of the additional photon at an angle of $\sim 30°$ to the electron, is shown in fig. 4b.

We next discuss the five other events (D–H) in which one of the two electron candidates fails at least one of the strict selection criteria described in table 1.

– The forward electron candidate of event D is associated with a track measured in the vertex detector as pointing to a coil of the magnet at a place where its thickness is ~0.5 radiation lengths. Three tracks, measured in the forward drift chambers, point to the energy cluster. One of them passes cut 2 but fails cut 6. Several MTPC clusters satisfy cut 4, but cuts 3 and 5 are never simultaneously satisfied (table 3). This configuration is consistent with the hypothesis of an electron initiating an electromagnetic shower in the magnet coil. The mass value listed in table 2 has been calculated under this hypothesis. Event D belongs to the data sample collected in 1982 and has been previously published [12].

The central electron candidate of event E passes all strict cuts but 4 and 5 (table 4). Its associated C5 cluster has a charge of only 2.4 mip and is 9 ± 2 mm away from the track impact. The occurrence of such a configuration in the present sample is compatible with the cut efficiencies listed in table 1 and event E is consistent with an electron pair hypothesis. We note however that the forward electron is accompanied by

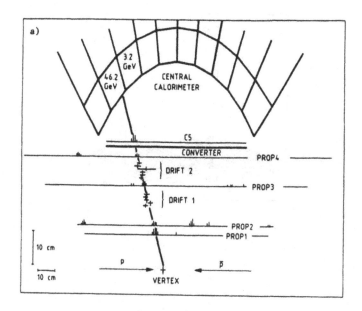

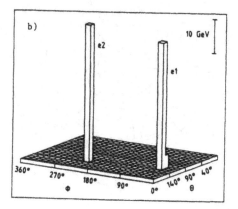

Fig. 3. (a) Longitudinal view of event A in the plane containing the central electron. In each of the four proportional chambers of the vertex detector (PROP 1–4) and in the preshower chamber C5 located behind the tungsten converter, signals are indicated whenever a coincidence in space was observed between the anode wire and the inner and outer cathode strips. The measurements from the two drift chambers (DRIFT 1 and 2) are indicated as crosses with sizes corresponding to the uncertainty on the charge division measurement. Energies measured in the electromagnetic cells facing the electron track are indicated, when non zero. (b) The cell transverse energy distribution for event A in the (θ, ϕ) plane.

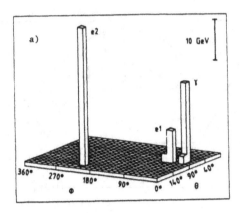

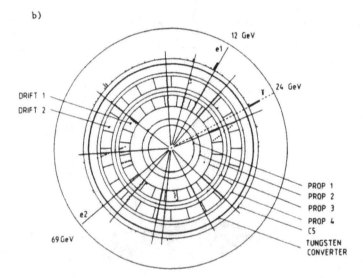

Fig. 4. (a) The cell transverse energy distribution for event C in the (θ, ϕ) plane. Electron e1 and the photon (γ) are observed in the central region, electron e2 in the forward region. (b) The transverse view of event C. Signals from the proportional and drift chambers are indicated as dots. Electron e1 and the photon (γ) are observed in the central region and associated with C5 signals (indicated by heavy lines proportional to pulse height). Electron e2 is observed in the forward region (not covered by C5).

Table 3
Electron parameters (forward regions).

Cut (see Table I)	Event	A	C	D	E
	θ (degrees)	142.2	150.3	155.8	148.3
	φ (degrees)	218.6	219.9	324.3	173.4
	E (GeV)	70.4 ± 1.6	68.5 ± 1.6	99.2 ± 6.0	58.1 ± 1.7
	I (mip)	0.9	0.9	1.9	0.9
1	Eadj (GeV)	0	0	3.0	0
	H (%)	0.4	0.7	0.4	0.3
2	Δφ(a,b) (mrad)	2	2	7	8
	N	2	2	2	2
3	Δx (mm)	11	29	81 [a]	60
4	Q (MTPC) (mip)	> 64	14	19	> 42
5	Δx (mm)	1.4	5.0	40.5	1.0
	Δy (mm)	1.8	4.7	8.8	2.1
6	$\dfrac{|p^{-1}-E^{-1}|}{\sigma(p^{-1}-E^{-1})}$	0.40	0.15	29.50 [a]	0.09

[a] This value fails the corresponding cut.

another particle having a measured momentum of 2.3 GeV/c and hitting the same calorimeter cell. The mass value listed in table 2 has been corrected accordingly. In minimum bias events the probability that a particle with a measured momentum >2.3 GeV/c hits a given calorimeter cell of the same θ is only 0.2%.

- One of the central electron candidates of event F passes all strict cuts but cut 3. The measured ratio of the signals from the two light guides of the impact cell is 0.71 ± 0.02 instead of 0.88 ± 0.02 as predicted from the track impact. Nothing suspicious has been found in the behaviour of the calorimeter cell from the monitoring of the stability of its calibration using light flasher and Co^{60} source measurements. We have also checked that the ratio between the light transmitted by the two light guides has the expected distribution in minimum bias events. Event F could be compatible with an electron pair hypothesis if a neutral particle (for example a hard bremsstrahlung photon) had entered the calorimeter cell very near its edge (~70 mrad away from the electron) causing many shower particles to cross the associated light guide. However the absence of a C5 cluster facing this region would imply that the photon did not convert in the 1.5 radiation lengths thick converter.

- The electron candidates of events G and H which do not pass the strict cuts are both observed in the central region. The latter fails cut 4 because it happens

to fall in the small region (Δφ = 18°) where C5 is non operational. We ignore this fact in the present discussion. Both fail cut 3 for the following reason: additional energy (~3 GeV) is observed in neighbour cells, inconsistent with lateral and longitudinal leakages of a shower initiated by an electron of the measured energy. In both cases this additional energy has an important component in the hadronic compartments: it is therefore difficult to ascribe it to radiative effects. We observe no track pointing to these cells, and no C5 cluster facing them. In the case of minimum bias events superimposed at random on large transverse momentum identified electrons we find that the probability of observing similar configurations is about 0.1%. We have also checked, using the light flasher system. that cross-talk between neighbour cells is negligible.

The presence of additional energy in these events has been ignored when calculating the invariant mass (table 2 and fig. 2). From the above discussion we retain the following points:

(a) Eight events (A - H) are observed with at least one electron passing the strict identification criteria. Their masses cluster in the 90 GeV/c^2 region where the expected background is only 0.03 events.

(b) Five of these events (A - E) are either identified as, or perfectly compatible with, e^+e^- or $e^+e^-\gamma$ configurations.

(c) The three other events (F-H) have both clus-

Table 4
Electron parameters (central region).

Cut (see Table I)	Event	A	B	C	C	D	E	F	F	G	G	H	H	H
	θ(degrees)	117.4	61.2	45.9	81.8	79.0	126.4	44.3	132.4	98.1	126.8	104.8	115.0	67.4
	φ(degrees)	38.3	299.7	130.2	59.5	27.7	147.2	1.7	123.3	306.4	346.6	162.2	198.6	27.0
	E(GeV)	49.5 ±2.0	48.3 ±2.3	73.2 ±4.0	11.4 ±0.9	24.4 ±1.4	50.8 ±2.0	38.6 ±2.1	53.0 ±5.5	46.0 ±1.8	45.0 ±1.8	47.5 ±2.3	46.5 ±2.0	42.1 ±1.7
	I(mip)	1.2	1.2		0.8		1.7	1.4	1.7	1.3	1.3	1.4	0.9	1.8
1	Max(R_θ,R_ϕ) (cell sizes)	0.25	0.36	0.39	0.39	0.26	0.31	0.40	0.16	0.38	0.36	0.16	0.13	0.12
	H(%)	6.5	5.0	12.9	5.0	8.9	8.1	6.2	10.8	8.7	8.9	3.1	10.6	4.4
3	Δ(cell sizes)	0.22	0.46	0.21	0.43		0.35	0.33	0.26	0.24	0.50	0.36	0.27	0.44
	P(%)	35	28	11	6	67	15	69	0.001[a]	83	0[a]	28	0.002[a]	60
4	q_s(mip)	14.5	24.8	23.8	13.9	29	4.5	2.4[a]	5.3	20.4	16.5	105.6	-[a]	31.6
5	d(mm)	3.2	1.8	0.9	1.5	—(γ)	1.9	8.9[a]	5.6	1.2	0.2	2.6	-[a]	3.2

[a] This value fails the corresponding cut.

ters in the central region. In each of these events one cluster fails cut 3. Although we retain the interpretation that this cluster is in each case associated with an electron, its configuration is inconsistent with our present knowledge of the detailed response of the central calorimeter to high energy electrons. However, we shall repeat the measurements of relevance in a high energy electron beam before drawing any definite conclusion on the significance of these inconsistencies.

The presence in fig. 2b of a signal free of background contamination, and the difficulties encountered in interpreting events F–H in terms of electron pairs, have led us to consider the hypothesis that the sample of fig. 2b could be contaminated by a background peaking in the (W, Z^0) mass region, but not made of genuine electron pairs. W and Z^0 decay modes other than $e^\pm \nu$ and e^+e^- (for example into two hadron jets) could provide such a mechanism. However we find it difficult to retain such an hypothesis because the sample of fig. 2a contains only two extra events, compatible with the background expectation of 0.7 events, in the mass region above 80 GeV/c^2.

4. Conclusions. The most likely interpretation of the eight events in fig. 2b is that they all result from the decays $Z^0 \to e^+e^-$ or $Z^0 \to e^+e^-\gamma$.

However three of these events (F–H) are not completely consistent with this hypothesis. Event D contains an electron for which the energy is not accurately measured. We restrict therefore the following discussion to the sample of four events (A–C and E) which can be used with confidence in an evaluation of the Z^0 mass and width.

From these events we measure the mass of the Z^0 boson to be:

$$M_Z = 91.9 \pm 1.3 \pm 1.4 \text{ GeV}/c^2, \qquad (2)$$

where the first error accounts for measurement errors and the second for the uncertainty on the overall energy scale.

The rms of this distribution is 2.6 GeV/c^2, consistent with the expected Z^0 width [13] and with our experimental resolution of ~3%.

Under the hypothesis of Breit–Wigner distribution we can place an upper limit on its full width

$$\Gamma < 11 \text{ GeV}/c^2 \quad (90\% \text{ CL}), \qquad (3)$$

corresponding to a maximum of ~50 different neutrino types in the universe [14].

The standard $SU(2) \times U(1)$ electroweak model makes definite predictions on the Z^0 mass. Taking into account radiative corrections to $O(\alpha)$ one finds [13]

$$M_Z = 77\, \rho^{-1/2}(\sin 2\theta_W)^{-1} \mathrm{GeV}/c^2, \qquad (4)$$

where θ_W is the renormalised weak mixing angle defined by modified minimal subtraction, and ρ is a parameter which is unity in the minimal model.

Assuming $\rho = 1$ we find

$$\sin^2\theta_W = 0.227 \pm 0.009. \qquad (5)$$

However, we can also use the preliminary value of the W mass found in this experiment [15]

$$M_W = 81.0 \pm 2.5 \pm 1.3 \ \mathrm{GeV}/c^2.$$

Using the formula [13]

$$M_W = 38.5\,(\sin\theta_W)^{-1} \mathrm{GeV}/c^2, \qquad (6)$$

we find $\sin^2\theta_W = 0.226 \pm 0.014$, and using also eq. (4) and our experimental value of M_Z we obtain

$$\rho = 1.004 \pm 0.052, \qquad (7)$$

in agreement with the prediction of the minimal $SU(2) \times U(1)$ model, with the recent results of the UA1 experiment [4], and with the results of low energy neutrino experiments [16].

This experiment would have been impossible without the collective effort of the staffs of the relevant CERN accelerators. They are gratefully acknowledged.

We are grateful to the UA4 Collaboration for providing signals for the luminosity measurement.

We are deeply indebted to the technical staffs of the Institutes collaborating in UA2 for their invaluable contributions.

Financial supports from the Schweizerischer Nationalfonds zur Förderung der Wissenschaftlichen Forschung to the Bern group, from the Danish Natural Science Research Council to the Niels Bohr Institute Group, from the Institut National de Physique Nucléaire et de Physique des Particules to the Orsay group, from the Istituto Nazionale di Fisica Nucleare to the Pavia group, and from the Institut de Recherche Fondamentale (CEA) to the Saclay group are acknowledged.

References

[1] S.L. Glashow, Nucl. Phys. 22 (1961) 579;
S. Weinberg, Phys. Rev. Lett. 19 (1967) 1264;
A. Salam, Proc. 8th Nobel Symp. (Aspenäsgården, 1968) (Almqvist and Wiksell, Stockholm) p. 367.

[2] G. Arnison et al., Phys. Lett. 122B (1983) 103.

[3] M. Banner et al., Phys. Lett. 122B (1983) 476.

[4] G. Arnison et al., Phys. Lett. 126B (1983) 398.

[5] UA2 Collab., B. Mansoulié, presented at XVIIIth Rencontre de Moriond on Proton–antiproton collider physics (La Plagne, March 1983), to be published.

[6] A. Beer et al., The central calorimeter of the UA2 experiment at the CERN $\bar{p}p$ collider, to be published in Nucl. Instrum. Methods.

[7] M. Banner et al., Phys. Lett. 115B (1982) 59.

[8] M. Banner et al., Phys. Lett. 122B (1983) 322.

[9] C. Conta et al., The system of forward–backward drift chambers in the UA2 detector, to be published in Nucl. Instrum. Methods.

[10] K. Borer et al., Multitube proportional chambers for the location of electromagnetic showers in the CERN UA2 detector, to be published in Nucl. Instrum. Methods.

[11] M. Banner et al., Phys. Lett. 118B (1982) 203;
P. Bagnaia et al., Measurement of production and properties of jets at the CERN $\bar{p}p$ collider, CERN-EP/83–94 (July 1983), submitted to Z. Phys. C.

[12] UA2 Collab., P. Darriulat, presented at Third Topical Workshop on $\bar{p}p$ collider physics (Rome, January 1983), CERN 83–04 (May 1983), eds. C. Bacci and G. Salvini (CERN, Geneva, 1983) pp. 235, 236.

[13] For a review see: M.A.B. Bég and A. Sirlin, Phys. Rep. 88 (1982) 1 and references quoted therein (in particular Section 4, references 27–39);
see also W.J. Marciano and Z. Parsa, Proc. AIPDPF, Summer Study on Elementary particle physics and future facilities (Snowmass, CO, 1982) (AIP, New York, 1983) p. 155;
M. Consoli, S. Lo Presti and L. Maiani, Higher order effects and the vector boson physical parameters, University of Catania preprint, PP/738-14/1/1983.

[14] J. Ellis, M.K. Gaillard, G. Girardi and P. Sorba, Ann. Rev. Nucl. Part. Sci. 32 (1982) 443, and references therein.

[15] UA2 Collab., G. Sauvage, preliminary results presented at Intern. Europhysics Conf. on High energy physics (Brighton, UK, July 1983);
UA2 Collab., A.G. Clark, at 1983 Intern. Symp. on Lepton and photon interactions at high energies (Cornell, USA, August 1983).

[16] For a recent review see: M. Davier, Supp. J. Phys. (Paris) 12, C3 (1982) 471, and references therein.

CHAPTER 8
HIGH PRECISION STUDIES
OF THE ELECTROWEAK FORCE

From the Top Down (1983 - 1995)

8.1. RADIATIVE CORRECTIONS AND TESTS OF THE STANDARD MODEL

The work of t' Hooft, Veltman, and others showed that the GWS model could be renormalized (see Chapter 4). This means that the radiative corrections will be finite, and this expanded the definition of $\sin^2\theta_w$. The pioneering calculations of the actual radiative corrections were done by A. Sirlin, W. Marciano, M. Peskin, J. Rosner, and others [8.1]. [We reprint here, a paper (A) by A. Sirlin, who was among the first to calculate the radiative corrections to the mass of the W and Z particles.) This is the first of the precision corrections to the standard model. Early detailed analysis of the data and fits to the parameters are discussed in Ref. [8.2].

Since the heyday of the 1970s and the W and Z particle discovery, the task of the experimentalists has been largely to measure the parameters of the model with ever increasing precision and to search for deviations from the standard model. We will briefly discuss the status of this endeavor in this chapter.

Ever since the discovery of the W and Z bosons, there has been a remarkable agreement of experimental data with the standard model. Future tests of the model might possibly go in three directions:

1. Higher energy fundamental collisions to search for breakdown of the standard model or further confirmation in the discovery of the Higgs boson;
2. Precision measurements of some key quantities predicted in the standard model (such as the W, Z mass difference) or the search for new processes at low energy that do not fit into the model. The observation of a WNC process that changes hadronic flavor would indicate a deviation from the standard model that would have significant consequences. Sensitive searches for FCNC processes have been carried out for the past 20 or so years. These early results helped frame the standard model and led to the need for the GIM mechanism [see Chapters 3 and 4.]. The observation of such processes now could lead to a comparable breakthrough.

This summary takes us to 1996 − 13 years after the discovery of the W and Z particles. During this time, there was remarkable improvement in the precision of the basic parameters of the standard model, the discovery of the (possibly) last quark (the t quark), and further searching for FCNC beyond those searches described in Chapter 3.

The easiest way to see the success of the model is to plot $\sin^2\theta_w$ as a function of momentum transfer, Q^2, for a variety of experiments and measurements, as we show in Fig. 8.1. Note the remarkable agreement.

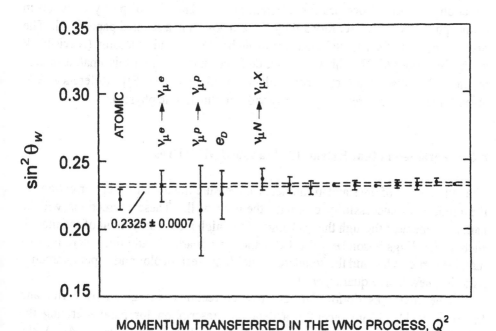

Figure 8-1. Plot of a recent set of measurements of $\sin^2\theta_w$ in different experiments and measurements as a function of the momentum transferred, Q^2, in the WNC process. (The range is from 10^{-6} to 10^4 GeV/c^2.)

8.2. PRECISION MEASUREMENT OF THE ELECTROWEAK PARAMETERS

The number of different types of measurements of the electroweak parameters is remarkable. As shown in Fig. 8.1, these range from atomic parity violation at very low Q^2 to the study of W and Z parameters. Currently, only the W, Z (i.e., $e^+e^- \to Z$ at LEP and SLC) and deep-inelastic neutrino scattering provide the most precise measurements. In the future, we expect improvements in all measurements. It is very important to study the electroweak parameters as a function of Q^2, since the radiation corrections vary with Q^2. The paper by P. Langacker gives an excellent introduction to these measurements [8.3].

In order to extend the study of the effects of the radiative corrections, several sets of new parameters have been proposed [8.4].

8.2.1. Atomic Physics Measurements

A key new type of measurement to study the standard model is that of parity violation in atomic processes for different atoms. The idea of parity violation in atomic physics was first set forward by Ya. Zel'dovich in a seminal paper [8.5]. The first observation of parity violation was made by Barkov and Zolotorev (Novisibirsk group) [see paper 6(E)]. This was a tour de force experiment for this small team and predicted the observation of parity violation at SLAC [see 6(F)]. Reference [8.6] gives some recent references for parity violation in atomic physics.

8.2.2. Parameters that Extend the Standard Model Tests

The ultimate test of the standard model is to measure the higher order corrections to the model and successfully calculate them as well. These corrections will be partially generated through the exchange of the high mass component of the model, such as the Higgs boson ($m_H < 1$ TeV) and the t quark. In addition, there may be massive particles beyond the standard model (*e.g.*, technicolor and supersymmetric particles, new heavy quarks, *etc.*).

The first prescription for radiative corrections was given by Sirlin and Marciano [8.1]. More recently, a detailed prescription for parameterizing the radiative corrections has been proposed by Altarelli, Peskin, and others [8.4]. A key method to measure these parameters is to carry out measurements of electroweak parameters at different energies and by different techniques (W,Z decays, atomic physics parity-violation measurements, low-energy neutrino measurements of $\sin^2 \theta_W$, *etc.*).

There is no completely universal prescription for these new parameters. We adopt the S, T, U parameterization in this chapter. At the tree level, the GWS model gives

$$\sin^2 \theta_W = \frac{e^2}{g^2} \text{ and } \frac{G_F}{\sqrt{2}} = \frac{g^2}{8 m_W^2} . \tag{8.1}$$

With the inclusion of radiative corrections, these definitions become

$$\sin^2 \theta_W = \frac{\pi \alpha / \sqrt{2} G_F}{m_W^2 (1 - \Delta r_W)} , \tag{8.2}$$

where Δr is due to the radiative corrections. At the tree level

$$\sin^2 \theta_W = 1 - \frac{m_W^2}{\rho_0 m_W^2} , \tag{8.3}$$

where $\rho_0 = 1$ is the standard model at tree level. Radiative corrections modify this to

$$\sin^2 \theta_W = 1 - \frac{m_W^2}{\rho_{mass} m_W^2} \quad , \tag{8.4}$$

where $\rho_{mass} = \rho_0 (1 + \delta\rho_{mass})$. Sometimes the $\rho_0 = 1$ case is used to define $\sin^2 \theta_W$ when the m_W and m_Z masses are obtained directly [8.4].

$$S = \Delta\rho \quad , \tag{8.5}$$

$$T = C_2^0 \Delta\rho + \frac{S_2^0 \Delta r_W}{(C_0^2 - S_0^2)} - 2S_0^2 \Delta k' \quad , \tag{8.6}$$

$$U = C_2^0 \Delta\rho + (C_0^2 - S_0^2) \Delta k' \quad , \tag{8.7}$$

where $\Delta\rho$ and $\Delta k'$ are given by

$$G_F = -\frac{1}{2}\left(1 + \frac{\Delta\rho}{\alpha}\right) \quad , \tag{8.8}$$

$$\frac{G_V}{G_A} = 1 - 4(1 + \Delta k') S_0^2 \quad , \tag{8.9}$$

where $S_0^2 = \sin^2 \theta_W$ before QED corrections. The parameters S, T, U are therefore related to $\Delta\rho, \Delta k', \Delta r_W$. For very large t quark mass, m_t, all of the corrections are dominated by $G_F M_t^2$. There is also a contribution to these parameters from the Higgs mass, but it is very small. The analysis in Ref. [8.3] gives the following for these parameters:

$$S = -0.29 \pm 0.46 \quad ,$$

$$T = -0.05 \pm 0.43 \quad ,$$

$$U = 0.37 \pm 0.93 \quad ;$$

thus the parameters are very poorly determined at present. So far there is no indication that these parameters deviate from zero.

8.3.   CONTRIBUTIONS FROM A MASSIVE t QUARK

In Section 8.2, we discussed the effects of radiative corrections on the parameters of the WNC, Z^0, and W bosons. The leading corrections likely arise from the t quark and the Higgs boson. Beyond this, most other models that give physics beyond the standard model will likely give some changes to the WNC radiative corrections.

The ultimate strategy for detecting these changes is to measure the S, T, U parameters (discussed in Section 8.2) precisely [8.4]. The leading corrections from the t quark could be evaluated after the t quark has been discovered [8.7]. Then any remaining contributions from the new physics could be identified, including the Higgs boson mass [8.8]

Technicolor models give rise to contributions to the parameter S of $\Delta S = 1/6\pi$ per new fermion doublet. Extra Z bosons and massive Majorana neutrinos give negative values of S. The parameter T can get contributions from massive Majorana neutrinos and novel Higgs models. Thus continued study of the electroweak parameters could reveal new physics – beyond the standard model. However at present, there is no reliable indication of such effects.

8.4.   PRESENT AND FUTURE SEARCHES FOR FCNCs

Another way to test the standard model and to search for physics beyond the standard model is to observe forbidden processes, such as FCNC processes.

The progress in the search for FCNC up to about 1986 was described by the author [reprinted here (B)]. The major development from the late 1960s (Chapter 3) was the use of charm and beauty particles to search for FCNC. While the limits have not really changed in the past 10 years, we expect important new searches in the next 10 years that we now describe.

In the search for tree-level FCNC, we seek a first order process, such as that shown in Fig. 8.2(A), compared with the expected second-order processes, such as those described by the diagrams of Fig. 8.2(B) and (C) (for $K^+ \rightarrow \pi^+ \nu\bar{\nu}$). Of course, effects from second order diagrams are observed, such as K^0–$\bar{K}^0$ and B^0–$\bar{B}^0$ mixing. These diagrams are shown in Fig. 8.3. In Fig. 8.4, we summarize the current limit on the FCNC decay modes as a function of the year the search was made [8.9].

8.4.1.   Prospects for Detecting FCNCs at New Facilities

Flavor factories, which use intense sources of strangeness, beauty states, and charm, as well as leptonic flavors such as the τ, have led in the past (and will in the future) to new and extremely sensitive searches for FCNC. In addition, the study and possible origin of CP violation will be probed at these facilities.

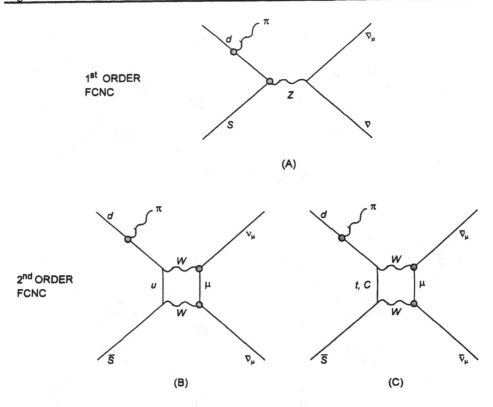

Figure 8.2. Diagrams for FCNC at (A) the tree level, and (B) and (C) by higher order corrections.

After the discovery of $B^0 - \overline{B}^0$ mixing, an intense study of possible B factories was started around the world [8.10]. The driving force behind this interest is the possibility of detecting CP violation in the b quark system.

In Section 8.4.2, we will highlight the possible future experiments that could extend the FCNC search to ever more sensitive levels. We will not discuss processes that change leptonic flavor, since the mass of the neutrino plays an important role in such processes in the standard model and perhaps reduces the sensitivity to new interactions for nearly massless neutrinos. We believe that a new round in the search for FCNC, especially for the heavy flavors such as charm and beauty, is of great importance.

8.4.2. Flavor Factories and Hadron Machine Sources of Heavy Flavors: Rare B Decays

The key to searching for FCNC or to studying flavor systems is machines – flavor factories, which provide intense sources of flavor. There are two main routes to flavor factories: (1) e^+e^- colliders and (2) hadronic production. Below we list some comparisons that we have made of these different approaches.

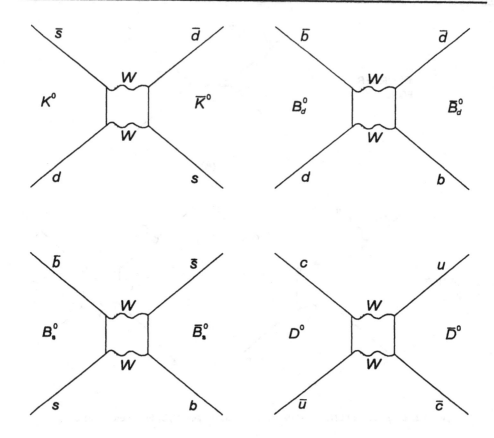

Figure 8.3. Diagrams that describe the process of flavor oscillations or mixing for (A) $K^0 - \overline{K}^0$, (B) $B_d^0 - \overline{B}_d^0$, (C) $B_s^0 - \overline{B}_s^0$, and (D) $D^0 - \overline{D}^0$. (A)-(C) have been detected; (D) has not been observed.

1. e^+e^- Collider Option: This option uses the increased cross section at a resonance, such as the ϕ, $\psi(3S)$, and others (Cornell, KEK, and SLAC are all constructing B factories). This method leads to a very clean environment for the study of the final states. However, since the cross section is relatively small, this current state of the art ($\mathcal{L} \gtrsim 3 \times 10^{33}$ cm^{-2} s^{-1} for a $B\overline{B}$ factory) will give, at most, 10^8 B particles to study.

2. The Hadronic Production Option: High-energy hadronic collisions produce copious numbers of flavors (charm, beauty particles, *etc.*). This is especially true for $B\overline{B}$ production with very high-energy machines, such as the LHC. In Table 8.1 we provide a comparison of the rate of production of $B\overline{B}$ pairs at Fermilab (TEV I and II) and at the LHC being constructed at CERN. We also provide some of the other properties of the particles, such as mean decay distance and mean momentum (taken from Ref. [8.9]).

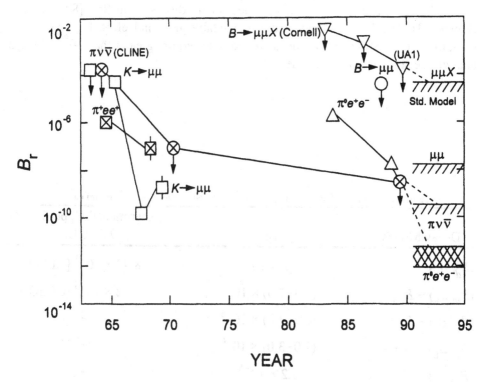

Figure 8.4. Current limits on FCNC decay modes, and the expectation of one model for tree level FCNC [8.9].

Table 8.1. Important parameters of beauty production in various hadronic experimental configurations [8.11].

	TEV II Fixed Target	TEV I Collider	LHC Collider
Interaction rate	$10^7 - 10^8/s$	$10^5/s$	$10^7/s$
$\sigma(pN \rightarrow B\bar{B})$	10 nb	20 μb	200–500 μb
$B\bar{B}/10^7$ s	$10^7 - 10^8$	4×10^8	$2 \times 10^{11} - 5 \times 10^{11}$
$\sigma(b\bar{B})/\sigma_T$	1/2,250,000	1/2,500	1/500 – 1/200
Multiplicity	≈ 15	≈ 45	"few" 100
$\langle P_b \rangle$	1.43 GeV/c	38 GeV/c	51 GeV/c
$\langle P_B \rangle$	118 GeV/c	22 GeV/c	43 GeV/c
$\langle P_\mu \rangle$	32 GeV/c	13 GeV/c	36 GeV/c
Median B decay length	8 mm	1.5 mm	3 mm
Mean B decay length	16 mm	4.7 mm	13 mm

The clean probes of FCNC in B decays are discussed in Ref. [8.9] and are shown in Table 8.2, which is adapted from the table given in Ref. [8.12]. As can be seen, many interesting decay modes are likely to require greater than 10^8 B decays and will be studied at the LHC detectors.

Table 8.2. Estimates of the branching fractions for FCNC B decays in the standard model.[*]

Decay Modes	Br	Experimental Upper Limits (90% CL)
$(B_d, B_u) \rightarrow X_s \gamma$	4.2×10^{-4}	8.40×10^{-4} [CLEO]
$(B_d, B_u) \rightarrow K^* \gamma$	$(4.0-7.0) \times 10^{-5}$	0.92×10^{-4} [CLEO]
$(B_d, B_u) \rightarrow X_d \gamma$	$(0.5-3.0) \times 10^{-5}$	$-$
$(B_d, B_u) \rightarrow \rho + \gamma$	$(1.0-3.0) \times 10^{-6}$	$-$
$(B_d, B_u) \rightarrow X_s e^+ e^-$	1.2×10^{-5}	$-$
$(B_d, B_u) \rightarrow X_s \mu^+ \mu^-$	6.7×10^{-6}	5.0×10^{-5} [UA1]
$(B_d, B_u) \rightarrow K e^+ e^-$	4.4×10^{-7}	5.0×10^{-5} [PDG]
$(B_d, B_u) \rightarrow K \mu^+ \mu^-$	4.4×10^{-7}	1.5×10^{-4} [PDG]
$(B_d, B_u) \rightarrow K^* e^+ e^-$	3.7×10^{-6}	$-$
$(B_d, B_u) \rightarrow K^* \mu^+ \mu^-$	2.3×10^{-6}	2.3×10^{-5} [UA1]
$(B_d, B_u) \rightarrow X_s \nu \bar{\nu}$	6.6×10^{-5}	$-$
$(B_d, B_u) \rightarrow K \nu \bar{\nu}$	5.2×10^{-6}	$-$
$(B_d, B_u) \rightarrow K^* \nu \bar{\nu}$	2.0×10^{-5}	$-$
$B_S \rightarrow \gamma \gamma$	2.0×10^{-8}	$-$
$B_S \rightarrow \tau^+ \tau^-$	1.8×10^{-7}	$-$
$B_S \rightarrow \mu^+ \mu^-$	8.3×10^{-10}	$-$
$B_S \rightarrow e^+ e^-$	2.0×10^{-14}	$-$

[*]For $m_t = 150$ GeV and $f_B = 200$ MeV. Note that the CKM-suppressed decays, given in rows 3 and 4, depend on $|V_{dt}|$, and the numbers correspond to $|V_{dt}| = 0.007$. Experimental upper limits are also listed. (Based on [8.12].)

8.4.2. Rare K Decay Studies

Ambitious experimental programs are underway at BNL and KEK to continue the search for FCNC K decays to a very low level. The level of sensitivity that the experiments hope to reach is given in Table 8.3. These new experiments will use more intense K_L^0 beams if used in a dedicated mode. In addition, ϕ factories will provide an intense source of K_S^0 mesons. The experience gained at BNL and KEK will be invaluable in order to access the achieved background levels for the rare K decay searches.

Rare K decay searches have been crucial to the development, which started with the search for $K^+ \rightarrow \pi^+ e^+ e^-$ in 1963 (see Chapter 3). A similar urgency exists today for the search for various decays, such as

$$K^+ \rightarrow \pi^+ \nu \bar{\nu} \ , \tag{8.10}$$

$$K_L^0 \rightarrow \pi^0 e^+ e^- \ , \tag{8.11}$$

as well as the further study of the decays

$$K^+ \rightarrow \pi^+ e^+ e^- \ , \tag{8.12}$$

$$K^+ \rightarrow \pi^+ \gamma\gamma \ , \quad \pi^0 \gamma\gamma \ . \tag{8.13}$$

In Table 8.3, we give some information about on-going studies and searches for rare K decays (experiments are labeled by the experiment number at the laboratory) [8.11].

It is clear that the search for tree-level FCNC in B, charm, and K decays is far from over, and in many cases is just starting. Thus some of the most important tests of the standard model will occur in the next 10 years at the new $e^+ e^-$ flavor factories, the LHC, and the intense K factories.

Table 8.3. Recent progress in rare kaon decays.[*]

Decay Mode	Experiment	Result	Sensitivity	Comments
$K^+ \to \pi^+ \nu \bar{\nu}$	BNL E787	$< 3.0 \times 10^{-8}$	10^{-10}	Detector works well
$K^+ \to \pi^+ \mu^+ \mu^-$		$< 2.3 \times 10^{-7}$		
$K^+ \to \mu^+ \nu \, \mu^+ \mu^-$		$< 4.1 \times 10^{-7}$		
$K^+ \to \pi^+ \mu^+ e^-$	BNL E777	$< 2.1 \times 10^{-10}$	$< 1.5 \times 10^{-10}$	Limited by beam halo
	BNL E865		$< 2.0 \times 10^{-12}$	
$K_L \to \mu^\pm e^\pm$	BNL E780	$< 1.9 \times 10^{-9}$		Limited by beam halo
$K_L \to e^+ e^-$		$< 1.2 \times 10^{-9}$		
$K_L \to \pi^0 e^+ e^-$		$< 3.2 \times 10^{-7}$		Will be continued ($\sim 10^{-10}$)
$K_L \to \mu^\pm e^\pm$	BNL E791	$< 2.2 \times 10^{-10}$	$\sim 2.0 \times 10^{11}$	Limited by accidentals from K_L decays
$K_L \to e^+ e^-$		$< 3.2 \times 10^{-10}$		
$K_L \to \mu^\pm e^\pm$	KEK E137	$< 4.0 \times 10^{-10}$	$\sim 2.0 \times 10^{-11}$	
$K_L \to e^+ e^-$		$< 5.4 \times 10^{-10}$		$K_L \to \pi^0 e^+ e^-$ ($\sim 10^{-10}$)
$K_L \to \mu^\pm e^\pm$	BNL E871		$\sim 3.0 \times 10^{-12}$	
$K_S \to \pi^+ \pi^- \pi^0$	FNAL E621	$< 1.5 \times 10^{-7}$	$\sim 3.0 \times 10^{-9}$	Expected rate $\sim 1.2 \times 10^{-9}$ (CP violation component)
$K_L \to \pi^0 e^+ e^-$	FNAL E371	$< 4.2 \times 10^{-8}$	$\sim 1.0 \times 10^{-8}$	
	CERN NA31	$< 4.0 \times 10^{-8}$		
$K_L \to \pi^0 e^+ e^-$	KEK E162		10^{-10}	
$K_L \to \pi^0 e^+ e^-$	FNAL P799		$10^{-10} - 10^{-11}$	

[*]Note that some results are preliminary while others have been published.

REFERENCES

8.1. A. Sirlin, *Rev. Mod. Phys.*, **50**, 573 (1978); W. J. Marciano and A. Sirlin, *Phys. Rev.*, **D22**, 2695 (1980); also see article (A).

8.2. U. Amaldi *et al.*, *Phys. Rev.*, **D36**, 1385 (1987); G. Costa *et al.*, *Nucl. Phys.*, **297B**, 244 (1988). Deep inelastic scattering was also considered by G. L. Fogli and D. Haidt, *Z. Phys.*, **C40**, 379 (1988).

8.3. P. Langacker, in *Discovery of Weak Neutral Currents: The Weak Interaction Before and After* (Proc., Santa Monica, CA 1993), A. K. Mann and D. B. Cline, eds. (AIP Conference Proceedings 300, AIP, New York, 1994), p. 289.

8.4. M. E. Peskin and T. Tacheuchi, *Phys. Rev.*, **D46**, 381 (1992); G. Altarelli and R. Barbieri, *Phys. Lett.*, **253B**, 161 (1991); G. Altarelli, R. Barbieri, and S. Jadach, *Nucl. Phys.*, **369B**, 3 (1992).

8.5. Ya. B. Zel'dovich, *Sov. Phys. JETP*, **6**, 1184 (1957).

8.6. M. A. Bouchiat and C. C. Bouchiat, *Phys. Lett.*, **46B**, 111 (1974); also *J. Phys. (Paris)*, **35**, 899 (1974); D. N. Stacey, in *Atomic Physics 13*, H. Walther, T. W. Hänsch, and B. Neizert, eds. (AIP, New York, 1992), p. 46; I. B. Khriplovich, *Commun. At. Mol. Phys.*, **23**, 189 (1989); N. Fortson and L. L. Lewis, *Phys. Rep.*, **113**, 289 (1984); D. M. Meekhof *et al.*, *Phys. Rev. Lett.*, **71**, 3442 (1993).

8.7. CDF Collaboration, F. Abe *et al.*, *Phys. Rev. Lett.*, **74**, 2626 (1995); D^0 Collaboration, S. A. Bachi *et al.*, *Phys. Rev. Lett.*, **74**, 2632 (1995).

8.8. See for example J. Ellis *et al.*, *Phys. Lett.*, **333B**, 118 (1994).

8.9. D. B. Cline, in *Discovery of Weak Neutral Currents: The Weak Interaction Before and After* (Proc., Santa Monica, CA 1993), A. K. Mann and D. B. Cline, eds. (AIP Conference Proceedings 300, AIP, New York, 1994), p. 175.

8.10. B^0 mixing was discovered by the UA1 group at CERN and the ARGUS group at DESY in 1987–1988.

8.11. See A. Ali, in *Discovery of Weak Neutral Currents: The Weak Interaction Before and After* (Proc., Santa Monica, CA 1993), A. K. Mann and D. B. Cline, eds. (AIP Conference Proceedings 300, AIP, New York, 1994), p. 437.

8.12. Some of these results are taken from *Discovery of Weak Neutral Currents: The Weak Interaction Before and After* (Proc., Santa Monica, CA 1993), A. K. Mann and D. B. Cline, eds. (AIP Conference Proceedings 300, AIP, New York, 1994).

RADIATIVE CORRECTIONS IN THE STANDARD MODEL

A. Sirlin
New York University, New York, N. Y. 10003

ABSTRACT

Recent developments in the analysis of radiative corrections in the standard model are discussed. Particular emphasis is given to the effect of these corrections on the theoretical predictions of intermediate boson masses.

INTRODUCTION

The basic aim of these studies is the verification of the Standard Model of Electroweak Interactions (SM), at the level of its quantum corrections.

Loop effects are important in many processes. For instance, an old problem where they play a crucial role is the analysis of universality of the weak interactions.[1] In the present context this problem can be formulated as follows: is the phenomenologically derived Kobayashi-Maskawa (K-M) matrix unitary? The first two entries of this matrix are obtained from the superallowed Fermi transitions, μ decay and $\Delta Q = \Delta S = 1$ semileptonic processes. Ignoring radiative corrections, one finds

$$|U_{ud}|^2 + |U_{us}|^2 > 1 \tag{1}$$

by about 3.2%, hardly the property of a unitary matrix! Clearly, large loop effects are needed that differentiate β and μ decays so that the largest coefficient $|U_{ud}|^2$ can be altered. In the SM there exists an asymptotic theorem that indicates that this indeed may occur. It states that in semileptonic processes mediated by $W^{\pm}$ the large m_Z behaviour of the $O(\alpha)$ corrections is given by[2]

$$\Delta P = \overset{o}{P}\left[\frac{3\alpha}{2\pi}(1+2\bar{Q})\ln(\frac{m_Z}{\mu}) + ...\right] \tag{2}$$

where $\overset{o}{P}$ is the zeroth order probability expressed in terms of the μ-decay coupling constant G_μ (defined in the local V-A theory from the muon lifetime), μ is an unspecified mass scale characteristic of the process, $\bar{Q}$ is the average charge of the quark isodoublet and the ellipses represent subdominant contributions. The coefficient of the leading term in Eq(2) is not affected by the strong interactions. For semileptonic processes, with the usual charge assignments one has $\bar{Q} = 1/6$; applying Eq(2) to μ-decay: $\bar{Q} = (Q^{(\mu)} + Q^{(\nu_\mu)})/2 = -1/2$ so that there are no large logarithms in the latter process. Detailed calculations based on Current-Algebra and Short-Distance Expansion techniques gives for the decay probability of the superallowed Fermi transitions[1]:

$$\Delta P = \overset{o}{P}\left[\frac{3\alpha}{2\pi}\ln(\frac{m_Z}{2E_m}) + ...\right] \tag{3}$$

where E_m is the end-point energy of the positron and the ellipses denote less important terms. The argument of the logarithm in Eq.(3) is typically $\simeq 2 \times 10^4$ and, as a consequence, $|U_{ud}|^2$ is decreased by $\simeq 3.7\%$. One finds, when radiative corrections are included: $|U_{ud}| = 0.9735 \pm 0.0016$. Combining with the recent value $|U_{us}|^2 = 0.221$[3]: $|U_{ud}|^2 + |U_{us}|^2 = 0.997 \pm 0.004$. Remembering $|U_{ub}|^2 < 10^{-4}$, consistency with unitarity of the K-M matrix is achieved!

Historically, it is interesting to recall that the early calculations of radiative corrections[4] were carried out in the framework of the Fermi theory, where the effective ultraviolet cutoff was infinite rather than m_Z. In particular, in the case of the local V-A theory, the difference in ultraviolet behavior embodied in Eq (2) manifested itself in a dramatic manner: while the QED corrections to μ decay were finite, those affecting β decay were divergent to 0 (α)! This has been one of the many features of the old theory which suggested the need to modify its space-time structure at short distances and the possible existence of the intermediate vector boson.[5]

Recently a new generation of problems have emerged: they involve the theoretical predictions of m_W, m_Z and their interdependence and the determination of $\sin^2\theta_W$. Several theoretical groups have studied these problems.[6] I will follow the approach of a program carried out in collaboration with W.J. Marciano. Its methodology is based to a large extent on the Current Algebra Formulation of Radiative Corrections[1] and a "Simple Renormalization Framework"[7].

SIMPLE RENORMALIZATION FRAMEWORK

This is a method to renormalize the theory at the S-matrix level. The basic idea can be readily described by considering the vector boson mass terms and their interactions with quarks and leptons:

$$L_M = \frac{\lambda_0^2 g_0^2}{4} W_\mu^\dagger W^\mu + \frac{\lambda_0^2 (g_0^2 + g_0'^2)}{8} Z_\mu Z^\mu \qquad (4a)$$

$$L_{int} = -\frac{g_0 g_0'}{(g_0^2 + g_0'^2)^{\frac{1}{2}}} A_\mu J_\gamma^\mu - (g_0^2 + g_0'^2)^{\frac{1}{2}} Z_\mu J_Z^\mu - \frac{g_0 (W_\mu^\dagger J_W^\mu + h.c.)}{2^{\frac{1}{2}}} \qquad (4b)$$

where the zero subscripts denote unrenormalized parameters, g_0 and g_0' are the gauge couplings associated with $SU(2)_L$ and $U(1)$ and $\lambda_0 = 2 < 0|\phi_0|0>$ is the vacuum expectation value of the scalar field. Writing $g_0 = g - \delta g$, $g_0' = g' - \delta g'$, $\lambda_0 = \lambda - \delta\lambda$, we need three constraints to determine $\delta g, \delta g'$ and $\delta\lambda$. We require that the renormalized version of the masses, namely $m_W^2 \equiv \lambda^2 g^2/4$, $m_Z^2 \equiv \lambda^2 (g^2 + g'^2)/4$ be the physical masses in the presence of the radiative corrections. Further we demand that $gg'/(g^2 + g'^2)^{\frac{1}{2}} = |e|$ be the conventionally defined charge of the positron. This determines δg, $\delta g'$, $\delta\lambda$ in terms of various self energies and charge renormalization counterterms. The method has some simple features illustrated below:

Tree Level	In This Method
$\cos\theta_W = m_W/m_Z$	Remains exact
$e = g\,\sin\theta_W$	Remains exact
$\dfrac{G_\mu}{2^{\frac{1}{2}}} = \dfrac{g^2}{8m_W^2}$	Corrected in $O(\alpha)$

THEORETICAL PREDICTION OF m_W, m_Z.

The first important application is to study the corrections to muon decay.[7] It is convenient to define G_μ by

$$\frac{1}{\tau_\mu} = G_\mu^2 \ P\left[1+ \frac{\alpha}{2\pi}\left(\frac{25}{4} - \pi^2\right)\left(1+ \frac{2\alpha}{3\pi}\ln(\frac{m_\mu}{m_e})\right)\right] \tag{5a}$$

$$P = \left[1-8\frac{m_e^2}{m_\mu^2}\right]\left[1+ \frac{3}{5}\frac{m_\mu^2}{m_W^2}\right]\frac{m_\mu^5}{192\pi^3} \tag{5b}$$

which, aside from the very small term $(3/5)(m_\mu^2/m_W^2)$, are the relevant formulae of the local V-A theory.[4] Evaluation of the corrections in the $SU(2)_L \times U(1)$ theory leads to

$$\frac{G_\mu}{2^{\frac{1}{2}}} = \frac{g^2}{8m_W^2(1-\Delta r)} \tag{6a}$$

where $\Delta r = \Delta r^{(1)}+\Delta r^{(2)}+\Delta r^{(3)}$ denote the $O(\alpha)$ correction; $\Delta r^{(i)}$ $(i=1,2,3)$ represent contributions from the bosonic, leptonic and hadronic sectors, respectively. It is important to note that Eq(6a) connects the fundamental parameters of the old and new theories. One finds that $\Delta r^{(1)}$ is a complicated function of $\sin^2\theta_W$ and $\xi \equiv m_H^2/m_Z^2$; it turns out to be small: $\simeq -0.22\%$ for $\xi=0$, 0.30% for $\xi=1$ and 1.18% for $\xi=100$. On the other hand

$$\Delta r^{(2)} = \frac{2\alpha}{3\pi}\left[\sum_{l=e,\mu,\tau} \ln\left(\frac{m_Z}{m_l}\right) - \frac{5}{2} + \frac{3}{8}\frac{(2s^2-1)}{s^4}\ln c^2\right] \tag{6b}$$

$$\Delta r^{(3)} = -e^2 \mathrm{Re}\ \pi_{\gamma\gamma}^{r(h)}(m_Z^2) + \frac{3\alpha}{4\pi}\frac{(2s^2-1)}{s^4}\ln c^2 \tag{6c}$$

where $s^2 \equiv \sin^2\theta_W$, $c^2 \equiv \cos^2\theta_W$, $\pi_{\gamma\gamma}^{r(h)}(m_Z^2)$ is the hadronic contribution to the renormalized vacuum polarization function evaluated at m_Z^2. For simplicity, in Eq.(6c) we have neglected terms of $O(m_f^2/m_W^2)$ but the complete expressions are known[8,9]. The large logarithms make $\Delta r^{(2)}$ and $\Delta r^{(3)}$ dominant. Their existence was anticipated in Ref. 10. The origin of the mass singularities in Eqs.(6b,c) can be traced to the definition $g \equiv e/\sin\theta_W$ and the fact that e is evaluated at $q^2=0$ (exceptional invariant momentum).

Numerically, one finds:

$$\Delta r = 0.0696 \pm 0.0020 \qquad \left[\begin{array}{l} \sin^2\theta_W = 0.217 \\ m_H = m_Z \\ m_t = 36 \text{GeV} \end{array} \right. \qquad (6d)$$

where the error arises from uncertainties in the hadronic contributions. The results are not sensitive to small variations in $\sin^2\theta_W$, m_H, m_t. For this reason we may take Eq(6d) as the "standard value" for Δr. Recently, these muon decay calculations have been confirmed analytically to $O(\alpha)$[11]. What about corrections of $O(\alpha^2)$ and higher? Although complete calculations do not exist, renormalization group arguments and the theorems on cancellation of mass singularities have been used[12] to show that terms of $O(\alpha^n \ln^n m)$ $n \geq 2$ as well as $O(\alpha^2 \ln m)$ (m is a generic fermion mass and we consider here the limit of small m) are contained in

$$\frac{G_\mu}{2^{\frac{1}{2}}} = \frac{g^2}{8m_W^2} \frac{1}{\left[1 - \Delta r + e^4 \text{Re } \pi_2^{(r)}(m_Z^2) \right]} \qquad (7a)$$

The last term in the denominator represents the fourth order contribution to the vacuum polarization and is very small for present applications. Neglecting such term we find:

$$m_W = \left(\frac{\pi\alpha}{2^{\frac{1}{2}} G_\mu} \right)^{\frac{1}{2}} \frac{1}{\sin\theta_W} \left[\frac{1}{1 - \Delta r} \right]^{\frac{1}{2}} = \frac{38.65 \pm 0.04 \text{ GeV}}{\sin\theta_W} \qquad (8a)$$

and we recall that $\cos\theta_W$ has been defined by

$$m_Z = m_W / \cos\theta_W \qquad (8b)$$

For GUT predictions a particularly useful definition is $\sin^2\hat{\theta}_W(m_W)$, defined by modified minimal subtraction (MS) (i.e. the counterterms are chosen to subtract the terms proportional to $(n-4)^{-1} + (1/2) \cdot (\gamma - \ln(4\pi))$ and the 't Hooft mass scale μ is set equal to m_W). The analytic relation to $O(\alpha)$ between $\sin^2\hat{\theta}_W(m_W)$ and $\sin^2\theta_W$ is given in Ref. 13. Numerically,

$$\sin^2\theta_W = 1.006 \sin^2\hat{\theta}_W(m_W) \text{ for } m_H \simeq m_Z \qquad (9)$$

To predict m_W and m_Z from Eqs (8a,b), $\sin\theta_W$ is needed. Most precise values are extracted from ν_μ deep inelastic scattering and the e-D asymmetry. In the case of $R_\nu = \sigma_T(\nu_\mu + N \rightarrow \nu_\mu + X) / \sigma_T(\nu_\mu + N \rightarrow \mu^- + X)$, the asymptotic theorem of Eq. (2) tells us that $\sigma_T(\nu_\mu + N \rightarrow \mu^- + X)$ is renormalized by a factor $1 + (\alpha/\pi) \ln(m_Z^2/\mu^2)$, not present in the neutral case. Detailed studies show that this leads to an effective value $\rho^2(\nu_\mu; h) = 0.983$ which, together with other effects, lowers $\sin^2\theta_W$ by $\approx 5\%$.[14] Using the 1981 data this led to[14,15]

$$\sin^2\theta_W = 0.217 \pm 0.014 \ (R_\nu) \qquad (10a)$$

$$\sin^2\theta_W = 0.218 \pm 0.020 \ (e\text{-}D) \qquad (10b)$$

Inserting this value in Eqs (8a,b) led to the 1981 prediction of m_W, m_Z which is compared in Table I with recent UA1 and UA2 data (Ref.22).

Table I. Comparison of recent m_W, m_Z data and 1981 Theoretical Predictions.

	UA1 Electron Data	UA2 Electron Data	Weighted Ave. Incl. μ Data	1981 Prediction
m_W(GeV)	80.9±1.5±2.4	83.1±1.9±1.3	82.1±1.7	$83.0^{+2.9}_{-2.7}$
m_Z(GeV)	95.6±1.4±2.9	92.7±1.7±1.4	93.0±1.7	$93.8^{+2.4}_{-2.2}$

THE m_W, m_Z INTERDEPENDENCE

But now we can take advantage of the recent experimental information about m_W, m_Z! It is convenient to write Eqs(8a,b) in the compact form

$$m_W = A/\sin\theta_W; \quad m_Z = m_W/\cos\theta_W \qquad (11a)$$

$$A = \left(\frac{\pi\alpha}{\sqrt{2}G_\mu}\right)^{\frac{1}{2}} (1-\Delta r)^{-\frac{1}{2}} = 37.281\text{GeV}(1-\Delta r)^{-\frac{1}{2}} \qquad (11b)$$

with "standard values" $\Delta r=0.0696 \pm 0.0020$ and $A=38.65 \pm 0.04$ GeV. We can now compute $\sin\theta_W$ from m_W, m_Z. For example, using $m_W=82.1 \pm 1.7$ GeV (see Table I) and $\sin\theta_W = 38.65\text{GeV}/m_W$ we find $\sin^2\theta_W = 0.222\pm0.009$. Using $m_Z=93.0\pm1.7$ GeV and $\sin(2\theta_W)=77.30\text{GeV}/m_Z$ we obtain $\sin^2\theta_W=0.222\pm0.011$. The weighted average of both determinations is:

$$\sin^2\theta_W = 0.222 \pm 0.007 \qquad \text{(from } m_W, m_Z) \qquad (12)$$

We can also eliminate $\sin\theta_W$ between the two formulae in Eq(11a) and obtain useful relations which describe the m_W, m_Z interdependence:[16,17]

$$m_Z = m_W/(1-A^2/m_W^2)^{\frac{1}{2}} \qquad (13a)$$

$$m_W = \frac{m_Z}{\sqrt{2}}\left[1+(1-4A^2/m_Z^2)^{\frac{1}{2}}\right]^{\frac{1}{2}} \qquad (13b)$$

$$\Delta r = 1 - \frac{(37.281\text{GeV})^2}{m_W^2(1-m_W^2/m_Z^2)} \qquad (13c)$$

For $m_Z=93.8 \pm 2.5$ GeV and $A=38.65$ GeV one predicts $m_Z-m_W = 10.8 \pm 0.6$ GeV. About 1 GeV, i.e. 10% of the mass difference for given

m_Z, arises from the quantum corrections![17]

One would like to have simple criteria to signal the possible need to modify the theory. Consider the quantity

$$\frac{1}{(1-\Delta r)^2} = \left[\frac{m_W}{37.281\,\text{GeV}}\right]^4 \left[1 - \frac{m_W^2}{m_Z^2}\right]^2 \qquad (14)$$

which represents the correction to the μ decay rate when expressed in terms of the fundamental parameters g and m_W. The "standard" value for this quantity is 1.1552. Although this is not extremely close to 1, all the large contributions arising from the radiative corrections have been taken into account; therefore it would be very difficult to explain large departures from this value by invoking additional loop effects. As an example, let us suppose that $m_W = 81.0$ GeV and $m_Z = 95.0$ GeV with great precision. Then we would have $(1-\Delta r)^{-2} = 1.66$, which would strongly indicate the need for a change of the theory at the tree level.

POSSIBLE EXOTIC CONTRIBUTIONS

If the deviations are small, they could signal new contributions to Δr; in this sense this quantity may become a window for exotic phenomena! One possible way to change Δr, involving large m_t, has been ruled out by the recent discovery of the top quark. Other interesting mechanisms to change Δr include: 1) the possible effect of new fermion generations and 2) large m_H.

It is convenient to write $\Delta r = (\Delta r)_{st} + \delta$ where $(\Delta r)_{st}$ refers to the "standard" value (cf. Eq(6d)). If the "up" and "down" masses m_i and m_j of an additional quark generation satisfy $m_i^2/m_j^2 = 1+\epsilon$ ($\epsilon \ll 1$), then for large m_i/m_W, m_j/m_W we have asymptotically

$$\delta = -\frac{\alpha}{4\pi}\frac{c^2}{s^4}|U_{1j}|^2\frac{(m_i-m_j)^2}{m_W^2} \qquad (15a)$$

an expression which is related to the well known Veltman correction[18] to the ρ parameter. A detailed study of the effect of higher generations is given in Ref. 9. Some of the highlights are depicted in Figs. 1 and 2 and Table II.

The general conclusion is that an additional generation can give contributions to Δr ranging from large negative to small positive values. For $m_Z = 93.80$ GeV it can at most increase $m_Z - m_W$ by ≤ 130 to 180 MeV. On the other hand, as indicated by Eq(15a) and depicted in Fig.2, if $m_i - m_j \gg m_W$ the negative contribution can cancel $(\Delta r)_{st}$ and therefore decrease the theoretical prediction for $m_Z - m_W$.

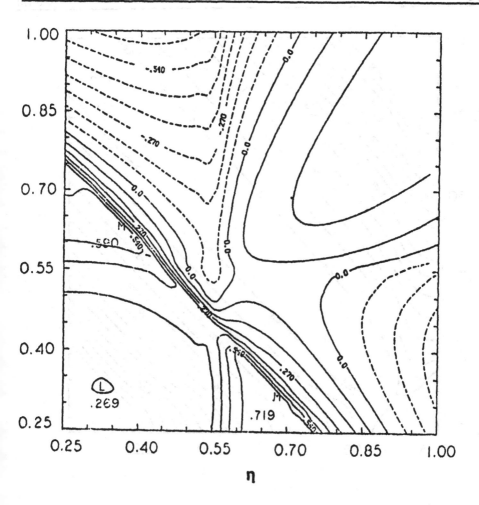

Fig.1. Contribution to Δr (in%) from an additional quark generation (Small Scale). Here $\zeta=m_i/m_W$,$\eta=m_j/m_W$, where m_i,m_j are the "up" and "down" masses, and mixing has been neglected. Solid and dashed curves are positive and negative equal level lines and the M's are upper bounds.

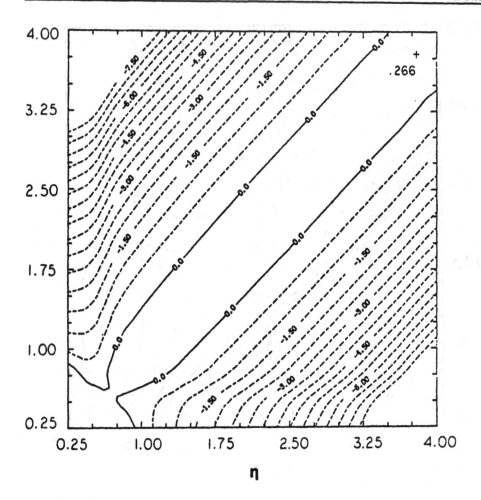

Fig. 2. Same as Fig. 1, in a larger scale.

Table II. Upper bounds to the contribution to Δr from an additional fermion generation according to whether or not the "up"
quark is restricted to be heavier and the neutrino massless.

	Heavier "up" Quark	No Restriction on Quarks
Massless Neutrinos	0.754×10^{-2}	0.883×10^{-2}
Massive Neutrinos	0.900×10^{-2}	1.029×10^{-2}

For large m_H, one has asymptotically

$$\delta = \frac{11}{48\pi} \; \frac{\alpha}{s^2} \; \ln\left(\frac{m_H^2}{m_Z^2}\right) + \ldots \tag{15b}$$

For $m_H^2/m_Z^2 < 100$, the increase in $\Delta r < 0.0088$, not a large effect. In this connection, it is worthwhile to point out that there are theoretical scenarios which lead to small upper bounds for m_H.[19,20,21] In particular, the theoretical assumptions and arguments of Ref.21 lead to $m_H \lesssim 125$ GeV.

If large deviations from the predictions are found, one obvious possibility is to consider

$$m_W^2/(m_Z^2 \cos^2\theta_W) = \rho \tag{16a}$$

rather than Eq.(8b). Eq.(16a) describes a more general class of $SU(2)_L \times U(1)$ theories with an additional fundamental parameter ρ arising from a more complicated Higgs sector. Combining Eq(16a) with $m_W = A/\sin\theta_W$:

$$\rho = \frac{m_W^2}{m_Z^2(1-A^2/m_W^2)} \tag{16b}$$

It is interesting to point out that the neutral current data imposes a useful constraint on ρ, even if we allow for exotic phenomena such as a fourth generation with $m_i - m_j >> m_W$ or large m_H. At the tree level ρ coincides with the renormalization factor $\rho_{NC}(\nu;h)$ of neutral current amplitudes in ν_μ-hadron scattering. Thus $\rho = \rho_{NC}(\nu;h)$. $(1+\epsilon_{NC})$ where $\epsilon_{NC} = O(\alpha)$ is known to be very small in the standard model. It is not difficult to show that if ρ is calculated via Eq(16b) with $A = (A)_{st} = 38.65$ GeV, ϵ_{NC} is not affected by the large hypothetical effects described by Eqs(15a,b).[17] Thus, one expects ϵ_{NC} to be small in any case. From ν_μ and $\bar\nu_\mu$ data, $\rho_{NC}(\nu;h) = 1.02 \pm 0.02$ which is a useful constraint on ρ.

It seems appropriate to end the discussion with a comparison of the theoretical expectations of the SM with information currently derived from the UA1 and UA2 experiments.[22] This is summarized in Table III.

Although the experimental information is generally in good agreement with the predictions, clearly smaller error bars are needed! It will be most interesting to see whether future experiments verify the SM at the level of its quantum corrections or signal the need to modify the theory in a significant way.

Table III.   Comparison of UA1 and UA2 results with theoretical expectations.

	UA1 (Electron Data)	UA2 (Electron Data)	SM with $\sin^2\theta_W=0.217\pm0.014$
m_W (GeV)	$80.9\pm1.5\pm2.4$	$83.1\pm1.9\pm1.3$	$83.0{-2.7}^{+2.9}$
m_Z (GeV)	$95.6\pm1.4\pm2.9$	$92.7\pm1.7\pm1.4$	$93.8{-2.2}^{+2.4}$
m_Z-m_W (GeV)	$14.7\pm2.1\pm0.4$	$9.6\pm2.5\pm0.2$	10.8 ± 0.5
Δr	$0.252\pm0.070\pm0.044$	$-0.025\pm0.211\pm0.032$	0.0696 ± 0.0020
ρ	$0.928\pm0.036\pm0.016$	$1.025\pm0.051\pm0.009$	1
$\sin^2\theta_W=\left(\dfrac{38.65\text{GeV}}{m_W}\right)^2$	$0.228\pm0.008\pm0.014$	$0.216\pm0.010\pm0.007$	0.217 ± 0.014
$\sin^2\theta_W=1-m_W{}^2/m_Z{}^2$	0.284 ± 0.034	0.196 ± 0.047	0.217 ± 0.014

This work was supported in part by the National Science Foundation under Grant No. PHY-8116102.

REFERENCES

1. A. Sirlin, Revs. Mod. Phys. $\underline{50}$, 573 (1978).
2. A. Sirlin, Nucl. Phys. B$\underline{196}$, 83 (1982).
3. H. Leutwyler and M. Roos, Ref. TH.3830-CERN(1984).
4. R.E. Behrends, R.J. Finkelstein and A. Sirlin, Phys. Rev. $\underline{101}$, 866(1956); S.M. Berman, Phys. Rev. $\underline{112}$, 267 (1958); T. Kinoshita and A. Sirlin, Phys.Rev. $\underline{113}$,1652(1959); S.M. Berman and A. Sirlin, Ann.Phys.(N.Y.)$\underline{20}$, 20 (1962); M. Roos and A. Sirlin, Nucl. Phys. B$\underline{29}$, 296 (1971).
5. A. Sirlin, Phys. Rev. Letters $\underline{19}$, 877 (1967).
6. Detailed reports and lists of references can be found in "Radiative Corrections in SU(2)$_L$xU(1)", edited by B.W. Lynn and J.F. Wheater, World Scientific Publishing Co. (1984) and the reviews of M.A.B. Bég and A. Sirlin, Phys. Rep. $\underline{88}$, 1(1982) and K. Aoki et al., Prog.Theor. Phys.Suppl. $\underline{73}$, 1(1982).
7. A. Sirlin, Phys. Rev. D$\underline{22}$, 971(1980).
8. W.J. Marciano and A. Sirlin, Phys. Rev. D$\underline{22}$, 2695(1980)
9. S. Bertolini and A. Sirlin, NYU preprint TR3/84 and Nuclear Physics B (to be published).
10. W.J. Marciano, Phys. Rev. D$\underline{20}$, 274 (1979).
11. M. Consoli, S. Lo Pestri and L. Maiani, Nucl. Phys. B$\underline{229}$, 474 (1983).
12. A. Sirlin, Phys. Rev. D$\underline{29}$, 89 (1984).
13. S. Sarantakos, W. Marciano and A. Sirlin, Nucl. Phys. B$\underline{217}$,84 (1983); this paper contains a detailed study of radiative corrections to v-lepton scattering.
14. W.J. Marciano and A. Sirlin, Nucl. Phys. B$\underline{189}$, 442 (1981).
15. C.H. Llewellyn Smith and J.F. Wheater, Nucl. Physics B$\underline{208}$, 27 (1982).
16. A. Sirlin, contribution to first work cited in Ref. 6, p. 155.
17. W.J. Marciano and A. Sirlin, Phys.Rev. D$\underline{29}$, 945 (1984).
18. M. Veltman, Nucl. Phys. B$\underline{123}$, 89 (1977).
19. N. Cabibbo, L. Maiani, G. Parisi and R. Petronzio, Nucl. Phys. B$\underline{158}$, 295 (1979).
20. D.J.E. Callaway, Nucl. Phys. B$\underline{233}$, 189 (1984).
21. M.A.B. Bég, C. Panagiotakopoulos and A. Sirlin, Phys. Rev. Letters $\underline{52}$, 883 (1984).
22. E. Radermacher, CERN-EP/84-41 (1984).

The Continuing Search for Flavor Changing Weak Neutral Current Processes

The early search for weak neutral current processes with a change of hadronic flavor led to the GIM mechanism and the prediction of charm. Present and future searches with very rare K decays, charm decays and B meson decays will increase in sensitivity and will provide constraints on various theoretical models. K meson factories and (B) "factories" at hadron colliders are needed to continue the search at an ever-increasing level of sensitivity.

1. INTRODUCTION

Ever since the discovery of the W and Z bosons there has been a remarkable agreement of experimental data with the standard model.[1] Future tests of the model will possibly go in more than one direction: higher energy fundamental collisions to search for breakdown of the standard model or further confirmation in the discovery of the Higgs boson; precision measurements of some key quantities predicted in the standard model such as the W, Z mass difference or the search for new processes at low energy that do not fit into the model. The observation of a weak neutral current process that changes hadronic flavor would indicate a deviation from the standard model that would have significant consequences. Sensitive searches for flavor changing weak neutral current processes (FCWNC) have been carried out for the past 20 or so years.[2,3] These early results helped frame the standard model and led to the need for the GIM mechanism.[4] The observation of such processes now could lead to a comparable breakthrough.

In this Comment we will review the present search for FCWNC and highlight the possible future experiments that could extend the search to ever more sensitive levels. We will not discuss processes that change leptonic flavor since the mass of the neutrino plays an important role in such processes in the standard model and perhaps reduces the sensitivity to new interactions for nearly massless neutrinos. We believe that a new round in the search for FCWNC, especially for the heavy flavors such as charm and bottom, is of great importance.

2. THE EARLY SEARCH FOR FCWNC IN K DECAYS

In the very early days of the study of K decays little attention was paid to processes such as

$$K^+ \to \pi^+ e^+ e^-$$

$$\to \pi^+ \nu \bar{\nu}$$

$$K^0 \to \mu^+ \mu^-$$

Only in the early 1960s were sufficient K decays available to enable a sensitive search for these decays.

The first explicit searches for flavor changing weak neutral currents occurred in 1963 with the search for the process[2]

$$K^+ \to \pi^+ e^+ e^- \tag{1}$$

followed by searches for[3]

$$K^+ \to \pi^+ \nu \bar{\nu} \tag{2}$$

$$\to \pi^+ \mu \mu \tag{3}$$

and

$$K^0_L \to \mu^+ \mu^- \tag{4}$$

In the past 20 years the limits on these decay modes have been reduced another 1–2 orders of magnitude and no explicit example of a flavor changing neutral current process has been detected. However, reactions (1) and (4) have been detected and are presumed to be due to a combined charged current reaction and electromagnetic effect. Observation of reaction (2) would provide clear evidence for flavor changing neutral currents, but is also expected in the standard model as a higher-order process.[3]

The present observation on charmed particles and beauty particles have failed to observe effects due to flavor changing weak neutral currents as well; however, the limits are much less restrictive than those for strangeness changing reactions.[5] (For the K decay situation, see Table I.)

3. FLAVOR CHANGING WEAK INTERACTIONS AND Z^0 DECAYS

The discovery of the K mesons in 1948 involved the first observation of strangeness and at the same time the observation of a strangeness changing weak decay. Until a theory of weak interactions was developed after the discovery of parity violation, the significance of the suppressed decay rates of $\Delta S = 1$ processes was not fully appreciated.[1] The success of the Fermi model is illustrated in Fig. 1 where the limits on many processes not allowed by the standard model are shown. We now know that the weak charged current is mediated by $W^{\pm}$ exchange that couples predominantly to the channels

$$W^+ \to u + \bar{d}$$

$$W^+ \to c + \bar{s}$$

$$W^+ \to t + \bar{b}$$

where u, d, c, s, t and b represent the quark states with six flavors of quark with $Q = -1/3$ for d, s, b and $Q = +2/3$ for u, c, t quarks. Processes such as

$$W \to u + \bar{s}$$

$$\to c + \bar{d}$$

$$\to c + \bar{b}$$

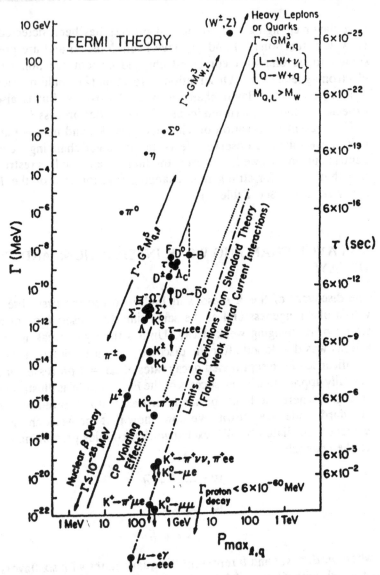

FIGURE 1 A summary of the widths of the decays of many allowed weak processes compared with the simple expectations of the Fermi or standard model of weak interactions. Also shown are the limits on other processes not allowed in the standard model.

are allowed by kinematics but are suppressed by small mixing angles. Thus the charged current flavor changing interactions are nearly diagonal in the mixing matrix that describes these transitions. This suppression is presumably due to dynamical effects.

The decays of Z^0 are expected to be very different due to the fundamental suppression of FCWNC with processes such as[4]

$$Z^0 \to d + \bar{d} \tag{5a}$$

$$\to s + \bar{s} \tag{5b}$$

$$Z^0 \to c + \bar{c} \tag{5c}$$

$$\to b + \bar{b} \tag{5d}$$

being fully allowed but processes such as

$$Z^0 \to s + \bar{d} \tag{6a}$$

$$\to c + \bar{u} \tag{6b}$$

$$\to b + \bar{d} \tag{6c}$$

$$\to t + \bar{c} \tag{6d}$$

being suppressed in the standard model.[4] In principle the search for the direct decays of Z^0 into these latter processes would be a very sensitive technique observed in the FCWNC process. However, the identification of these processes in the presence of other Z^0 decays at a sensitive level is very formidable.

The GIM mechanism was invented to explain the suppression of FCWNC and predicts that the Z^0 decays (5a)–(5d) will have a negligible branching ratio compared to processes (6a)–(6d).[4] However, higher-order processes to be discussed later can give rise to the forbidden decays (6a)–(6d).[6]

Phrased in another way, there is no first-order weak transition for the quark transitions

$$c \rightarrow u$$

$$t \rightarrow u$$

$$t \rightarrow c$$

$$s \rightarrow d$$

$$s \rightarrow b$$

$$b \rightarrow d$$

We will now review the present limits on the search for such transitions in the decays of strange, charm and bottom particles.

4. SEARCH FOR CHARM CHANGING WEAK NEUTRAL CURRENTS—PRESENT LIMITS

The first searches for charm changing weak neutral currents were carried out with neutrino beams where the reactions searched for were

$$\nu_\mu + N \rightarrow \mu^- + \overline{D}^0 + \cdots$$

$$\rightarrow D^0$$

$$\rightarrow \mu^- + \nu + \cdots$$

$$\nu_\mu + N \rightarrow \nu_\mu + D^0 + \cdots$$

$$\rightarrow \mu^+ + \nu + \cdots$$

$$\nu_\mu + N \rightarrow \mu^- + D^0 + \cdots$$

$$\rightarrow \mu^+ \mu^-$$

which give rise to same sign dimuons, wrong sign leptons and trimuon events in neutrino interactions.[1] The resulting limits, although poor compared to K decays, did further indicate the success of the GIM model.[1,4]

The charm case is an interesting example of the interplay of the search for direct FCWNC reactions such as

$$D^0 \rightarrow \mu^+ \mu^- \text{ (i.e., } c \rightarrow u + \mu^+ + \mu^-)$$

$$\nu_\mu + u \rightarrow \nu_\mu + c$$

and the indirect search through the mixing of the D^0 and $\overline{D}^0$ mesons. Both reactions must be studied in order to draw conclusions concerning the right-handed and left-handed coupling of the Z^0 to FCWNC processes. For a first order $D^0-\overline{D}^0$ transition through an intermediate Z^0 exchange, the virtual coupling could be either left-handed (g_L^c) or right-handed (g_R^c). The transition of a D^0 into a $\overline{D}^0$ can be characterized by two lifetimes (Γ_+, Γ_-) and two masses (m^+, m^-) in analogy to the K^0, $\overline{K}^0$ (K_s^0, K_L^0) system.[7] We define

$$\Gamma = \frac{\Gamma_+ + \Gamma_-}{2}; \quad \delta\Gamma = \Gamma_+ - \Gamma_-$$

$$\delta M = M_+ - M_-$$

The transition into a specific channel can be written

$$\frac{\Gamma(D^0 \rightarrow \overline{D}^0 \rightarrow \mu^- + \cdots)}{\Gamma(D^0 \rightarrow \mu^+ \cdots)} = \frac{\delta m^2 + 1/4 \delta\Gamma^2}{2\Gamma^2 + \delta m^2 - (1/2) \delta\Gamma^2}$$

It is generally expected that

$$\delta\Gamma^2 \ll \delta m^2$$

for D^0 (and B^0) systems and therefore we can write

$$\frac{\Gamma(D^0 \rightarrow \mu^- + \cdots)}{\Gamma(D^0 \rightarrow \mu^+ + \cdots)} \simeq \frac{\delta m^2}{2\Gamma^2 + \delta m^2} = r$$

Γ is directly measured from the D^0 lifetime and δm^2 can be deduced from the number of wrong sign leptons produced in a "beam" of D^0 mesons. Experimentally it is not this simple since most processes give a mixture of D^0, $\overline{D}^0$, $D^{\mp}$ final states. Nevertheless, recent estimates give a value for r of[8]

$$r < 1.2 \times 10^{-2} \text{ (90\% confidence level)}$$

A very general expression for the mass difference δm is given by

$$\delta m = \sqrt{1/2} \, GM_D f_D^2$$

$$\times \left[(g_L^c - g_R^c)^2 + \frac{1}{3}g_L^{c2} + g_R^{c2} + 4g_L^c g_R^c \left(\frac{M_D}{M_C + M_U} \right) \right]$$

where f_D is related to the vector

$$\langle 0| \, \overline{c} \, \gamma_\mu \, \gamma_s \, u \, |\overline{D}^0\rangle = if_D P_\mu^D / \sqrt{2} M_D$$

[f_D is usually estimated to be $O(200 \text{ MeV})$]. From the present limit on D^0–$\overline{D}^0$ mixing we can derive[8]

$$|\delta m| < 5.0 \times 10^{11} \text{ s}^{-1}$$

(using $\tau_{D^0} = 3 \times 10^{-13}$ s). Without further assumptions about the coupling of Z^0 to $u\overline{c}$ and $\overline{u}c$, the g_L^c and g_R^c components cannot be separated. In the case of the direct FCWNC decays

$$\frac{\Gamma(c \rightarrow u + \mu^+ + \mu^-)}{\Gamma(c \rightarrow s + \mu^+ + \nu_\mu) + \Gamma(c \rightarrow d + \mu^+ + \nu_\mu)}$$

$$= \frac{1}{4}[g_L^{c2}] + g_R^{c2}] [1 - 4 \sin^2\theta_W + 8 \sin^4\theta_W]$$

We note that some combinations of g_L^c and g_R^c could give a small (δm) and considerably larger branching ratios for the direct $c \rightarrow u + \mu^+ + \mu^-$ decays.

In the case of a pure left-handed process ($g_R^c = 0$) these formulas simplify to a function of g_L^c only and we find

$$|\delta M| = \sqrt{1/2}\, GM_D f_b^2 [(4/3) g_L^{c2}]$$

$$\text{giving } (g_L^c)^2 < 6 \times 10^{-7}/f$$

where f is a parameter estimated to be of $O(1)$.[7] In this case the branching ratio for the FCWNC decay of charmed particles can be directly estimated to be

$$\frac{B(c \rightarrow u + \mu^+ + \mu^-)}{B(c \rightarrow \mu^+ + \cdots)} \simeq \frac{1}{4} \times 6 \times 10^{-7} \sim O(10^{-8})$$

Presumably, a direct measurement of the $c \rightarrow u + \mu^+ + \mu^-$ branching ratio that is well above this value would indicate a strong deviation from either the GIM model for the Z decay or some other new physics. The present limits are far from this level, leaving room for a new experimental discovery.

The most restrictive direct search for charm changing FCWNC comes from the search for[9]

$$D^0 \rightarrow \mu^+ \mu^-$$

by the EMC group working at CERN. This group searched for D^0 decay by muon interactions to produce a large sample and is studying 3μ final states. The limit achieved so far is[9]

$$B(D^0 \rightarrow \mu^+ \mu^-) \leq 3.4 \times 10^{-4} \text{ (0\% C.L.)}$$

Note that this limit is far greater than the simple prediction obtained above from considerations of D^0–$\overline{D}^0$ mixing and assuming only left-handed currents.

We note that there is considerable room for improvement since the cross section for charm production by μ beams is of $O(10 \text{ nb})$ at 250 GeV, whereas the cross section for charm production in hadronic interaction is of $O(10-100 \text{ } \mu b)$ at FNAL and SPS energies. Unfortunately there are large backgrounds in these hadronic interactions as well at the low mass of the D^0. Nevertheless, a dedicated experiment with a hadronic calorimeter spectrometer might improve these limits by several orders of magnitude.

5. FUTURE RARE K DECAY SEARCHES—K MESON FACTORIES

An ambitious experimental program is underway at Brookhaven National Laboratory and KER (Japan) to continue the search for FCWNC K decays to a very low level.[10] The level of sensitivity that the experiments hope to reach is given in Table I. These new experiments will use more intense K^+ and K_L^0 beams as well as larger solid angle detectors with a great degree of sophistication to reduce background levels.[10]

There are also proposals for new accelerators dedicated to providing intense $K^\pm$, K_L^0 beams. These so-called meson factories have been proposed for Los Alamos (LAMPF II) and TRIUMPH.[11]

TABLE I

Limits on FCWNC K decays

Mode	Present Limits** K^+ Decays	Next Level To Be Searched
$K^+ \to \pi^+ e^+ e^-$	$(2.7 \pm 0.5) \times 10^{-7}$	
$K^+ \to \pi^+ \mu^+ \mu^-$	$< 2.4 \times 10^{-6}$	
$K^+ \to \pi^+ \nu\bar\nu$	$< 1.4 \times 10^{-7}$	$\sim 10^{-10}$
$K^+ \to \pi^+ e^\mp \mu^\pm$	$< 5 \times 10^{-9}$	$\sim 10^{-11}$
	K_L^0 Decays	
$K_L^0 \to \mu^+ \mu^-$	$9.1 \pm 1.9 \times 10^{-9}$	
$K_L^0 \to e^+ e^-$	$< 2 \times 10^{-7}$	
$K_L^0 \to e^\pm \mu^\mp$	$< 6 \times 10^{-6}$	$\sim 10^{-11}\text{--}10^{-12}$*
$K_L^0 \to \pi^0 e^+ e^-$	$< 2.3 \times 10^{-6}$	
$K_L^0 \to \pi^0 \mu^+ \mu^-$	$< 1.2 \times 10^{-6}$	
$K_L^0 \to \pi^0 \nu\bar\nu$	—	

* Branching ratio levels as small as 10^{-14} might be reached at a kaon factory such as LAMPF II proposed at Los Alamos National Laboratory.
** Limits taken from Particle Data Group, Dev. Mod. Phys. **56**, S1 (1984).

We note that the Fermilab $\bar{p}$ source target could also produce intense $K^{\pm}$, K_L^0 beams if used in a dedicated mode.[12]

The experience gained at BNL and KEK will be invaluable in order to access the achieved background levels for the rare K decay searches. If these levels are small enough then it would be useful to increase the $K^{\pm}$, K_L^0 flux as proposed at a K meson factory.

6. $B\bar{B}$ PRODUCTION AT $\bar{p}p$ COLLIDERS—HADRON COLLIDERS AS B MESON FACTORIES

Recent experimental evidence from the UA1 experiment at CERN leads us to believe that the cross section for $b\bar{b}$ production has reached the level of $O(3\ \mu b)$ at 540–630 GeV/c^2 center of mass energy.[13,14] We will briefly review this evidence here, since the results indicate a new "source" of b quarks possibly suitable for extending the search for b FCWNC to very sensitive levels where new physics could be detected.

The production of $B\bar{B}$ pairs can be observed through the decay of the Bs into final states with two leptons of the same electric charge. Such events have been detected in the UA1 experiment and lead to a determination of the cross section for $B\bar{B}$ production in the central region.

The events that are most characteristic of $B\bar{B}$ production are those with muons that have nearby jet activity. One such event is shown in Fig. 2. Table II gives the most recent breakdown of the dimuon events into hadronically isolated and nonisolated, same charge or opposite charge. The nonisolated events are directly associated with heavy flavor production such as $b\bar{b}$, $c\bar{c}$. From these events, using the nonisolated sample, the UA1 group obtains the following cross section[14]:

$$\sigma_{\bar{B}}B \simeq 2.4 \pm 0.2\ \mu b \quad \text{(using the ISAJET program}$$
$$\text{for detection efficiency)}$$

$$\text{Ratio} \frac{\text{Same Sign}}{\text{Opposite Sign}} = 0.46 \pm 0.1$$

These estimates are limited to $|\eta| < 2.5$ and $P_{\perp_\mu} > 5$ GeV/c due to the detector cuts (where η is the pseudo-rapidity).

bb̄ EVENT

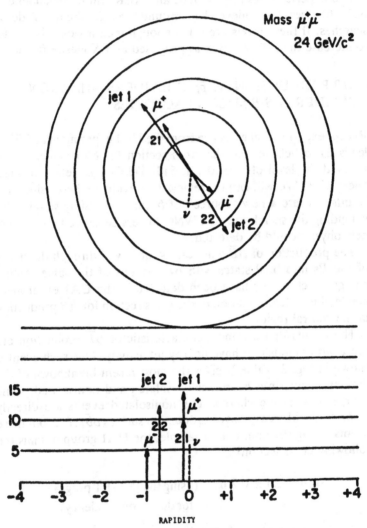

FIGURE 2 A schematic representation of a bb̄ event produced in a p̄p collision at CERN and detected with the UA1 detector by a μμ decay final state. The figure shows the transverse momenta of the muons and jets as well as the rapidity of the particles and jets.

TABLE II

Breakdown of UA1 dimuon events into same charge and opposite charge categories (Refs. 13 and 14)

	Isolated	Nonisolated
$\mu^+\mu^-$	44	106
$\mu^{\pm}\mu^{\pm}$	7	55

Therefore the UA1 group concludes that the cross section for $B\bar{B}$ production is of the order of several μb at $\sqrt{s} > 630$ GeV, which is almost three orders of magnitude larger than the limits for $B\bar{B}$ production at the FNAL/SPS energies.

The large $B\bar{B}$ production cross section is in good agreement with expectations from theory and phenomenology.[14] The angular distributions of the jets and the dilepton events and the transverse momentum characteristics are also in good agreement with the $B\bar{B}$ production hypothesis.

7. PROSPECTS FOR SEARCHING FOR B FLAVOR CHANGING NEUTRAL CURRENTS AT A SENSITIVE LEVEL

The search for FCWNC of b quark provides a new, sensitive technique to test the standard model. Again, as in K decays, it will be necessary to search for either B^0–$\bar{B}^0$ mixing or specific decays of B mesons that occur through flavor changing neutral currents. We first review the present limits from various experiments.

The B^0 system can be composed of two quark states B_d^0 ($b\bar{d}$) and B_s^0 ($b\bar{s}$). Thus either system can mix. In principle the B_d^0 and B_s^0 systems can mix as well but this is expected to be further suppressed due to the mass difference between strange and d quarks. In the standard model the mixing of these systems comes from the diagrams shown in Figs. 3(a) and 3(b). Also, in the standard model, the B_d^0 transition is expected to be very small and thus an experimental search for B_d^0 is very sensitive to the existence of FCWNC. In Fig. 4 we show the current limits on B_d^0 mixing from CESR and PEP experiments.[15]

The standard model for B_s^0 mixing gives a much larger, possibly

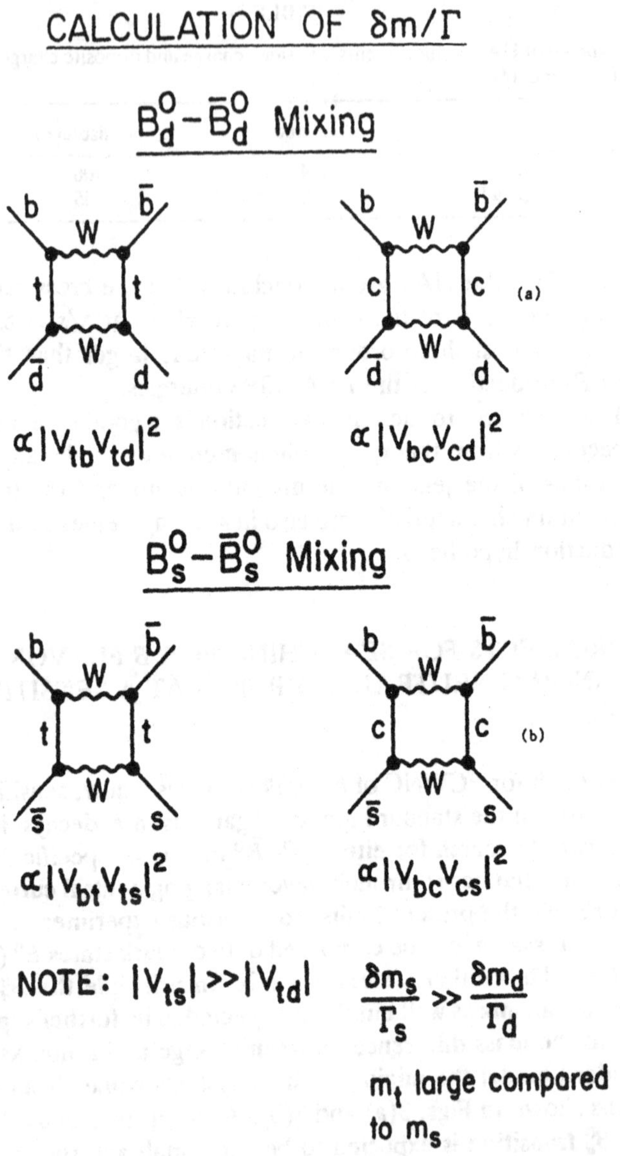
CALCULATION OF δm/Γ
$B_d^0 - \bar{B}_d^0$ Mixing
b
$\bar{b}$
W
t
t
W
$\bar{d}$
d
$\propto |V_{tb}V_{td}|^2$
b
$\bar{b}$
W
c
c
(a)
W
$\bar{d}$
d
$\propto |V_{bc}V_{cd}|^2$
$B_s^0 - \bar{B}_s^0$ Mixing
b
$\bar{b}$
W
t
t
W
$\bar{s}$
s
$\propto |V_{bt}V_{ts}|^2$
b
$\bar{b}$
W
c
c
(b)
W
$\bar{s}$
s
$\propto |V_{bc}V_{cs}|^2$
NOTE: $|V_{ts}| >> |V_{td}|$
$\frac{\delta m_s}{\Gamma_s} >> \frac{\delta m_d}{\Gamma_d}$
m_t large compared to m_s

FIGURE 3 Box diagrams used to estimate the mixing of B_s^0 and B_d^0 states.

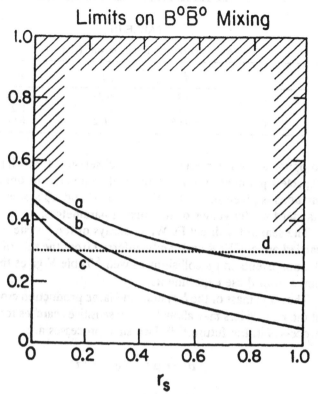

FIGURE 4 Current limits on B^0 mixing from Cornell and SLAC experiments where $\gamma d(s) = \Gamma \, (B^0_d(s) \rightarrow l^- x / \Gamma, \, B^0_d(s) \rightarrow l^+ x)$ is the mixing term. The hatched area is excluded to 90% confidence. The upper limits are given for various values of the production of d and s states (a) (0.35, 0.1); (b) (0.375, 0.15); (c) (0.4, 0.2). The upper limit from the CLEO experiment is also shown.

maximal effect.[16] Recent UA1 data on same sign and opposite sign dimuons can be interpreted as arising from maximal B^0_s mixing (Table II). This system is probably not very useful to further the search for FCWNC, however.

Within the standard model it is possible to estimate the mixing of B^0_d and B^0_s mesons, by calculating the $\delta m/\Gamma$ from the diagrams shown in Fig. 3.[16,17] The value of the K–M matrix (current estimates are given in Table III) lead to a much larger mixing effect in the B^0_s system than the B^0_d system. This effect may have been observed in the UA1 experiment where a large ratio of same sign

TABLE III

$K-M$ matrix and B_d^0/B_s^0 mixing (Ref. 17)

	d	s	b
u	0.97	0.23	<0.04
c	−0.22	0.96	0.2
t	~0.06	0.2	0.98

to opposite sign dimuons has been observed (Table II).[13,14] The expected parameters for the ratio of same sign to opposite sign dimuons is given in Table IV for the B_s^0 and B_d^0 systems, for the standard model values of the mixing parameters.

The search for direct FCWNC decays of the b system has been carried out at CESR and can be inferred from the rate of low mass dimuon events in $\bar{p}p$ collisions as well.[5] Table V gives the current limits from these experiments.

The large mass of the b system and large production cross section at the $\bar{p}p$ colliders may allow for very sensitive searches for FCWNC processes in the future.[18,19] Two such processes are

$$B^0 \rightarrow \mu^+\mu^- \text{ and } \tau^+\tau^-$$

$$B^+ \rightarrow K^+\mu^+\mu^-$$

In order to search for such processes at hadron colliders the event signature must be extremely clean and well defined. Several recent studies have shown that the CERN or FNAL $\bar{p}p$ colliders can produce $\sim 10^8$ B decays per year of actual running times.[18] The

TABLE IV

Mixing measurement $\dfrac{N(\mu^+\mu^+ + N(\mu^-\mu^-)}{N(\mu^+\mu^-) + N(\mu^-\mu^+)} = 2\Delta/1 + \Delta^2$

$$\Delta_{d,s} = \frac{\delta M_{d,s}/\Gamma_{d,s})^2}{2 + (\delta M_{d,s}/\Gamma_{d,s})^2}$$

expect $(\delta M_s/\Gamma_s) > 2$ $\qquad \Delta \rightarrow 1 \qquad (B_s^0)$

$\qquad (\delta M_d/\Gamma_d) \ll 1 \qquad \Delta \ll 1 \qquad (B_d^0)$

TABLE V

Mode	Limit B Decay	Confidence Level	Reference
$B \rightarrow l^+l^-x$	$< 3 \times 10^{-3}$	90%	5
$B^0 \rightarrow l^+l^-$	$< 3 \times 10^{-4}$	90%	5
D Decay			
$D^0 \rightarrow \mu^+\mu^-$	$< 3.4 \times 10^{-4}$	90%	9
$D \rightarrow \mu^+\mu^-x$	$< 10^{-2}$		1

SSC could produce several orders of magnitude more decays per year.[19] Thus these machines will be B meson factories if the correct detectors can be used to match the specific signatures of the decays. The rates of B mesons are several orders of magnitude larger than that from the LEP machine using Z^0 decays as well.[19]

There is a great challenge to continue the search for FCWNC to lower levels using the B decays. Table VI gives a partial list of

TABLE VI

Rare decays of B mesons and the physics implications of the observations

Process	Estimated Branching Ratio	Remarks—Interesting Physics
$B \rightarrow u + \cdots$	$< 10^{-2}$	Measures $|U_{bu}|^2$ important for CP
$B \rightarrow \tau + \nu_\tau$	$\sim 1/3 \times 10^{-4}$ (fb/200 MeV)^2	CP violation tests
$B^0_s \rightleftarrows \bar{B}^0_s$	~ 1	K-M Matrix—Possible CP violation if fourth generation exists
$B^0_s \rightleftarrows \bar{B}^0_d$	0.3 to $<< 0.1$	Possible test of CP violation $(\alpha\gamma) = (N^{++} - N^{--})/N^{+-} \neq 0$ if fourth generation exists
$B^\pm \rightarrow K^\pm \Psi + \cdots$	$\sim 1.5\%$	Possible large CP violation $B(B^- \rightarrow L^- \Psi) \neq B(B^+ \rightarrow K^+ \Psi)$
$B^0_{s/d} \rightarrow \mu^+\mu^-$	$\sim 10^{-8}$	Flavor changing neutral currents
$B^\pm \rightarrow K^\pm \mu^+\mu^-$	$\sim (10^{-5}$–$10^{-8})$	GIM breaking flavor changing neutral currents

the interesting B decays that might be observed with such a B meson factory. In a real sense the study of rare B decays will likely be as powerful as the study of rare K decays.

A suggestion has been made that a "pure" B_d^0 beam might be obtained from diffractive Σ_b^+ B_d^0 production at the $\bar{p}p$ colliders.[18,20] In Table VII we estimate the limits on B FCWNC processes that might eventually be achieved at the B meson factories at hadron colliders.

8. PROSPECTS FOR A SENSITIVE SEARCH FOR FCWNC PROCESSES IN THE FUTURE

The search for flavor changing weak neutral currents during the past 20 years has yielded very important null results. The fabrication of the standard model rests partially on these results.[4] We may now anticipate that the continued search for such processes will yield an equally important breakthrough if observed to be above the expectations of the standard model. It appears that a good place to look for new effects will be in D^0-$\overline{D}^0$ and in the b quark processes, in the former case, because one is dealing with a charge $+2/3$ quark for the first time, which could behave dif-

TABLE VII

Possible sensitivity for FCWNC processes using heavy flavor production at hadron colliders

Process	Estimate of the Possible Limit That Might Be Reached	Hadron Machine
B_s^0-$\overline{B}_s^0$ mixing	$\sim 10^{-3}$*	$S\bar{p}pS$/TeVI
	$\sim 10^{-3}$*	SSC
D^0-$\overline{D}^0$ mixing	$\sim 10^{-4}$**	$S\bar{p}pS$/TeVI
	$\sim 10^{-4}$**	SSC
$B^+ \rightarrow K\mu^+\mu^-$	$\sim 10^{-6}$-10^{-7}	$S\bar{p}pS$/TeVI
	$\sim 10^{-8}$-10^{-9}	SSC
$B^0 \rightarrow \tau\bar{\tau}$	$\sim 10^{-5}$*	SSC
$D^0 \rightarrow \mu^+\mu^-$	$\sim 10^{-5}$-10^{-6}**	$S\bar{p}pS$/TeVI

* Background limits assuming microvertex detectors.
** Rate limits.

ferently than the well studied K^+, K^0 system (i.e., the S, $\bar{S}$ quarks). The most promising processes most likely involve b quarks due to the large mass and the relation with the very heavy mass t quark in the standard model ($M_t > 25$ GeV). As described above, the B_s^0 will most likely not be a sensitive system due to large contributions expected in the standard model. Most likely t quark processes will be insensitive to new physics, as well.

Of course there could still be a small first-order coupling of Z to the FCWNC process as illustrated in Fig. 5(a). Presumably this will eventually show up in the decays of Z^0 as well as the decays mediated by Z^0 particles. A number of authors have speculated on the possible mechanisms that could yield significant FCWNC for b processes. We will not attempt to list all of their contributions in this Comment but simply point to two possibilities:

1. Heavy Higgs bosons could give rise to FCWNC in some models.[21]
2. It has been stated that any theory that relates the mixing angles of the quark–quark transition to the masses of quarks will necessarily give FCWNC processes. An example of this are the so-called "horizontal" gauge theories.[22] An example of a horizontal gauge symmetry breaking process is shown in Figs. 5(b), 5(c) and 5(d).

A very sensitive search for B_d^0–$\bar{B}_d^0$ mixing is most likely the most sensitive test of FCWNC processes. Unfortunately it may be difficult to exclude contamination from the much larger B_s^0–$\bar{B}_s^0$ mixing in such experiments. The possible direct mixing of the B^0, D^0 states via horizontal gauge bosons is illustrated in Fig. 5(d).

It is more likely that specific searches for B decays, such as

$$B^+ \rightarrow K^+ \mu^+ \mu^-$$

$$B^0 \rightarrow \mu^+ \mu^-$$

$$B^0 \rightarrow \mu^+ e^-$$

could be the most sensitive experiments in the next round of searches. Certainly the search for FCWNC K decays at present accelerators and K meson factories will be of great importance as well.

B FLAVOR CHANGING RARE DECAY PROCESSES

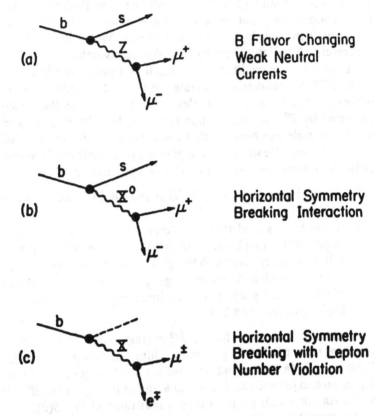

(a)   B Flavor Changing
       Weak Neutral
       Currents

(b)   Horizontal Symmetry
       Breaking Interaction

(c)   Horizontal Symmetry
       Breaking with Lepton
       Number Violation

FIGURE 5 Speculative processes that could induce flavor changing weak neutral currents. (a) Coupling through a small direct effect in Z^0 exchange. (b) Exchange of a horizontal gauge boson that violates FCWNC. (c) Horizontal gauge symmetry breaking that also violates lepton number. (d) Horizontal gauge symmetry breaking bosons that can give rise to D^0 and B^0 mixing. (e) A mechanism in the standard model to give a FCWNC process through t quark exchange.

However, there are processes within the standard model that can also give rise to the same final states as the FCWNC processes. See, for example, the process depicted in Fig. 5(e) that gives rise to the process

$$B \to K \mu^+ \mu^-$$

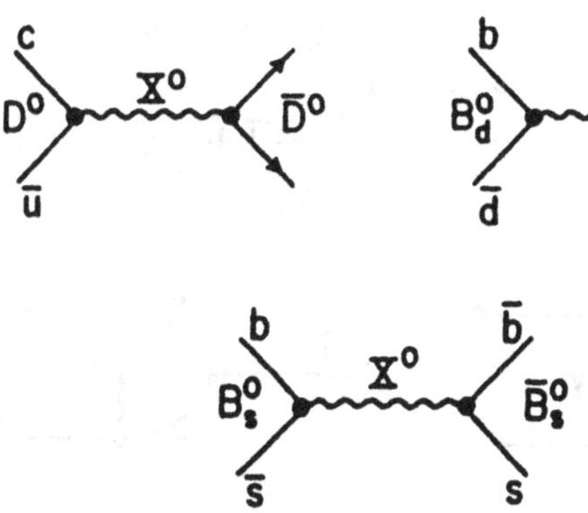

(d)

Standard Model Mechanism for
$B \rightarrow K\mu^+\mu^-$ Decay

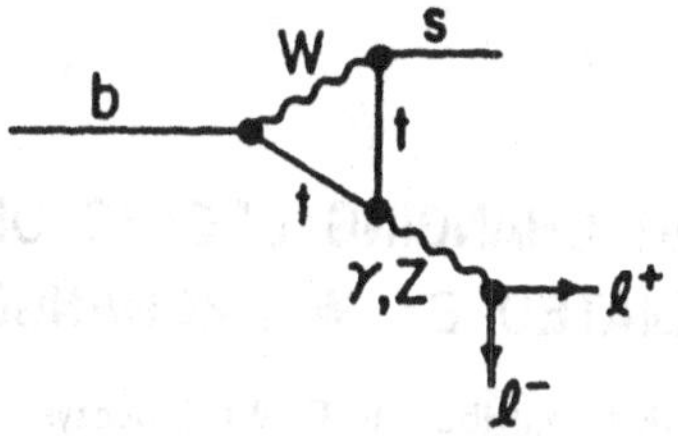

(e)

FIGURE 5 (continued)

within the standard model. Thus it is important to calculate the branching level at which these processes that are allowed by the standard model occur.

Finally, there are also induced FCWNC decays of the Z^0 boson from diagrams shown in Fig. 6. The search for these decay modes is of great importance in the study of rare Z^0 decays at the SLC and LEP e^+e^- colliders.[23] The experimental challenge is to first detect one of these processes and then to determine whether it is due to a correction term within the standard model or to a fundamental FCWNC process possibly illustrated in Fig. 5. We believe

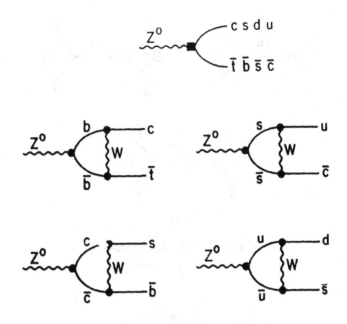

FLAVOR CHANGING DECAYS OF Z^0
MEDIATED BY W EXCHANGE

FIGURE 6 Diagrams that contribute to FCWNC decays of the Z^0 boson in the standard model.

that the next decade will see an intense search for these processes and that hadron colliders offer a new intense source of charm and *B* mesons with which to carry out this search.

References

1. See, for example, papers in "50 Years of Weak Interactions," Proceedings of the Wingspread Meeting, 1984, eds. D. Cline and G. Riedasch, University of Wisconsin (1985).
2. U. Camerini, D. Cline, W. Fry and W. Powell, Phys. Rev. Lett. **13**, 318 (1964).
3. See, for example, J. Ellis and J. Hagelin, Nucl. Phys. B **217**, 189 (1983) and references therein.
4. S. L. Glaskow, J. Iliopoulos and L. Maiani, Phys. Rev. D **2**, 1285 (1970).
5. For a recent limit on B decays into flavor changing weak neutral currents, see P. Avery *et al.*, Phys. Rev. Lett. **53**, 1309 (1984).
6. See, for example, G. G. Volkov *et al.*, Sov. J. Nucl. Phys. **37**, 39 (1983) and references therein; M. A. B. Beg and A. Sirlin, Phys. Rep. **88**, 1 (1982).
7. F. Buccella and L. Oliver, Nucl. Phys. B **162**, 237 (1980).
8. A. Benvenuti *et al.*, CERN EP/85-11 June (1985).
9. J. J. Aubert *et al.*, CERN EP/85-24 (1985).
10. A partial list of the various experiments at BNL is M. P. Schmidt and W. M. Morse, BNL–Yale collaboration, "A Search for the Flavor Changing Neutral Current Reaction $K_L^0 \to \mu e$ and $K_L^0 \to e^+ e^-$"; M. E. Zeller spokesman BNL–Seattle–Yale Collaboration, "Search for the Rare Decay Mode $K^+ \to \Pi e^+ e^-$," ACS Experiment E777.
11. A proposal to construct LAMPF II, a high intensity proton machine, has been submitted to the Department of Energy by the Los Alamos National Laboratory. There are similar considerations in progress for machines in Canada and Europe.
12. There are two possible sources of high intensity K beams at FNAL using the antiproton source target or using the high intensity booster to produce K beams.
13. G. Arnison *et al.*, Phys. Lett. **155B**, 442 (1985).
14. These data were presented by the UA1 group at The New Particles Conference, Madison, May 1985; Bari and Kyoto International Conferences.
15. T. Schaal *et al.*, Phys. Lett. **160B**, 188 (1985) and references therein.
16. See, for example, A. Ali and C. Jarlskog, Phys. Lett. **144B**, 266 (1984).
17. M. Kobayashi and T. Maskawa, Prog. Theor. Phys. **49**, 652 (1973); also A. Carter and A. I. Sanda, Phys. Rev. D **23**, 1567 (1981).
18. D. Cline *et al.*, "A Study of the Possibility of Searching for Rare B Decays at the $\bar{p}p$ Colliders," Florence–Wisconsin preprint, Madison, Dec. (1985).
19. J. Cronin *et al.*, Snowmass 84 Proceedings, page 162 (1985).
20. P. D. Collins and T. P. Spiller, J. Phys. Gen. Phys. **10**, 1667 (1984) and references therein.
21. L. L. Chau, Phys. Rep. **95**, 1 (1983); R. H. Cahn and H. Harari, Nucl. Phys. B **176**, 135 (1980); G. L. Kane and R. Thun, Phys. Lett. **94B**, 513 (1980).
22. See, for example, R. Barbieri *et al.*, Phys. Lett. **74B**, 344 (1978).
23. See, for example, K. Hikasa, Phys. Lett. **149B**, 221 (1984).

CHAPTER 9
BACK TO THE FUTURE WITH THE HIGGS BOSON

The God Particle Again (1995 – 2010)

9.1. THE CURRENT STATUS OF THE TESTS OF THE ELECTROWEAK FORCE THEORY

During the period between the discovery of the W and Z particles (1983) and now (1996), there have been two types of important observations:

1. Precision tests of the model and production (partially covered in Chapter 8), including the prediction and observation of the t quark with a mass near 180 GeV [9.1],[9.2].
2. Searches for deviations from the model (such as FCNC) and further confirmation, such as the discovery of the Higgs boson.

The t quark has been observed by the CDF and D^0 groups at FNAL using the $\bar{p}p$ collider [9.1],[9.2]. The extraction of this signal from the background appears to have been very difficult but successful. Nevertheless, it has been observed, and the mass is in excellent agreement with the expectations of the electroweak theory using the precision measurements (Chapter 8).

To illustrate just how well the standard model works, we reproduce some tables from the latest study of the so-called "LEP Electroweak Working Group" [9.3]. In Table 9.1, they show the results of different measurements at different machines and compare them with the standard model predictions; in Table 9.2, they provide a combined fit to all of the data to obtain the best value of $\sin^2\theta_W$ and of the t quark mass. It should be noted that the best error on $\sin^2\theta_W$ is now $\pm 2.4 \times 10^{-4}$ (compared with 0.1 twenty years ago) − a remarkable accomplishment. Note also that the predicted t quark mass is in excellent agreement with the measurements at FNAL [9.1],[9.2]. These fits have assumed a Higgs boson mass in the range of 60–1000 GeV/c^2. (The current limit on the Higgs mass from LEP is about 60 GeV.) Current attempts to extract a best fit to this mass are probably only reliable to a few hundred GeV, although there is some debate over this issue.

It is clear that to an incredible level of precision, the electroweak theory works very well. This is one of the major motivations of the effort to discover a Higgs boson. It was the driving force behind the now-cancelled Superconducting Super Collider and will play an extremely important role in the LHC machine at CERN. In this chapter, we will concentrate on the LHC discovery potential, but we should not forget that the LEP II e^+e^- machine now starting to work at CERN can observe a Higgs boson with mass up to about 90 GeV/c^2.

9.2. THE MECHANISM OF ELECTROWEAK SYMMETRY BREAKING AND THE HIGGS BOSON

In the mid-1990s, Peter Higgs set out to try to understand the concept of spontaneous symmetry breaking described by condensed matter theorists. (In the article reprinted

Table 9.1. Summary of measurements included in the combined analysis of standard model parameters. (Adapted from [9.3].)

	Measurement with Total Error	Systematic Error	Standard Model
LEP			
Line-shape & lepton asymmetries:			
m_Z (GeV)	91.1884 ± 0.0022	0.0015	91.1882
Γ_Z (GeV)	2.4963 ± 0.0032	0.0020	2.4973
σ_h^0 (nb)	41.488 ± 0.078	0.077	41.450
R_ℓ	20.788 ± 0.032	0.026	20.773
$\sin^2\theta_{eff}^{lept}$ ($\langle Q_{FB}\rangle$)	0.2325 ± 0.0013	0.0010	0.23172
SLD			
$\sin^2\theta_{eff}^{lept}$ (A_{LR})	0.23049 ± 0.00050	0.00015	0.23172
$\bar{p}p$ and νN			
m_W (GeV) ($\bar{p}p$)	80.26 ± 0.16	0.13	80.35
$1 - m_W^2/m_Z^2$ (νN)	0.2257 ± 0.0047	0.0043	0.2237

Table 9.2. Results of a fit to all available data. (Adapted from [9.3].)

	LEP + SLD + $\bar{p}p$ and νN Data
m_t (GeV)	$178 \pm 8^{+17}_{-20}$
$\alpha_S(m_Z^2)$	$0.123 \pm 0.004 \pm 0.002$
χ^2/d.o.f.	$28/14$
$\sin^2\theta_{eff}^{lept}$	$0.23172 \pm 0.00024^{+0.00007}_{-0.00014}$
$1 - m_W^2/m_Z^2$	$0.2237 \pm 0.0009^{+0.0004}_{-0.0002}$
m_W (GeV)	$80.346 \pm 0.046^{+0.012}_{-0.021}$

here by P. Higgs, there is a very candid discussion of this sequence of events. Please refer to it for the appropriate references; also see Ref. [9.4].) He succeeded brilliantly, and his work was a key concept to the Weinberg–Salam model (see Chapter 5). By 1971, when the theory had been shown to be renormalized by t' Hooft (see Chapter 4), the concept of a real particle associated with the symmetry breaking was advanced. Ever since then, there has been some effort to find such a particle, starting with the very-low-mass region of ≤ 100 MeV. Since the mid-1970s, there has been one search or another in progress to detect the Higgs boson. One of the shortcomings of this theory (at least to an experimentalist) is the lack of a specific prediction for the Higgs boson mass and, therefore, the first searches started with the very-low-mass region that could be detected in the π and K decays, e.g.,

$$K^+ \to \pi^+ + H^0 \ . \tag{9.1}$$

At higher mass, processes like

$$e^+ + e^- \to Z + H^0 \tag{9.2}$$

have been sought (see Ref. [9.5] for early searches). The current limit on the Higgs boson mass is about 60 GeV/c^2.

While the mass of H^0 is not predicted by the theory, the method of decay is well understood. Very approximately, for an H^0 with mass

$$m_{H^0} < 2m_W \ , \tag{9.3}$$

the dominant decays will be two quarks or heavy leptons, and for

$$m_{H^0} > 2m_W \quad \text{or} \quad 2m_Z \ , \tag{9.4}$$

the W^+W^- and Z^0Z^0 decays will be dominant. This prediction is incorporated into the present and future searches for the H^0. It is generally agreed that if the H^0 mass is below approximately 90 GeV, it can be detected at the LEP I and II e^+e^- colliding-beam machine at CERN. Going above this mass range will require higher energy pp colliding-beam machines for the discovery.

9.3. THE PROSPECTS FOR DISCOVERY OF THE HIGGS BOSON AT THE LARGE HADRON COLLIDER AT CERN

The first realistic attempts to observe the Higgs boson in the multi-GeV mass range have been made at LEP [9.5]. The LHC at CERN is being built partially to discover the H^0. The pp beam energies will be 7 TeV on 7 TeV. Of these two major detectors, we will focus on the potential of the CERN compact muon solenoid (CMS) detector to discover the H^0 particle. The author (DBC) and his group are part of this collaboration. The CMS detector for the LHC is shown schematically in Fig. 9.1, and Ref. [9.6] gives information about both the CMS and ATLAS detectors being constructed for the LHC.

The detection of the H^0 with mass less than 2 m_W will be extremely difficult, with the most promising decay modes,

$$H^0 \rightarrow \gamma\gamma \ ,$$

$$H^0 \rightarrow \mu^+\mu^-\mu^+\mu^- \ ,$$

being the ones that are most likely to stand out above background.

In the mass range above 2 m_W, the Z^0Z^0 decay mode should be easy to detect. Figure 9.2 shows the expected signal in this mass range, indicating a clear discovery potential, as the reader can see easily. Thus, with the advent of the LHC, it will be possible to search for the Higgs boson over the entire mass range up to about 1 TeV [9.6]. If it is not seen in this range, it is unlikely to exist, at least as a well-defined particle.

It is possible that the theory is more complex, including possible supersymmetric particles and several different Higgs bosons – even charged ones [9.7]. While this is an extremely interesting possibility, it is beyond the scope of this book. However, a good introduction to this topic can be found in Ref. [9.7].

9.4. THE UNIVERSAL IMPORTANCE OF THE WNC IN NATURE

During the intervening years, we have learned that the WNC plays a key role in the Universe. It drives the explosions of supernovas and the subsequent production of all of the heavy elements required for life on our planet. In addition, the formation of neutron stars or black holes from the supernovas in the Universe are related to the WNC forces. Perhaps the most exciting possibility is that WNC could provide the central mechanism for the basic structure of the DNA double helix.

9.4.1. The Weak Neutral Current and Biology

For more than one century, there has been evidence for the chiral, or handedness, nature of life forms on Earth. *The origin of this effect is one of the great mysteries of life.* Pasteur was among the first to point this out (1884), and the universal nature of chiral symmetry breaking in DNA and RNA is now very well-established for all life forms. The following quote, reproduced here because of its clarity, is from one of Richard Feynman's lectures [9.8]. In this quote, Feynman describes an experiment in which sugar is reproduced by chemical means. This sugar, which was produced

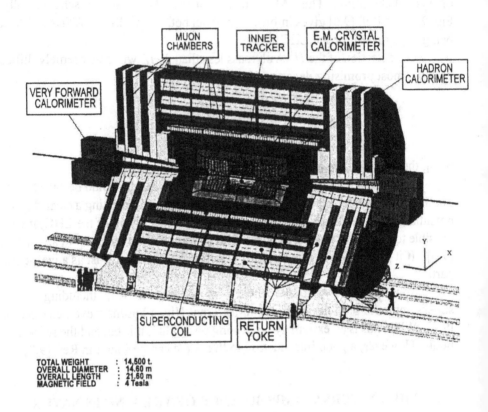

Figure 9.1. Three-dimensional view of the CMS detector (from [9.6]).

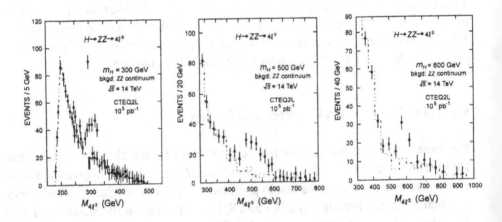

Figure 9.2. The four-lepton mass distributions.

Figure 9.2. The four-lepton mass distributions.
in the laboratory with exactly the same chemical composition, is compared to natural sugar, and certain differences are found to exist.

> "This is a most remarkable fact, and it seems at first sight to prove that the physical laws are not symmetric for reflection. However, the sugar that we used that time may have been from sugar beet; but sugar is a fairly simple molecule, and it is possible to make it in the laboratory out of carbon dioxide and water, going through lots of stages in between. If you try artificial sugar, which chemically seems to be the same in every way, it does not turn the light. Bacteria eat sugar: **if you put bacteria in the artificial sugar water it turns out that they only eat half the sugar, and when the bacteria are finished and you pass polarized light through the remaining sugar water you find it turns to the *left*.** The explanation of this is as follows. Sugar is a complicated molecule, a set of atoms in a complicated arrangement. If you make exactly the same arrangement, but with left as right, then every distance between every pair of the atoms is the same in one as in the other, the energy of the molecules is exactly the same, and for all chemical phenomena not involving life they are the same. **But living creatures find a difference.** Bacteria eat one kind and not the other."

So bacteria only eat right-handed sugar: this is an example of a chiral substance [9.8].

With the discovery of parity violation in charged current reactions in 1956 and of WNCs in 1973, two universal symmetry-breaking processes were uncovered that may have affected the structure of DNA and RNA. The main problem is that this interaction represents an extremely small symmetry-breaking effect ($\varepsilon/kT \sim 10^{-17}$); however, there are plausible non-linear mechanisms that could have amplified this small symmetry-breaking process up to the full symmetry breaking that is observed in life forms [9.9]. We note that there is a long-standing controversy as to whether or not these non-linear effects are actually large enough to have caused the selection of the chirality of the biomolecules.

This type of chiral symmetry-breaking effect could have aided, or perhaps even affected this transition. The key parameter is the time to affect the transition and is given by the parameter τ, which is related to the inverse of $\gamma N \varepsilon^2$, where γ is the replication time of the system [9.10], N the number of biomolecules, and ε the chiral symmetry-breaking parameter ($\varepsilon = 10^{-17}\ kT$ for WNC or charged currents). This is an example of field non-equilibrium thermodynamics.

If we consider the totality of biomolecules in the oceans of the earth, we estimate the number to be

$$N \sim 10^{23}$$

and, assuming a replication time of $\gamma \sim 1 \ hr^{-1}$, we find that $\tau > 10^7$ years for $\varepsilon = 10^{-17}/kT$. This is a plausible period of time for life to have formed on Earth, given the amplification mechanisms pointed out above [9.10],[9.11].

Thus, over very long periods of time, small effects can be magnified. *It is remarkable that all current and detailed calculations of energy force for different biomolecules (including the WNC interaction level) find that the lowest energy level occurs for the ones actually observed in Nature.*

Recently, Y. Liu, a graduate student at UCLA, and I have studied this problem [9.12]. We have carried out simulations of the possible transition of the system from the unbroken symmetry to the broken symmetry resulting from WNCs. We agree with the previous works and have tried to address some of the criticisms of these earlier works in our simulation, including the effects of large random noise at the critical transition point in the system.

There are many examples in Nature where small effects can be magnified to make large effects over sufficient time. One example is found in the annealing of a metal, where the very slow cooling can lead to a near crystalline structure. Another example is the buildup of the polarization or direction of the spin of an electron in a storage ring owing to the direction of the magnetic field. While this effect is very tiny during each revolution of the electron around the machine, after several hours (at 100-million revolutions per hour) a large effect ($\sim 90\%$ of the spins in one direction) builds up. The buildup of the chiral structure of the biomolecules of life on Earth could have followed a similar path, but over millions of years instead of several hours.

Modeling this theory, we have exploited a bi-stable dynamic system. It can transit to one of two states when the system passes through the transition point in the presence of a small bias (similar to WNC interactions), smaller in magnitude than the rms value of the fluctuations (white Gaussian noise). Such a system can be simply described by a first-order stochastic equation as

$$d\frac{\alpha}{dt} = -A\alpha^3 + B(\lambda - \lambda_c)\alpha + Cg + \varepsilon^{1/2} f(t) \ ,$$

where α is the amplitude of the symmetry breaking solution, λ the control parameter, g is the interaction or bias symmetry-breaking selector, λ_c the symmetry-breaking transition point, $\varepsilon^{1/2}$ is the rms value of fluctuation (noise), and $f(t)$ is the normalized fluctuations (noise).

$$\lambda = \lambda_0 + \gamma t \ ,$$

where λ_0 is the initial value of λ, γ is the evolution rate, and t is the evolution time.

First, we took a series of calculations with fixed g (< 0 in all calculations) and $\varepsilon^{1/2}$, and varied γ to investigate the effect on the transition rate. It is quite clear that for a given ratio, $\varepsilon^{1/2}/g$, the selectivity depends on the evolution rate, γ, or on the evolution time, $T_{tr} \sim 1/\gamma$. Also, we notice that for a different $\varepsilon^{1/2}/g$, the selectivity can be the same with different values of γ. The slower the process evolves, the higher the selected probability. If the environment is very noisy, the $\varepsilon^{1/2}$ is far larger than $|g|$, and the symmetry breaking in the bio-system is reduced.

Second, we examined the effects of varying g and found that the g coupling (mixing) with $\varepsilon^{1/2}$ has considerable effect on the system around the transition region. In this area, the effect of g seems to be amplified and, thus, gives a bigger contribution to the process at the far regions before and after the transition area. Hence, either a bigger g or $\varepsilon^{1/2}$ does not make much contribution to the system, and the system tends to be stable. Reference [9.12] presents the results of around 40,000 trajectories and gives a $P_- = 88\%$ chance that the process favored by g will emerge dominant even though the rms values of the random chiral influences (white Gaussian noise) are 10 times larger. However, such sensitivity can not be realized if the system does not evolve through the critical point. To save computing time and for easier graphical visualization, we set all parameters larger than those in the bio-system. Future simulations will attempt to explore the much smaller bio-system parameters.

In connecting these calculations to the life forms in Nature, two conjectures may be made:

1. Before the organic molecules formed, the WNC effect existed in the Universe. The chirality of the organic molecules, such as DNA (RNA) and protein, were automatically selected; then these selected molecules underwent their reproduction and evolved into modern life on Earth. When the evolved molecules became stable, their chirality could not be changed independently of the noise level.

2. Because of the large noise in the environment, the organic molecules initially displayed a chiral form almost equal to the WNC effect. However, during some period, the noise became relatively smaller and the evolution rate sped up, finally leading to the life forms that display the chirality present on Earth.

The fact is that there are many unclear factors that should be investigated, but the WNC could be a key factor in evolutionary processes.

The final word is not in on this issue of whether the WNC can contribute. It should be noted that experimental studies are going on at various places, and these studies could determine if the structure of the DNA building blocks can arise from the weak interaction. In the words of the late and great Russian physicist Ya. Zel'dovich who, along with the chemist C. Franck, in some ways initiated these ideas, wrote in one of his last works [9.10],

"We can thus conclude only that an analysis of this question *does not* rule out the possibility that the chirality of biomolecules was determined by the asymmetry of the weak interactions."

9.4.2. The Importance of the Weak Neutral Current in Astrophysics

We now turn to one of the great mysteries of astrophysics: What makes a massive star at the end of its life explode into a supernova, such as the one observed on Valentine's Day in 1987 in the Large Magellanic Cloud (Galaxy)? A supernova explosion today is the closest we can get to the holocaust of the Big Bang.

One of the major issues concerning supernovas is how they actually explode. It appears now that WNC processes are crucial to the explosion process. J. Wilson (LLNL) is the only "simulator" to demonstrate an actual explosion. His code requires high-energy neutrinos that were produced through the WNC process.

Soon after the discovery of WNC, D. Freedman (MIT) suggested (1975) that the WNC force might somehow provide pressure from the neutrinos produced in the core of the process. This was an important step to our current understanding. Unfortunately, more detailed calculations by D. Schramm (Chicago) and others failed to confirm this idea. For another 10 years, this problem remained. However in the mid-1980s, another idea was born. It was suggested that the WNC interaction could be responsible for the production of two of the neutrino flavors, the μ and the τ, which exit the supernova with relatively high energies. The other neutrinos were rapidly diminished in energy by interactions with electrons. The μ and τ neutrinos might be able to carry energy into the shock front to somehow keep the explosion going.

The incredible idea that emerged was that the high density of neutrinos would cause annihilation of the neutrinos and antineutrinos by their interaction, becoming electron and positron pairs, which then deposited their energy through the electromagnetic interaction into the front of the explosion.

There is little doubt that WNC is of great importance in the explosion of a supernova, the source of most of the heavy elements in the Solar System.

REFERENCES

9.1. CDF Collaboration, F. Abe *et al.*, *Phys. Rev. Lett.*, **74**, 2626 (1995).

9.2. D^0 Collaboration, S. Abachi *et al.*, *Phys. Rev. Lett.*, **74**, 2632 (1995).

9.3. A Combination of Preliminary LEP Electroweak Measurements and Constraints on the Standard Model, CERN report CERN-PPE/95-172 (November 1995).

9.4. P. W. Higgs, *Phys. Lett.*, **12**, 132 (1964); *Phys. Rev.*, **145**, 1156 (1966); F. Englert and R. Brout, *Phys. Rev. Lett.*, **13**, 321 (1964); T. W. Kibble, *Phys. Lett.*, **155**, 1554 (1967); G. S. Guralnik, C. R. Hagen, and T. W. B. Kibble, *Phys. Rev. Lett.*, **13**, 585 (1964).

9.5. ALEPH Collaboration, D. Decamp *et al.*, *Phys. Lett.*, **246B**, 306 (1990); DELPHI Collaboration, P. Abreu *et al.*, *Nucl. Phys.*, **342B**, 1 (1990); L3 Collaboration, B. Adeva *et al.*, *Phys. Lett.*, **248B**, 203 (1990); OPAL Collaboration, M. Z. Akraway *et al.*, *Phys. Lett.*, **253B**, 511 (1991).

9.6. ATLAS Collaboration Technical Proposal, CERN/LHCC 94-93 (1994); CMS Collaboration Technical Proposal, CERN/LHCC 94-38 (1994).

9.7. For reviews on supersymmetric theories, see P. Fayet and S. Ferarra, *Phys. Rep.*, **32**, 249 (1977); H. P.Nilles, *Phys. Rep.*, **110**, 1 (1984); H. Haber and G. Kane, *Phys. Rep.*, **117**, 75 (1985); R. Barbieri, *Riv. Nuovo Cimento*, **11**, 1 (1988).

9.8. R. Feynman, *The Character of Physical Laws* (the Cornell Lectures) (MIT Press, Cambridge, 1965).

9.9. See for example, A. Salam, *J. Mol. Evol.*, **33**, 105 (1991).

9.10. Ya. B. Zel'dovich and A. S. Mikhailov, *Sov. Phys. Usp.*, **30**, 11 (1987).

9.11. D. D. Kondepudi and G. W. Nelson, *Nature*, **314**, 438 (1985).

9.12. D. B. Cline, Y. Liu, and H. Wang, *Origins of Life.*, **25**, 201 (1995).

SBGT AND ALL THAT

Peter Higgs
Department of Physics, University of Edinburgh,
Edinburgh. EH9 3JZ, U.K.

ABSTRACT

I give a personal account of the genesis, about twenty years
ago, of spontaneously broken gauge theories.

Contrary to the custom at this conference, I want first of all
to disclaim priority for some of the concepts to which my name is
commonly attached in the literature. For this exaggerated view of
my originality I have to thank the late Ben Lee, who at the 1972
High Energy Physics Conference at Fermilab plastered my name over
almost everything concerned with spontaneous symmetry breaking.
"Higgs fields", for example, are just the scalar fields of a linear
sigma model, which was discussed in 1960 by Gell-Mann and Lévy[1] but
had been introduced three years earlier by Schwinger[2]. And "the
Higgs mechanism" was first described by Philip Anderson[3]: perhaps it
should be called "the ABEGHHK'tH.... mechanism" after all the people
(Anderson, Brout, Englert, Guralnik, Hagen, Higgs, Kibble, 't Hooft)
who have discovered or rediscovered it! However, I do accept
responsibility for the Higgs boson; I believe that I was the first
to draw attention to its existence in spontaneously broken gauge
theories[4].

My interest in spontaneous symmetry breaking was stimulated in
1961 by the work of Nambu[5]. His idea was to generate the mass
splittings within hadron multiplets (and even the masses themselves)
by spontaneous breaking of the appropriate internal symmetries. The
field theoretic models which he studied were inspired by the BCS
theory of superconductivity, so the scalar fields which developed
vacuum expectation values were associated with fermion pairs rather
than elementary bosons. I found this programme very attractive; it
seemed to me that the internal symmetries of particle physics would
be a little less mysterious if they were genuine (i.e. unbroken)
symmetries at the level of Lagrangian field theory.

However, there was an obstacle to the realization of Nambu's
programme; this was the Goldstone theorem[6,7]. It states that if a
manifestly Lorentz invariant local field theory exhibits spontaneous symmetry breaking, it will contain massless spin-zero bosons,
the non-existence of which is rather easy to establish experimentally!
Thus it seemed as if spontaneous symmetry breaking was not enough;
there would have to be explicit breaking as well to give mass to
would-be Goldstone bosons, such as the pion in Nambu's original
model. This rather spoiled the elegance of the theory.

Over the next few years a debate developed about whether the
Goldstone theorem could be evaded. Anderson[3] pointed out that in a

superconductor, where the broken symmetry is a local "gauge" symmetry, the Goldstone (plasmon) mode becomes massive due to the gauge field interaction, whereas the electromagnetic modes are massive (Meissner effect) despite gauge invariance. He concluded that "the Goldstone zero-mass difficulty is not a serious one, because we can probably cancel it off against an equal Yang-Mills zero-mass problem". However, he did not discuss explicitly any relativistic model: since Lorentz invariance was a crucial ingredient of the Goldstone theorem, he had not provided a convincing demonstration that it could be evaded. Meanwhile, the Goldstone theorem had fallen into the hands of axiomatic field theorists, who proceeded to prove it with impeccable rigor.

In March 1964 Abraham Klein and Ben Lee published a note[8] which provided the first clue to how the theorem could be evaded. They analysed the structure of the ground state expectation value of the commutator of a symmetry current with one of a multiplet of scalar fields – the object which was central to the proof of the theorem – in a "relativistic" description of a condensed matter system such as a superconductor, where the rest frame of the medium was identified by a timelike unit four-vector n. They had no trouble in showing that the occurrence of n as well as the four-momentum k in the Fourier transform of this function allowed the theorem to be evaded. They speculated that it might be possible to find a truly relativistic model in which the same happened. I read their note while I was convalescing after an illness and it cheered me up considerably, but I could not see how to construct such a model.

Three months later, at the end of June, a reply to Klein and Lee from Walter Gilbert appeared. He pointed out that in a relativistic field theory with a Lorentz invariant vacuum as its ground state there is no medium whose rest frame could provide a special four-vector n. Therefore the Goldstone theorem could not be evaded in this way. I got very angry when I read Gilbert's paper, because I didn't want to believe it but I saw no way to bring back the banished vector n.

But a few days later, early in July, it suddenly struck me that I had known for years a relativistic field theory involving just such a vector; it is called quantum electrodynamics! And the feature of this theory which permits the appearance of such a vector in the formalism, without loss of Lorentz invariance of the physical content, is gauge invariance. Thus I had found a loophole in Gilbert's argument; gauge theories could evade the Goldstone theorem.

Now most people would probably not have thought of quantum electrodynamics as a theory involving such a vector. But I had long been a follower of Schwinger in preferring the Coulomb gauge formalism, which involves a choice of inertial frame n for the gauge, to the more commonly used covariant formalism, in which the photon propagator contains unphysical polarizations. More specifically, in parallel with my interest in spontaneous symmetry breaking I had been interested in an apparently unrelated problem, the connection between gauge invariance and zero-mass quanta. Schwinger[10] had

shown that gauge invariance alone does not prevent the photon from
being massive. He had invented a model in 1+1 dimensions in which
this occurs, but he had no example of it in 3+1 dimensions. For me
at this time,the psychologically most important thing which I had
learned from Schwinger's papers was the most general form in Coulomb
gauge of the photon propagator; it suggested to me how the vector $\underline{n}$
would occur in other vacuum expectation values in a gauge theory,
such as the one considered by Klein and Lee.

   So far, all I had established was that Goldstone bosons need
not occur when a <u>local</u> symmetry is spontaneously broken in a
physically Lorentz invariant theory. Obviously, the next thing to
do was to find out what <u>did</u> happen in such a theory. Since the
Goldstone phenomenon was known[6] to occur in <u>classical</u> scalar field
theories, the simplest model to study was clearly the locally
symmetric version of Goldstone's classical model. Linearizing the
field equations, I saw at once that the relativistic version of the
Anderson mechanism did indeed occur; the Goldstone mode provided
the third polarization of a massive vector field. The other mode of
the original scalar doublet remained as a massive scalar.

   I quickly wrote a short paper, "Broken Symmetries, Massless
Particles and Gauge Fields", which described how gauge theories may
evade the Goldstone theorem, and submitted it to Physics Letters.
It was received on 27 July and published on 15 September[11]. Before
writing up the work on what is now known as the Higgs model I spent
a few days searching the literature to see whether it had been done
before. I thought that Schwinger, in particular, might well have
done something of the kind years earlier and I might have overlooked
it. When I had satisfied myself that he hadn't, I wrote a second
short paper, "Broken Symmetries and the Masses of Gauge Bosons",
and submitted it too to Physics Letters. It was rejected. In his
letter the Geneva editor, Jacques Prentki, wrote that it was not the
kind of result which called for rapid publication in Physics Letters
but that a fuller account of the work might be suitable for Il Nuovo
Cimento.

   I was indignant – I thought my discovery was important! My
colleague Euan Squires, who spent the month of August 1984 at CERN
in the Theory Division, later told me that people there just didn't
see the point of what I had done. In retrospect this is not surprising. In 1964 particle theory in Europe was dominated by S-matrix
theory and the doctrines of Geoffrey Chew. Quantum field theory was
out of fashion, and I had been rash enough to base my claims on the
linearized version of a <u>classical</u> field theory (invoking implicitly
the de Broglie relations). What relevance could this possibly have
to real particle physics ?

   Realizing that I had failed to sell my work sufficiently, I
rewrote the paper, adding some remarks on a model in which the
broken gauge symmetry was the fashionable hadronic SU(3), and submitted it to Physical Review Letters. This time it was accepted[4];
but the referee * asked me to add a comment on the relation of my

_____

*   After I had given this talk, Nambu told me that he was the
referee.

work to that of Englert and Brout[12], which had just been published.
(It was published on 31 August, the day that my paper was received
by Physical Review Letters. It had been received on 26 June, but
in 1964 the Brussels group did not send preprints to Edinburgh. As
far as I am aware, they still don't.) Englert and Brout summed
tree diagrams in perturbation theory to obtain the mass formula
for vector bosons in a general spontaneously broken non-Abelian
gauge theory: they considered models with composite scalar fields
as well as those based on elementary scalars, but they did not
explicitly dispose of the Goldstone theorem nor did they discuss the
residual scalar particles, as I did.

During the following October I had discussions with Guralnik,
Hagen and Kibble, who had discovered[13] how the mass of noninteract-
ing vector bosons can be generated by the Anderson mechanism, and
with Streater, who was involved in the more rigorous proofs of the
Goldstone theorem. But it was not until September 1965, when I
arrived in Chapel Hill on sabbatical leave at the invitation of
Bryce DeWitt, that I settled down to work out the details of my
Abelian model. The result of this work was my Physical Review
paper[14], which appeared as a preprint in December 1965. In the New
Year 1966 I received an invitation from one of the recipients of
that preprint, Freeman Dyson, to give a colloquium in March at the
Institute for Advanced Study. The previous summer, at the General
Relativity Conference in London, Stanley Deser had invited me to
give a talk at the joint seminar at Harvard sometime during my year
in the U.S.A., so I took the opportunity to arrange this for the
day following my Princeton talk.

At tea before my Princeton talk on 15 March 1966 Klaus Hepp
told me that there must be an error in my work, since Kastler,
Robinson and Swieca had just proved the Goldstone theorem by C*-
algebraic methods – the ultimate in rigor! Nevertheless I survived
the questions of the Princeton axiomatists. Encouraged by this
experience, I was ready for a rather different style of discussion
the next day at Harvard. Years later, when I met Sidney Coleman
again, he told me that he and his colleagues "had been looking forward to some fun tearing to pieces this idiot who thought he could
get round the Goldstone theorem". Well, they did have some fun,
but I had fun too!

By then I had already spent some time fruitlessly trying to
construct a realistic model. The trouble was that, like so many
people at that time, I was too preoccupied with the breaking of
hadronic (flavour) symmetries: I was aware that leptonic symmetries
had been proposed by various people, but I had not appreciated their
significance. Shelly Glashow, in his Nobel lecture[15], said of
Goldstone, Kibble and myself: "These workers never thought to
apply their work on formal field theory to a phenomenologically
relevant model. I had had many conversations with Goldstone and
Higgs in 1960. Did I neglect to tell them about my $SU(2) \times U(1)$
model, or did they simply forget ?" I should explain that I first
met Glashow in 1960 at the first Scottish Universities' Summer
School in Physics where he was a participant and I was a member of
the executive committee with the duties of steward. I do not

recall hearing about the SU(2) x U(1) model there: perhaps I did but, in my defence, I would point out that my duties as steward kept me from taking part in all the discussions which continued far into the night (lubricated by wine which seemed to wander from my not very secure store) between Glashow, Cabibbo, Veltman and others. Glashow, later in the Nobel lecture, said a propos the failure of Bjorken and himself in 1964 to solve the problem of strangeness-changing neutral currents: "I had apparently quite forgotten my earlier ideas of electroweak synthesis". He must still have forgotten them in 1966, since he was at my Harvard seminar!

My own attempts to find a phenomenologically relevant model continued after my return to Edinburgh in August 1966. Contrary to Glashow's belief, I was so absorbed in model building that I neglected to follow up the more formal aspects of spontaneously broken gauge theories. Tom Kibble's 1967 paper[16] dealt with the problem of the Goldstone theorem in the Lorentz gauge and with the structure of non-Abelian theories.

What may have been my last informal contribution to this story also took place in 1967. On my way to Marshak's "Particles and Fields" Conference at Rochester I visited my friend Michael Fisher at Brookhaven. While I was there I joined in a discussion between Weinberg, Boulware and others on the masses of $1^{\pm}$ and 0^- mesons. Weinberg and I had been studying these masses in rather similar field theoretic models, and my contribution to the discussion, as I remember it, was to stress the desirability of generating the $1^{\pm}$ masses by spontaneous symmetry breaking, which unfortunately did not seem to be possible for SU(2) x SU(2) or SU(3) x SU(3) hadronic chiral symmetries. I have often wondered to what extent this discussion contributed to Weinberg's subsequent realization that (as he later described it in his Nobel lecture[17]) he "had been applying the right ideas to the wrong problem".

REFERENCES

1. M.Gell-Mann and M.Levy, Nuovo Cimento 16, 705 (1960).
2. J.Schwinger, Ann.Phys. (N.Y.) 2, 407 (1957).
3. P.W.Anderson, Phys.Rev. 130, 439 (1963).
4. P.W.Higgs, Phys.Rev.Letters 13, 508 (1964).
5. Y.Nambu and G.Jona-Lasinio, Phys.Rev. 122, 345 (1961); 124, 246 (1961).
6. J.Goldstone, Nuovo Cimento 19, 154 (1961).
7. J.Goldstone, A.Salam and S.Weinberg, Phys.Rev. 127, 965 (1962).
8. A.Klein and B.W.Lee, Phys.Rev.Letters 12, 266 (1964).
9. W.Gilbert, Phys.Rev.Letters 12, 713 (1964).
10. J.Schwinger, Phys.Rev. 125, 397 (1962); 128, 2425 (1962).
11. P.W.Higgs, Phys. Letters 12, 132 (1964).
12. F.Englert and R.Brout, Phys.Rev.Letters 13, 321 (1964).
13. G.S.Guralnik, C.R.Hagen and T.W.B.Kibble, Phys.Rev.Letters 13, 585 (1964).
14. P.W.Higgs, Phys.Rev. 145, 1156 (1966).
15. S.L.Glashow, Rev.Mod.Phys. 52, 539 (1980).
16. T.W.B.Kibble, Phys.Rev. 155, 1554 (1967).
17. S.Weinberg, Rev.Mod.Phys. 52, 515 (1980).